Amateur Radio INSIGHTS

# SDR 2.0

## Software Defined Radio

by

Andrew Barron, ZL3DW

This book has been published by Radio Society of Great Britain of 3 Abbey Court, Priory Business Park, Bedford MK44 3WH, United Kingdom **www.rsgb.org.uk**

Amateur Radio Insights is an imprint of the Radio Society of Great Britain

First Edition 2016

Revised 1st Edition 2017

Second Edition 2024



Cover design: Kevin Williams, M6CYB
Typography and design: Andrew Barron ZL3DW
Production: Mark Allgar, M1MPA

**Notes:**
The book refers to software and software defined radios created by many individuals and companies. All such software and intellectual property relating to hardware mentioned in this book are copyrighted by their respective authors.

The opinions expressed in this book are those of the author and are not necessarily those of the Radio Society of Great Britain. Whilst the information presented is believed to be correct, the author, the publishers and their agents cannot accept responsibility for consequences arising from any inaccuracies or omissions.

The author has no association with any equipment manufacturers or resellers. He has not received any discounts or incentives and paid full retail prices for the equipment featured in the book.

ISBN: 9781 9139 9566 9

Printed in Great Britain by Short Run Press Ltd. of Exeter, Devon

Any amendments or updates to this book can be found at: www.rsgb.org/booksextra

## TABLE OF CONTENTS

# ABOUT THE AUTHOR

Well, if you have managed to get this far you deserve a cup of coffee and a chocolate biscuit. It is not easy to digest large chunks of technical information. I hope you enjoyed the book and that it has made learning about SDR a little easier.

I live in Christchurch, New Zealand. I am married to Carol who is very understanding and tolerant of my obsession with amateur radio. She describes my efforts as "Andrew playing around with radios." We have two children, two cats, and some tropical fish. After agreeing that we would never, ever buy a pedigree cat we now have a $700 Birman and a tabby. One son works for a large food wholesaler and the other is a doctor working at Christchurch Hospital.

I am a keen amateur radio operator who enjoys radio contesting, chasing DX, digital modes, and satellite operating. But I am rubbish at sending and receiving Morse code. I write extensively about many aspects of the amateur radio hobby, writing occasional columns for several magazines and other publications. The 1st edition of this book was my 2nd amateur radio book. I have just completed my 20th amateur radio book.

Thanks for reading my book!

73 de Andrew ZL3DW.

## ACKNOWLEDGMENTS

Thanks to my wife Carol for her love and support and to my sons James and Alexander for their support and their insight into this modern world.

Thanks also to everyone who has ever helped with the design or production of software defined radios for radio hams and shortwave listeners, and to the hundreds of software developers who spend thousands of hours writing SDR software and then make it available to the community for free!

Finally, many thanks to you, for taking a chance and buying my book.

## ACRONYMS

Software defined radio, and radio in general, is full of acronyms and technical jargon. I have tried to expand acronyms the first time they are used. The glossary near the end of the book has explanations for most of the technical phrases and acronyms used in the book. I have assumed that the reader is familiar with common radio abbreviations like RF, HF, VHF, SSB, AM, FM, USB, and LSB. USB can mean 'upper sideband' or the 'universal serial bus,' on a computer. I have used a few Q codes. These are codes that were used by telegraph operators, way back before radio was invented. They are a popular shorthand with amateur radio operators. QRP indicates low power transmission. Less than 10 watts for SSB or 5 watts for digital text modes or CW. A QSO is a conversation between two amateur radio operators. QTH is your house or the place you are transmitting from. DX is a code that originally meant 'distant exchange.' It is used to mean a station that is far away or difficult to reach by radio. CQ is short for "seek you." It refers to a general call to anybody who wishes to respond.

# INTRODUCTION

Everyone is talking about software defined radios. But is an SDR right for you? This book explains the basics without getting too technical. It will help you to get the most out of your new software defined radio or decide whether you want to buy one. The use of software defined radio by amateur radio operators and shortwave listeners is rapidly growing in popularity as people become aware of their great features and performance. What used to be the domain of a few dedicated hackers and experimenters has now reached the 'mainstream' ham population. Virtually all new amateur radio transceivers and receivers intended for the hobby market have some level of SDR integration. Some are 'real' SDR designs and some are a hybrid mix of superheterodyne technology and SDR technology.

Things have changed since I wrote the first edition in 2016. This updates edition places more emphasis on the latest SDR and hybrid radios and introduces some new ideas, such as the design of the QDX transceiver from QRP Labs. All the basic theory remains the same. Some radios have disappeared and some new ones have joined the list.

For many hams, software defined radio conjures up images of circuit boards, tricky software, and black box radios without any knobs or buttons. But that is not how it is anymore. Most of the new SDR receivers and transceivers are just as easy to set up and use as any other ham radio transceiver or shortwave receiver. They have all the buttons, switches, knobs, and controls you are used to. These SDRs "with knobs" do not have to be connected to a personal computer (PC) for their operation. You might not even know that you are using an SDR. They are "the same on the outside, but different on the inside." Take a look at the Icom IC-7300 or the Elecraft KX4. They look just like many other amateur radio transceivers, but they are both software defined transceivers. There are new black box radios without knobs, as well. In some cases, you can buy the same radio with a conventional layout or as a black box version.

I got into software defined radio because I wanted one of those fancy bandscopes like the ones on the latest high-end transceivers like the Icom IC-756. My first SDR was a SoftRock kit on 9 MHz, wired into the receiver I.F. stage of an old transceiver. I liked it… a lot. I found it amazing to be able to see the signals on the band as well as hear them. The SoftRock experience was so positive that I bought a FLEX-1500 QRP radio. I used it to drive a 100 W linear amplifier made from the power amp block of a surplus Yaesu FT-301 transceiver. Within a couple of months, I was hooked on the SDR experience. I have moved on to one of the new 100-watt SDR transceivers. My old HF transceiver sits gathering dust. Now, when I use a conventional radio, I feel deprived of the bandscope. It feels like I am operating blind. The improved audio quality and ease of use are great, but for me, the best thing about using an SDR is being able to see the signals across the band.

With the SDR, I can set my receiver filter to exactly the right receiver bandwidth, avoiding unwanted noise. I can see, and often eliminate, any interfering signals. There are other benefits too, such as the ability to watch a net frequency on the 80m band while working an FT8 station on 20m. Contest and DX stations can monitor other bands for an opening without moving off their run frequency.

SDR receivers have opened up the VHF, UHF, and sometimes the SHF microwave bands to casual listeners. Previously you needed converters or very specialized receivers to hear signals on the high bands. Now receivers like the Icom IC-R8600, Malahit DSP2, and dozens of cheap SDR dongle receivers can receive from as low as 10 kHz all the way up to 2 or 3 GHz. You get the advantages of an SDR, including the bandscope, variable RF bandwidth, excellent noise reduction, and notch filters. Plus you often get CW, PSK, RTTY, or FT8 decoders, FM stereo for the FM broadcast band, or a variety of digital-voice decoders.

Software defined radios offer a new way to operate and some amazing new features, including remote operation, looking at several bands at the same time, and improved transmitter performance using new modulation techniques. In many cases, you can use your black box, or dongle radio, with several different software packages. It is like getting a brand-new radio just by loading a new computer program. A lot of SDR software is free, not shareware, or limited in some essential way. It is created under the 'open source' rules by amateur software developers or groups. These programs are fully featured and can be downloaded and used completely free of charge. Not only that, they are usually very well written and well supported.

You don't need a PC to operate the new generation of receivers and transceivers, but you will need a PC or SBC (single board computer) to use black box 'no knobs' software defined receivers and transceivers. If you have already connected your computer to a radio for digital mode operation, monitoring CW signals, logging stations, or remote control, you should have no difficulty adjusting to life with a software defined radio. It is different, and it does take a bit of getting used to. But operating a software defined radio quickly becomes just like using any other computer application.

Software defined radio is a logical extension of the digital signal processing (DSP) found in many amateur radio transceivers and some shortwave receivers. Digital signal processing in conventional superheterodyne receivers and transceivers is performed by specialized DSP chips in the radio. They are programmable devices running software or 'firmware,' that are specially designed to manage radio, audio, or more often, video signals. The first generation of radios with digital signal processing provided audio frequency (AF) DSP. It works by converting the audio output from the demodulator into a digital signal. The digital signal is manipulated to apply filtering and noise blanking and then it is converted back into an audio signal and passed on to the audio amplifier stage.

The biggest advantage was the introduction of manual and automatic notch filters. AF DSP was excellent for noise reduction although the DSP audio filters had a reputation for adversely affecting the audio quality. The main problem with AF DSP was that it occurred after the receiver AGC. Although you could filter strong interference signals out of the demodulated audio, the AGC action still desensitized the receiver making weaker, wanted signals disappear. This was a particular problem with PSK reception. Strong local signals would push down the level of weak DX stations we wanted to decode.

To combat this issue the DSP was moved to the I.F. stage. That way strong signals can be filtered out before they affect the receiver's sensitivity. With I.F. DSP, (intermediate frequency digital signal processing,) the I.F. output of the first or second mixer of the receiver is converted into a digital signal. Features like I.F. shift and I.F. width control are performed using firmware code running on the DSP chip. Often a bandscope is provided so that you can see a display of signals above and below the receiver's frequency. After the DSP stage, the digital signal is converted back into an analogue signal and passed on to the demodulator. Or the demodulation may occur in the DSP stage and the resulting digital audio signal is converted back to analogue audio and sent to the audio amplifier and speaker.

Software defined radios take the DSP process further by performing the analogue to digital conversion as early as possible in the receiver and as late as possible in a transmitter. Receiver performance is improved by eliminating analogue mixers and I.F. amplifiers which contribute noise and distortion products in traditional designs. Most new HF and VHF receivers now use direct digital sampling, where virtually the entire HF spectrum or the VHF range is sampled by an ADC (analogue to digital converter), close to the antenna. It is usually placed immediately after the antenna switching, bandpass filters, front-end preamplifiers, and attenuators. Some radios use Mirics or other 'receiver chips,' but these still use direct sampling at HF and VHF frequencies. Most ADCs cannot sample fast enough to allow direct sampling of UHF frequencies. Receivers for frequencies above 200 MHz usually employ a heterodyne stage with a hardware oscillator and mixer to convert the incoming frequencies down to a range that the SDR can process. For example, the LTC2208 ADC used in some radios, can sample up to 700 MHz. That allows a direct digital sampling receiver to receive signals up to 350 MHz.

Some recent HF transceivers employ 'hybrid SDR' or 'SDR technology.' They are conventional superheterodyne radios with an I.F. stage replaced by an ADC and an FPGA (field programmable gate array). The FPGA may only be performing the task previously performed by the second oscillator and mixer, down-converting the incoming frequency to a low 24 kHz or 36 kHz I.F. for connection to the input to the same Texas DSP chip used in earlier models.

This book covers the many things that an SDR can do better than your old technology receiver or transceiver. I think there is something for everyone.

Here are some of the features and benefits that you can expect from a software defined radio. Not all of these functions are available with all software defined radios, or software. But these nine items will give you a taste of what you can gain by buying an SDR.

1. Generally, SDRs have very good receiver performance. This is mostly due to the fundamental difference in technology. Direct sampling digital down conversion HF receivers have no hardware mixers so they don't suffer from the intermodulation distortion effects and 'birdies' that mixers and oscillators contribute to superheterodyne receivers. Of course, the manufacturers of conventional radios do go to great lengths to try to eliminate these problems by using excellent designs, quality components, and picking I.F. frequencies that place image signals outside the ham bands. But, great performance comes at the cost of increased complexity and a higher price.

2. The panadapter screen of a software defined radio displays the signals across a ham or shortwave band. A simple mouse click or finger tap tunes the radio to any signal that you can see. The panadapter shows other interesting things such as the type of modulation being employed, the transmitted signal's bandwidth, interfering signals, or the effect of poor-quality transmissions. CW key clicks, over-modulated digital mode signals, and linear amp splatter are easy to spot.

3. Some receivers use Internet capability to display the name or callsign of received stations directly on the panadapter or band-scope display. This is a great aid to shortwave listeners trying to identify broadcast stations, or to amateur radio stations wanting to identify contest or rare DX stations.

4. Tuning around the band with a conventional receiver you might easily miss a station making occasional calls. With an SDR, you will see the signal pop up on the waterfall or spectrum display. You can click to move directly to the station without having to tune up and down the band. This is fantastic during amateur radio contests because you can click on each station in turn as you work up or down the band.

5. With some software defined radios, you can listen to two receivers at the same time. You can use this to tune around a band while listening for a call on another band or frequency. Or to hear a DX station and also the stations they are working in the pileup. The panadapter can show both the DX station and the pileup of calling stations. This lets you pick a relatively quiet spot to transmit on or work out which way the DX station is progressing through the calling stations. You can easily see when the DX station is transmitting and time your call appropriately. Two receivers can let you monitor two repeater channels at once. This is useful but not as good as a scanning receiver.

6. Being able to monitor two or more bands at the same time can be useful. You can quickly see which HF bands are 'open.'

If you have arranged to work the same station on another band, you can see their signal come up on the new frequency, without moving off the old frequency. No more changing bands only to find the frequency is in use or you can't hear the other station. You can watch and hear your old frequency and your new frequency at the same time. You can watch a frequency for a 'Net' to begin, or for a DX station while working or listening to a different band.

7. Modern transceivers and receivers only require one USB cable to carry CAT (CI-V) commands and audio between the radio and your PC. No interface units or multiple cables audio cables. Black box radios use PC software to perform the DSP, control, and display functions. They can interface with external digital mode or other software with 'virtual' com and audio cables.

8. You won't need external audio equalizers either since the SDR, or SDR software, usually includes microphone and receiver audio equalizers.

9. The CW Skimmer program is very popular with CW DX stations and contesters. It can decode many Morse Code signals at the same time. A software defined radio can send the entire CW band segment, or even multiple bands, to CW Skimmer, so that all of the CW signals can be identified. With conventional receivers, only signals within the relatively narrow 3 kHz receiver bandwidth can be seen by the CW Skimmer application.

I have noticed that many articles in amateur radio magazines feature screenshots of SDR receiver displays. It seems that the people who are interested in the development of new digital modes, EME, propagation, shortwave signals, weather satellites, ham satellites, contesting, interference detection, and monitoring intruders on our bands are finding that software defined radios are very useful for their particular interest. An SDR receiver can be used as a spectrum analyser to observe different types of modulation or interference signals. You can even see the frequency response of the audio modulation on a single sideband signal. You can also use the panadapter to monitor your own transmitted signal to ensure that your transmission is clean and legal.

The ham radio and shortwave listening software defined radio world is developing so fast that much of the information online and many of the products for sale are already out of date. It can be quite confusing deciding what radio to buy. This book explains the merits of different types of SDR and there is a comprehensive list of the key parameters of many of the currently available models. I hope that this book will convince you to buy a software defined radio and will help you to get the most out of using one. One of the joys of software defined radio is that it is evolving rapidly. New features are being developed continuously. This is great for those who like to tinker with new technology and be at the bleeding edge of learning what is possible with these new radios. Others may be more interested in the competitive edge that using an SDR can provide.

# BUYING AN SDR

Firstly, you need to decide whether you are looking for a transceiver or a receiver and often your budget will be a major factor. Nearly all new amateur radio HF transceivers released since 2017 have some sort of SDR technology. There are a lot of SDR receivers on the market because that is where the biggest improvements lie. Most of them are black box or SDR dongle types, but increasingly some designs include physical controls and a built-in bandscope. The Icom IC-R8600 is an SDR receiver with coverage from 10 kHz up to 3 GHz. It uses direct sampling for the HF bands and hybrid technology from 30 MHz to 3 GHz.

Kit builders and those looking to connect a cheap bandscope to the I.F. output of a conventional receiver may prefer a QRP Labs kit or a black box receiver such as an SDRplay. The RTL dongles are a very cheap way to have a play with SDR and they are tremendously popular. I am not a fan because the performance of these devices is quite limited. The RTL dongles were designed primarily as a cheap way to receive digital TV signals and display the channels on portable computers. I believe that the FUNcube Dongle Pro+ USB dongle receiver is a better choice. They are a bit more expensive, but they have dedicated front-end filters with excellent SAW filters on the 2m and 70cm ham bands. The frequency coverage is amazing, and they were specifically designed to be a receiver for amateur operators and for schools involved with the FUNcube satellite project.

At the other end of the scale are the direct digital sampling 100-watt HF and VHF transceivers. The Icom IC-7300 has proved to be a 'game changer' for amateur radio worldwide. It was introduced in 2016 and it is probably still the best-selling amateur radio transceiver. Sales reached 100,000 units in May 2024. The IC-7300 has many features that were previously not available, or only available on much more expensive radios. These include the touchscreen bandscope, VOX, CW, and RTTY keyers. Plus an RTTY decoder, antenna tuner, off-air recording, plain English or Japanese menu rather than cryptic codes and numbers, and a one USB cable interface to your computer.

If you are buying a black box radio, it is important to evaluate the SDR software as well as the hardware because it is your interface with the radio. Some radios can use a range of software applications. Others are limited to a particular program that comes bundled with the radio.

If you are interested in VHF, UHF, or microwave frequencies, some radios work up into the GHz range. Many inexpensive SDR receivers cover immense frequency ranges that were simply out of reach before SDRs became available.

The Icom IC-9700 opened easy 'all mode' plus D-Star access to the 1.2 GHz (23cm) amateur band, as well as covering the 2m and 70cm bands. The IC-905 has extended that to include the 2.4 GHz, 5.6 GHz, and 10 GHz bands.

What you buy is up to you. The choices are a bit like buying a car. An RTL dongle is like a 'Go-kart.' It is technically a car, but it lacks some of the capability and features that you might prefer. My fear is that you might be disappointed with the performance of one of these little radios and that this might disillusion you about software defined radios in general. The older QSD (quadrature sampling detector) based receivers and transceivers are like a 'Model T.' Again, they have many of the features of a modern car but nowhere near the performance that is available from a newer model. Of course, many people love the cheap price of the Go-kart or they only need the capabilities of the Model T and that is fine. Owning either of them can be loads of fun. As with everything that you choose to buy for your ham shack or hobby room, the purchase of an SDR comes down to a careful analysis of what you want to achieve balanced by an equally careful analysis of your available funds. There are a myriad of different SDR types. USB dongles, small box QSD-based receivers and transceivers, SDR kits, VHF/UHF/SHF boards that use a mixer to get an I.F. frequency low enough for direct sampling, HF transceivers using direct digital sampling, SDRs with knobs and onboard DSP, superheterodyne radios with an IQ audio output from their DSP stages, and commercial or military radios that use digital sampling and DSP chips.

Check out the Catalogue of software defined radios near the end of the book. It contains basic details of some of the software defined radios that are currently available. Whether you want a receiver, a QRP transceiver, a radio for VHF, UHF, or microwave frequencies, or a 100W HF transceiver, you should be able to find a software defined radio that meets your needs. If not..., buy two!

## PANADAPTERS AND BANDSCOPES

I have mentioned panadapters and bandscopes several times without explaining what they are.

A **bandscope** is the spectrum and waterfall display fitted to I.F. DSP radios and the new SDRs with knobs. It shows a 'span' of frequencies on either side of the frequency that the receiver is tuned to. Icom and Yaesu call it the 'Centre' display.

A **panadapter** is the spectrum and waterfall that a black box SDR software displays. It shows a range of frequencies. You can tune the receiver 'slice' anywhere across the displayed bandwidth. Most radios with knobs can display a bandscope or a panadapter. Icom and Yaesu call the panadapter a 'FIX' (fixed) display.

The two names are often used interchangeably but they are actually different. To make things simple I often refer to both types as a 'spectrum and waterfall display.'

Both types normally show a spectrum display and a waterfall image. A spectrum display shows the signal strength of received signals across a range or 'span' of frequencies. It is similar to the display of a spectrum analyser.

A waterfall shows the signal strength of the same range of received frequencies over time. Both provide useful information. The spectrum shows the activity on the band and the bandwidth of the received signals. The waterfall shows the bandwidth of signals and captures a trace of signals that were present recently but have now stopped transmitting.

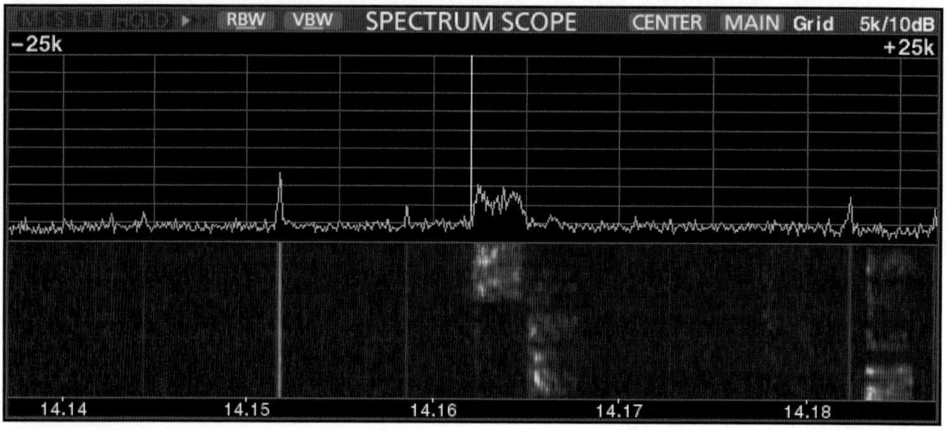


Figure 1: The CENTRE bandscope display

This bandscope or 'CENTRE' display has a span of ±25 kHz on each side of the VFO frequency. You can see the bandwidth of the three SSB signals and three unmodulated carrier signals. The station that was transmitting beside the wanted signal has stopped, but you can see the remnants of the transmission on the waterfall display.

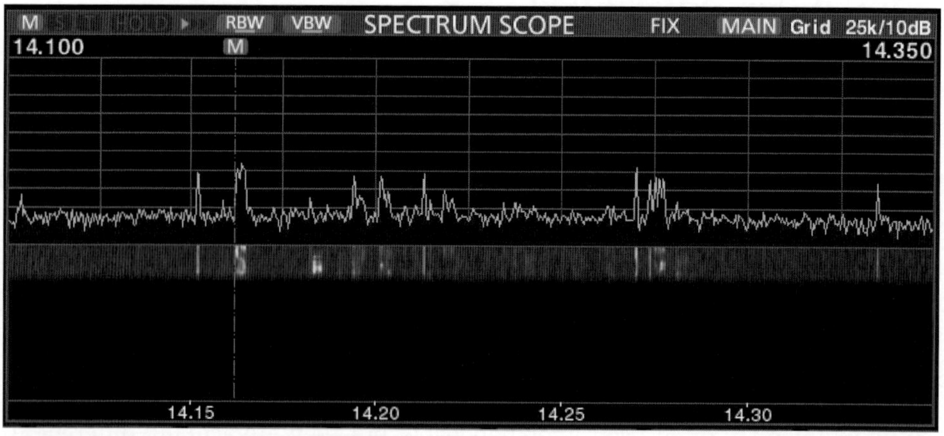


Figure 2: The FIX panadapter display

This panadapter is showing the same signal on a 14.100 to 14.350 MHz panadapter. On this radio, you can set the panadapter width anywhere between 5 kHz and 1 MHz. The panadapter 'FIX' display is great for contests because you can see each station and work up the band. It is also easier to identify contest stations that you have worked before. Some radios can even tag the signals with their callsigns.

## RECEIVERS

Receivers get more performance gains from changing to an SDR design than transmitters do. Even a casual listener can hear the improvement, especially on the HF shortwave bands. SDR receivers are less noisy, have a better dynamic range, have far better noise reduction, notch filters, and noise blankers, and they usually have a wider frequency coverage. You can set the receiver bandwidth to exactly match the transmitted signal so that you are not listening to band noise outside of the transmitted signal or compromising the fidelity of the signal with a receiver filter that is too narrow. The bandscope lets you see the signals across the band and measure their relative signal strength. Features like built-in recording and decoders are helpful as well. Radios with multiple receivers can let you monitor two or more channels at the same time.

SDRs are not as good in mobile applications. The bandscope is a major benefit, but it could be distracting while driving. Most SDRs are not great at high-speed scanning or for receiving trunk radio systems.

You need to be careful in your selection. The price is not necessarily a good indicator of the radio's performance. Check out the specifications online at the manufacturer's website and check for reviews on 'eHam reviews' http://www.eham.net/reviews/. It is not a good sign if there is sketchy or no data available online.

The updated 'Catalogue of software defined radios' starting on page 241, includes the basic specifications of some of the SDR receivers on the market. I have written 'Radio Today guides' for the Icom IC-R8600 and the Malahit DSP2. The Icom is a fantastic receiver, possibly the best non-professional receiver on the market, but it is quite expensive. The Malahit is a Russian design built under license in China. It carries both FCC and CE certifications, which is unusual for a Chinese clone radio. The DSP2 is also unusual because it is a stand-alone radio with a display, controls, an internal speaker, and a built-in battery.

Most small SDR receivers are black box models or SDR dongles that require a PC connection. The DSP is performed by SDR software running on your computer. The computer may be a desktop or laptop PC. Or, it could be a single-board computer such as a Raspberry Pi, or a cell phone.

Make sure that the receiver covers the bands that you want. Some SDR dongles only cover frequencies above 24 MHz. There are several specialist experimenters' boards with huge frequency ranges extending from around 10 MHz to as high as 6 GHz. Some have dual receivers, and some are 'sub-QRP' transceivers.

QSD designs with PC or USB soundcards are based on older technology but they still work very well. The RTL dongles that use 8-bit analogue to digital converters will have a high noise floor and a poor dynamic range.

14-bit direct sampling designs can have excellent performance. They are often just as good as 16-bit radios, but they usually have a lower sampling rate, which means reduced frequency coverage. An HF receiver using a 14-bit ADC might not include coverage of the 6m amateur band.

Radios advertised with 24-bit ADCs are usually hybrid receivers where the analogue to digital conversion is being performed at audio frequencies or less than 48 kHz. The number of ADC bits is much less important for hybrid architecture receivers. Some QSD receivers have 24-bit audio frequency ADCs, but 16-bit USB or computer soundcards are more common.

## TRANSCEIVERS

A few transmitters use features like adaptive pre-distortion to improve their linearity or more efficient modulation methods like CESSB (controlled envelope SSB), but generally, SDR transmitters have very similar performance to traditional superheterodyne models. The transmitters in SDR transceivers usually use the same class of technology as the receivers. If the radio uses a QSD receiver it will have a QSE transmitter. If an HF radio has a direct sampling receiver the transmitter will use digital up-conversion directly to the transmitter frequency. And if the receiver is a hybrid superheterodyne front end followed by an SDR, working at RF, I.F., or audio; the transmitter will usually generate a similar audio, I.F., or RF frequency which will be mixed up to the transmitted frequency.

There are a few more transceiver choices than there were when I wrote this book in 2016. Most of the major manufacturers now include direct sampling SDR HF transceivers in their product line. The two hybrid transceivers produced by Yaesu also have very good technical specifications.

There are two distinct types of SDR transceivers. They use different technologies because they come from different heritages. On one branch there are transceivers that stem from the SDR experimenters in the early 2000s. They were developed by interested amateurs who developed innovative radios and open-source software. Their initial radios were black box designs that used a PC to perform the DSP, control, and display. These morphed into a few 100 watt transceivers and many QRP transceiver and receiver designs. The latest transceivers in this group have SDR boards, usually with onboard DSP. They often include a computer inside the radio. Some use a small Windows computer or a Linux single-board computer, some use an STM32 microcontroller, and the Apache Labs ANAN G2 uses a Raspberry Pi4 'compute module.' Whether these radios have knobs, or not, they are characterized by the use of an SDR board plus an internal or external computer.

The other SDR transceiver type are the radios produced by companies that used to make conventional superheterodyne transceivers that have now "jumped on board" with the new technology. Their SDR models are based on their old radios with modifications that embrace the new technology.

These new radios look very like the old radios, other than having a larger screen. They have the same buttons, knobs, and functions, plus a few new features such as message keyers, decoders, touchscreens, and the ability to display a panadapter or a band scope. The radios are characterized by mainboards with integrated SDR chips. They generally have a microcontroller, or two, to manage the controls and display as the earlier radios had, and the DSP is performed by the same or similar DSP chips that were used in the earlier radios.

### SDRs with knobs

Icom led the way with the IC-7300 and IC-7610. Both are direct sampling SDRs. They were followed by the IC-9700 tri-band VHF/UHF all-mode radio. Then the IC-705 QRP radio for hiking, POTA, and SOTA. The IC-905 provides easy access to the SHF microwave bands. Yaesu released the FTdx101 and FTdx10 hybrid radios and followed that with the FT-710 their first direct sampling SDR. Elecraft has released the K4 direct sampling SDR in several configurations.

### SDRs with internal or external computers

Sixteen-bit, direct digital sampling (DDS) radios, with one or two analogue to digital converters represent the top of the heap for HF software defined transceivers. The Apache Labs ANAN and the FlexRadio Signature radios both use 'state of the art' direct digital sampling technology. However, their technical design and their sales philosophy differ. Your choice comes down to a comparison of the specifications, performance figures, and software features, balanced against the cost of each model.

The FlexRadio FLEX-8400 and FLEX-8600 'Signature series' radios have digital signal processing performed within the radio and this dramatically reduces the data load on the Ethernet connection between the radio and the computer. Because the digital signal processing is being done inside the radio, you can use less powerful devices like tablets and Netbooks to control the radio and display the receiver panadapters. FlexRadio's SmartSDR operating program comes bundled with the radio. Its capabilities are amazing! The addition of the Maestro controller and the introduction of 'M' models that include a 'Maestro' front panel, means that FlexRadio now offers both 'black box' and 'with knobs' models. You can use any model with Maestro, a tablet, an iPhone, or a macOS or Windows PC running SmartSDR.

The Apache Labs ANAN series radios have been developed in conjunction with the OpenHPSDR group of amateur radio and SDR enthusiasts. When you buy an Apache Labs radio, it does not ship with any PC software. Thetis the successor to Power SDR has been developed under the 'open source' rules and is made available to radio users for free. The latest ANAN G2 transceiver is available as a black box – no knobs model, or a full front panel model with a display and knobs. There is also a 1500-watt black box version. Computer power is supplied by a built-in Raspberry Pi4 compute module.

You can use any model with the piHPSDR program operation on the compute module, external piHPSDR software, or Thetis.

The approach taken by the Expert Electronics team for their Expert Electronics MB1 transceiver is closer to the Apache Labs design. The radio can be operated completely stand-alone with no connection to an external PC because it has a Windows 10 PC with an Intel i7-9700 CPU built into the box alongside the radio board. As well as running the SDR program, the internal PC can be used for other computing tasks such as your station log or digital mode software.

## QRP TRANSCEIVERS

There is a big range of QRP (low power) transceivers both with knobs and without. For the HF bands, I recommend a direct digital sampling radio with at least 5 watts of output power and either a USB or an Ethernet connection. There are quite a few options including the ELAD FDM-Duo, the Sun SDR2 pro, or the Xiegu X6100.

For VHF and above, there are 'sub-QRP' very low power transceivers like the PLUTO+, or HackRF One units. These UHF/SHF radios and boards are very low-power transceivers designed for experimenters. They have extremely wide frequency coverage. The two Pluto models are popular as drivers for 2.4 GHz power amplifiers being used to access the QO-100 satellite.

There is a trap for the unwary in buying any QRP radio. By all means, buy one if you want to operate a QRP station or if you want a low cost introduction to software defined radio. But, if you are planning on adding a 100-watt linear amplifier, you might find that the combined cost of the QRP radio plus the linear amplifier is more expensive than simply buying a 100-watt SDR transceiver.

## A SHOPPING CHECKLIST

Work out exactly what you want from an SDR. Do you want a cheap toy to play with or a serious amateur radio station? Do you want a kit to build? Is an unboxed circuit board OK, or do you want a fully cased model? How much of an experimenter are you? Finally, do you want a transceiver or just a receiver?

Generally, there is a strong correlation between price and performance, but you still have to be careful. Here are some points to consider.

- Some old models still carry high price tags.
- You may have to factor in freight costs, import duty, and taxes.
- Some websites offer radios that are not 'currently' available. You may have to pay a deposit and wait a month or two for the radio.
- Some websites look good but have very little technical information about their radios. You might be disappointed with your purchase.

Good manufacturer's websites have specification sheets, which include the most important features. Often you can download a user manual. There may be a Facebook group or a forum on https://groups.io.

- Radios with 8-bit analogue to digital converters will have reduced dynamic range compared to radios that use 14 or 16-bit ADCs. This might be OK on boards that cover VHF to SHF frequencies, but it will be a problem on the HF bands.
- Radios with no band-specific bandpass filters before the ADC may suffer from ADC gain compression, aliased image signals, and overload. Unlike conventional receivers, a software defined radio receiver is affected by all of the signals being sampled, right across the input spectrum, not just the signals that you can see on the panadapter.
- Does the radio use a QSD design, or does it use direct sampling? Direct sampling is better but QSD-based radios are usually cheaper.
- What is the ADC sampling frequency? For direct sampling HF radios, higher sampling rates are better because you get a wider frequency range.
- What is the interface to the PC? They are typically USB or Ethernet. Some very old designs used the obsolete FireWire interface.
- What software can be used with the black box radio? More than one program? Is it easy to use, and does it have the features you want?
- What frequencies does the radio cover? Buy a radio that covers the bands that you want to use.

Here are some more key facts:

- Don't buy a kit unless you want to build a kit. There are ready-built options at around the same prices.
- QRP transmitters are fine if you want to operate QRP. But by the time you add in the cost of an external 100 W power amplifier, it is usually more cost-effective, to simply buy a 100 W radio.
- Most cheap software defined receivers can only display one panadapter or bandscope. More expensive software defined radios can display two or more panadapters or bandscopes, which means that you can receive two (or more) bands at the same time.
- A radio with a 24-bit or 32-bit analogue to digital converter is based on a hybrid architecture or it might be a QSD design. The ADC is sampling at audio (less than 48 kHz), not directly sampling the incoming RF frequency. This usually indicates that the radio is a hybrid with a dedicated DSP chip. This is not bad. It is just a different method.

The Yaesu FTdx10 and FTdx101 use this architecture and they are placed near the top of the Sherwood receiver performance list.

- With direct sampling radios, "more bits is better." But as far as signal to noise performance and dynamic range is concerned, 14-bit ADCs are just as good as 16-bit ADCs. This is because the ENOB (effective number of bits) of a 16-bit ADC is pretty much the same as a 14-bit ADC.

- The ADC sampling rate is very important if you are buying a direct digital sampling HF radio. It governs the maximum bandwidth that can be passed to the DSP stage of the receiver. In some cases that is equal to the maximum amount of spectrum that can be displayed on the panadapter. In other cases, the panadapter can only display a section of the sampled spectrum.

- With QSD radios, the analogue to digital conversion is done at audio so the number of bits is less important. Most soundcards can manage 16 bits at a 96 ksps sample rate. Some soundcards create 24 bits and a few can sample at 192 ksps. Even if the computer or USB soundcard has a 24-bit analogue to digital converter, the displayed bandwidth and dynamic range will be much less with a QSD receiver than you can get using a direct sampling receiver.

- An advertised panadapter or bandscope bandwidth of 48 kHz, 96 kHz, or 192 kHz indicates that the analogue to digital conversion is being done at audio frequencies with a computer or USB soundcard. It usually indicates that the receiver uses a QSD design. This is not necessarily a problem, but it is 'good to know.'

# WHAT IS DIFFERENT ABOUT SDR?

The internal construction of software defined radios is different from the traditional superheterodyne receiver and transmitter architecture found in 'conventional' amateur radio or shortwave receivers and transceivers. When you look at the circuit board there are no adjustable tuned circuits in little cans, just a few large surface-mounted integrated circuits. Some components do look familiar though. You may be able to spot the receiver's bandpass filters and the low pass filters with their toroid cores in the transmitter output stage. The main TCXO oscillator and the transmitter power amplifier look just the same.

SDRs are a logical extension of the introduction of digital signal processing (DSP). Most recent ham radio transceivers and some shortwave receivers feature DSP. In a DSP equipped radio, the signal is converted from analogue to digital. DSP magic is performed using mathematics and software code and then the signal is converted back to analogue again so you can hear it from the speaker. Software defined radio is essentially the same. The main difference is how early the signal is converted to a digital signal. In an SDR receiver, the signal is converted to digital as close to the antenna as possible. The resulting digital signals can be manipulated, filtered, amplified, modulated, and demodulated without adding noise and distortion to the signal. The same thing occurs in an SDR transmitter. The speech audio signal from your microphone or digital mode tones is converted to digital bits and processed using DSP software. The digital signal is up-converted or generated directly at the final transmit frequency and converted back to an analogue signal just before the RF power amplifier and low pass filters. CW signals can be generated in the same way as in most SSB transceivers by SSB modulating a keyed audio tone, or they can be keyed directly at the RF output frequency by simply generating or not generating a carrier signal.

SDR receivers are relatively immune to intermodulation distortion (IMD). In a conventional receiver, large signals within the I.F. passband can mix with local oscillator signals in the mixers and I.F. amplifiers. This causes unwanted signals known as intermodulation products, which can interfere with the signals that you want to hear. Susceptibility to receiver IMD is one of the main differentiators between an 'average' receiver and a 'good' receiver. If you can't pick out individual stations working a busy contest and everything sounds mushy and distorted, then don't blame the contesters. Poor receiver IMD performance is the most likely reason. Intermodulation distortion is easy to measure in conventional superheterodyne receivers using the well-established 'two tone IMD' test method. Direct sampling software defined radios don't have mixers before the analogue to digital converter, so the causes and effects of intermodulation distortion are completely different. The IMD performance of an SDR receiver is less predictable than the linear rise of intermodulation products in a superheterodyne receiver as the level of the interference signals increases.

Large signals near the operating frequency may actually improve the receiver's performance.

Software defined radios don't necessarily perform better than conventional radios. Like conventional architecture radios, you generally pay more for radios with higher levels of performance. However, SDRs do provide many new features and continual development.

TABLE 1: CONVENTIONAL VS SDR RADIOS

| **Feature** | **Superheterodyne conventional radio** | **Software defined radio** |
|---|---|---|
| Bandscope (panadapter display) | Top-end radios only, small display | Nearly all SDRs, large display |
| Waterfall signal display | Top end radios | Nearly all SDRs |
| Bandscope/waterfall display of multiple bands at the same time | No | Some SDRs |
| Wideband bandscope/waterfall display of the entire HF spectrum or higher | No | Some SDRs |
| Listen to two signals on the same band | Yes on some transceivers | Many SDRs |
| Listen to signals on multiple bands at the same time | Yes, on some VHF/UHF radios but not usually on HF radios | Some SDRs |
| Listen to more than two signals on the same band | No | Some SDRs |
| Split transceiver operation | Yes on 'all mode' transceivers | Yes on 'all mode' transceivers |
| CAT (CI-V) control for PC applications | Yes – most HF transceivers | Yes – most HF SDR transceivers. |
| Change the look and functions of the radio by changing the software | No – but can use different CAT-based programs | Many SDRs |
| Transmitter pre-distortion to improve transmitter IMD performance | No | Some SDRs |
| Operate on different bands or modes using different computer devices simultaneously. | No | FlexRadio 6000 and 8000 series |
| Firmware or software updates to the control software and DSP | Yes – most HF transceivers | Yes |
| Transmit power level | Most are 100 W, some are high power or QRP | Most are QRP some are 100 W |

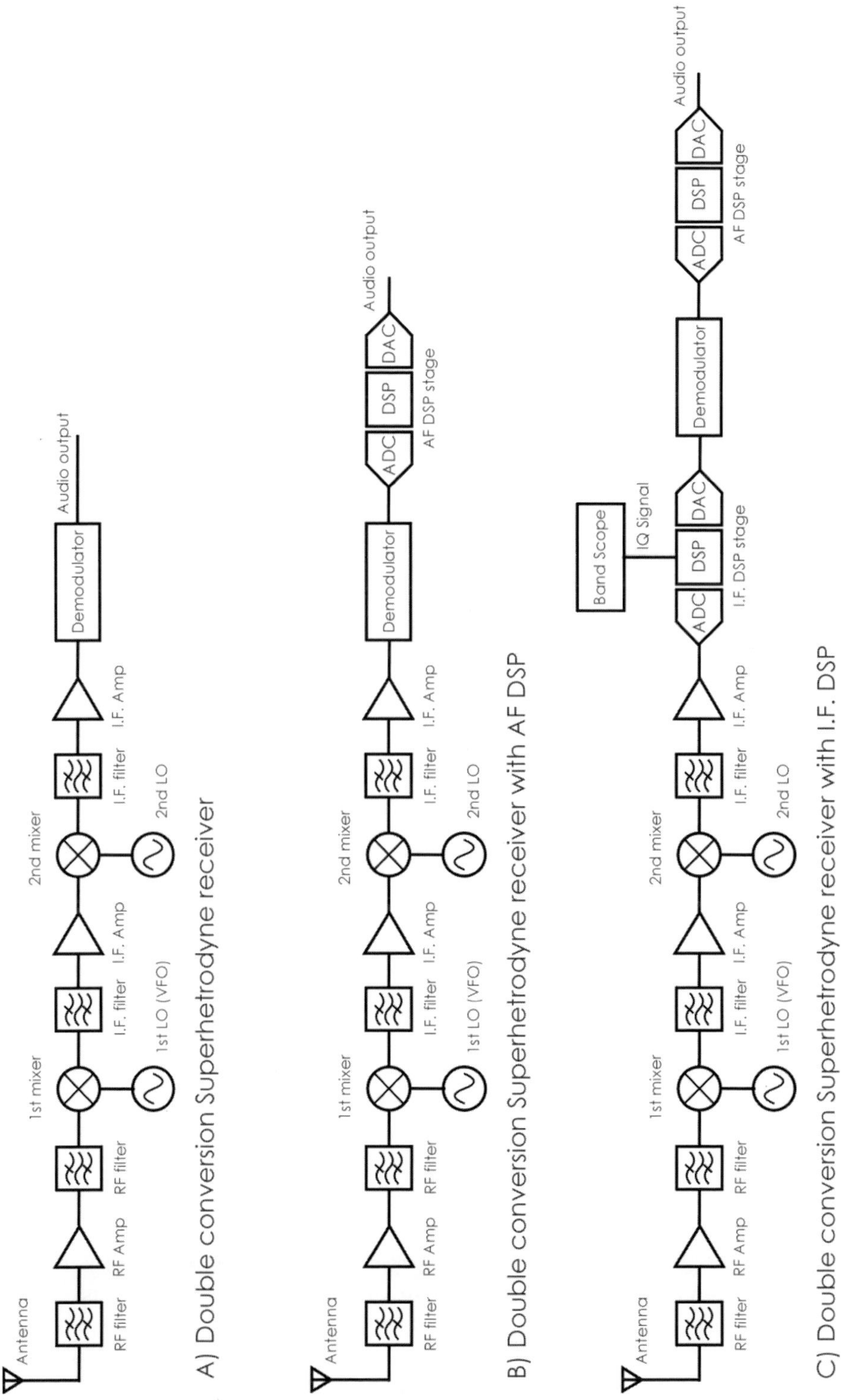


Figure 3: Progression of receiver technology (part 1)

Digital to analogue conversion and digital signal processing are not new. The block diagrams on the previous page show the progression of receiver technology toward software defined radio. Picture A is a typical double-conversion superheterodyne receiver. Picture B shows the introduction of AF DSP following the demodulator. Picture C shows the introduction of I.F. DSP.

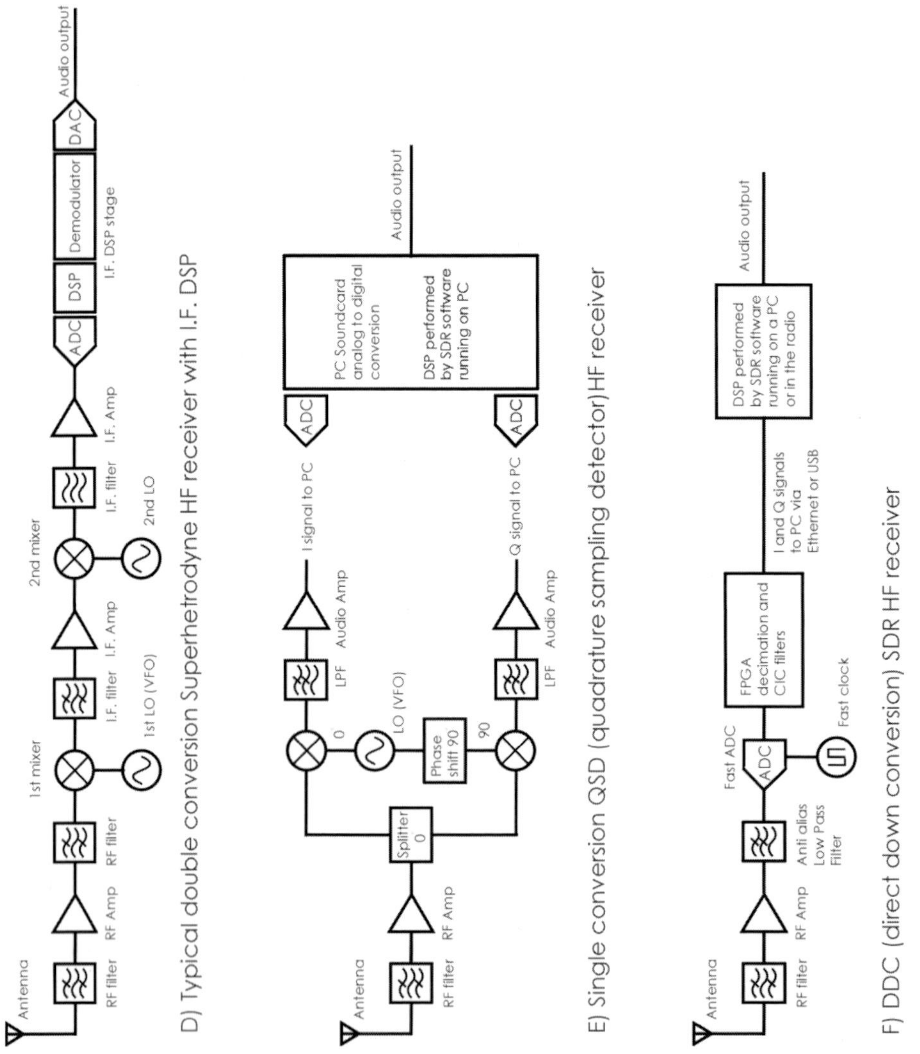


Figure 4: Progression of receiver technology (part 2)

These block diagrams show more progression of receiver technology from the conventional superheterodyne to software defined radio. Picture D is a typical superheterodyne receiver with I.F. DSP. In this case, the receiver demodulator is inside the DSP code block. Picture E shows a QSD-based software defined radio. Picture F shows a direct digital sampling software defined radio for the HF bands. It uses digital down conversion (DDC).

# WHAT COMPUTER SKILLS DO YOU NEED?

Operating any of the SDR radios that have knobs doesn't require any computer skills at all. Other than new features such as touchscreens, digital mode decoders, and plain English menus, they are very similar to earlier models from the same companies. You don't have to connect the radio to a computer to operate modes like SSB, AM, FM, or CW. But you can connect a computer for digital text modes like FT8, RTTY, or PSK.

The radios without knobs do need a computer connection to supply the control interface, panadapter display, and usually the DSP functions. Most amateur radio operators and 'shortwave listeners' use a computer for emails and accessing the Internet and many also use a computer in their ham shack or office as part of their radio activities. But strangely, some people recoil from the idea of operating their radio using a PC display, keyboard, and mouse. In reality, operating a radio in this way is not much different from browsing the Internet. 'Point and click' to listen to a station. Fine-tuning is done with the little wheel between the left and right buttons on the mouse and you 'click and drag' the image on the screen to change the size of filters. Your receiver or amateur radio transceiver becomes just another PC application and you quickly become used to using it. If you really must have a knob for tuning, there are options available. Control knobs can be purchased from WoodBox Radio or other suppliers and some people use 'Midi' controllers such as the Hercules DJ console. We have been able to use CAT commands to remote control our transceivers for many years and a lot of amateur operators and SWLs use PC software to send or receive digital modes, such as RTTY, PSK, FT8, and CW. If you already use a PC in the shack, you probably won't have any difficulty using a black box software defined radio.

My advice is to give it a real chance. Some folks try an SDR and give up after a few days. Like any new radio, an SDR does take some getting used to. Use the SDR like you would use your normal transceiver or receiver. If you are a contester, use it in a contest. If you are a CW operator, make a few CW QSOs. Experiment with the filters and settings. I found that I very quickly became addicted to the bandscope display. Being able to see the signals and precisely filter out interference was a revelation to me. If I ever did go back to using an old-style transceiver, I would be sure to add a bandscope.

Initially, while you are setting up your new radio and software, you do need to be able to run the installation program and you might have to download software from the Internet. If you want to run digital modes you will probably have to download a USB driver software and a 'virtual audio cable.'

These topics are covered later in the book. Once the software has been loaded, making the radio work is pretty easy and after that, it is all plain sailing! You just have to learn how to use the radio software.

I have included information on some popular SDR software applications, and some tips which will make using your radio even more enjoyable.

Most black box software defined radios use either a USB or an Ethernet connection. Generation 1 QSD receivers usually need audio cables to your PC soundcard as well as a USB cable for frequency control.

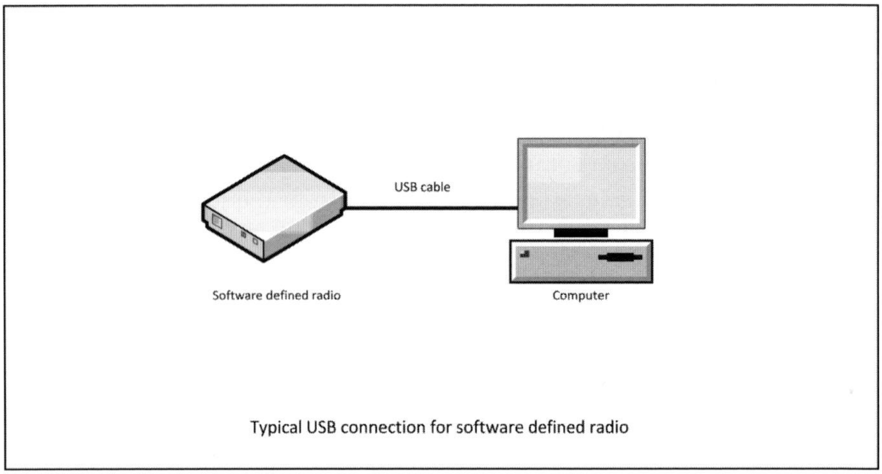


Figure 5: A USB connection to your computer

## USB connections

USB connections are often plug-and-play. You plug them into the USB port on your PC, start up the SDR software application and everything should be fine.

You often have to select a COM port number. The same as you would when configuring a digital modes program. The easiest way to find the correct COM port number is to open **Device Manager** on Windows and select the **Ports (COM & LPT)** section. Look for a COM port that disappears when you disconnect the cable between the radio and the computer and re-appears when you restore the connection. To use some receivers with SDR programs that were initially designed for different SDR hardware, you may have to download a file called extio.dll and copy it into the same directory on your computer as the SDR program files.

Modern receivers and transceivers with knobs are usually connected to your PC with a single USB cable. It carries audio to and from the computer and one or often two 'virtual' com ports. One com port is used for CAT (CI-V) control and the other is used for CW and RTTY keying, or other data. The single cable dispenses with the need for interface boxes containing isolation transformers and optocouplers, RS232 serial cables, and audio cables. You usually have to download and install a USB 'driver' program to make this work. The Yaesu driver suits most Yaesu models and the Icom driver suits most Icom models. You should only have to install the relevant USB driver once.

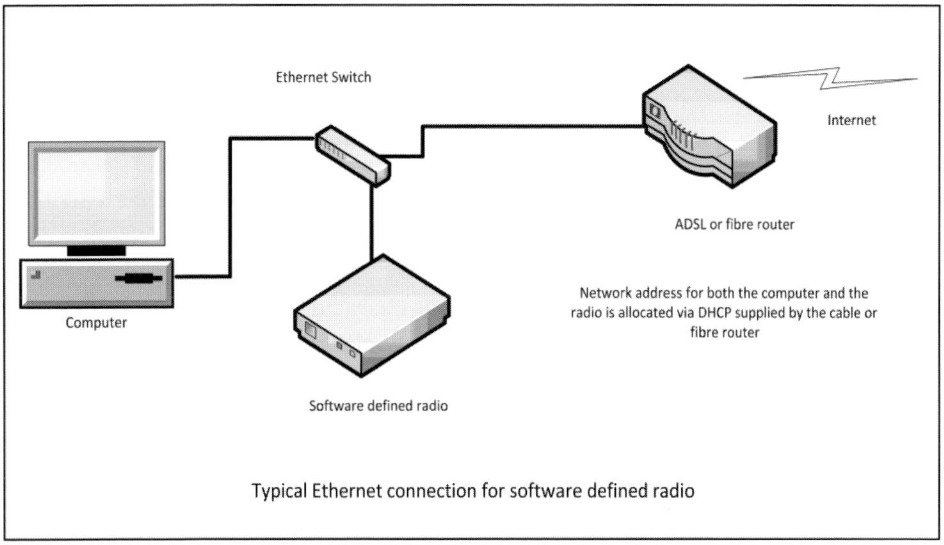


Figure 6: Ethernet connection is easier using a Gigabit Ethernet switch

### Ethernet connections

In most cases, your main Internet router is not located in the room where you have your radio(s). To use an SDR that has an Ethernet interface, you will probably have to buy a CAT5e cable and a Gigabit Ethernet Switch. These typically have four or six Ethernet ports and should cost less than $100. In this configuration, network addresses are allocated by your Internet router using DHCP. Some radios, like the IC-705 can be controlled over your WiFi network. If your PC is using a WiFi connection to your network or it has two Ethernet ports, you can connect the radio directly to your computer. This usually requires a static IP connection, which is more difficult to set up.

Generation one 'soundcard' type SDRs will need a stereo connection to the PC soundcard for the audio IQ signals. A mono 'mic input' won't work. You need two channels, one for the I signal and one for the Q signal. Frequency control of the radio is normally done over a USB cable. Some generation 1 SDRs, like the early SoftRock kits, use a crystal oscillator and don't have variable frequency control.

Some radios that have knobs can be used with the same SDR software running on your computer as the black box versions use. FlexRadios can use SmartSDR, and Apache Labs ANAN radios can use Thetis or piHPSDR. The Icom IC-R8600 receiver can use HDSDR or SDR Console V3, or connect to an external SDR receiver running its normal SDR software. The Xiegu X6100 can use HDSDR or connect to an external SDR running its normal SDR software. This capability gives you the "best of both worlds." You can operate the radio on its own with the normal controls, or from your computer with a large display and often additional features. For example, the FLEX-6600 and FLEX-8600 radios can display two panadapters on the front panel display, but up to four panadapters and four receivers on your PC.

# WHAT IS DIGITAL?

I guess most readers will be familiar with what I mean when I talk about digital and analogue signals. But just to make sure, I will include a quick refresher.

Sound waves that travel through the air and the radio waves that carry our conversations or music out to the world are examples of analogue waveforms. When sound is converted into an electrical signal by the microphone, it becomes an electrical analogue signal. An analogue signal is characterized by having a constantly changing instantaneous level. You can display an electrical analogue voltage signal on an oscilloscope.

*An analogue or analog signal is any continuous signal for which the time varying feature (variable) of the signal is a representation of some other time varying quantity, i.e., analogous to another time varying signal. https://en.wikipedia.org/wiki/Analog signal.*

In the case of the electrical signal from your microphone, the electrical signal is an analogue of the sound waves that the microphone picks up.

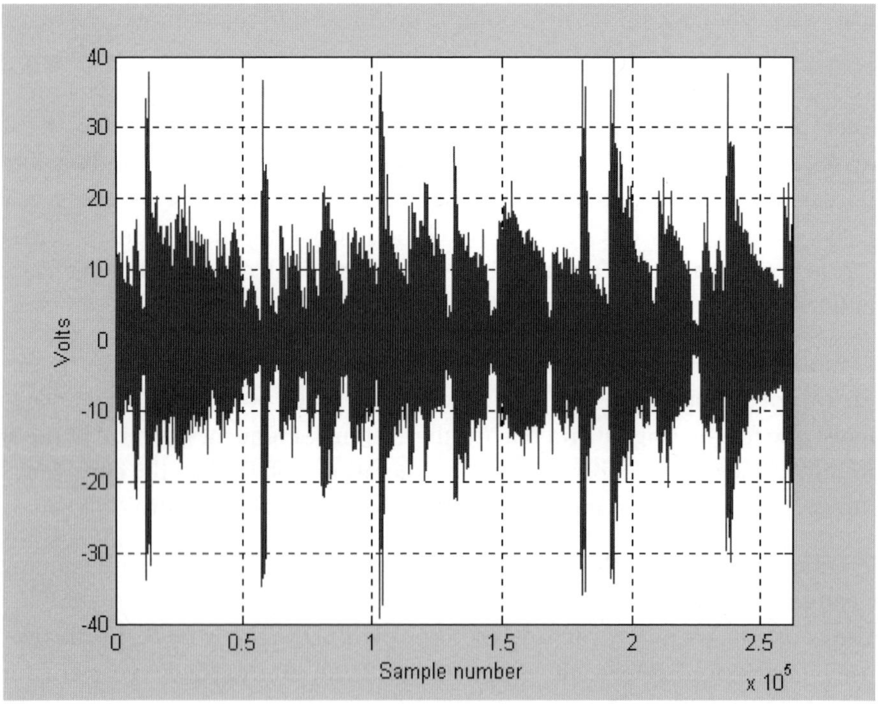


Figure 7: Analogue signal – an Oscilloscope trace of speech from a microphone

Image from www.rocketroberts.com

Digital signals are 'discrete and finite,' meaning there is a limited set of values that they can be. This could mean just two total possible integer values (0 or 1), or it could be a range of 256 or more individual values.

Digital signals are always discrete. That does not mean they are careful about what they say, it means that each digital number is fixed for a known period. Analogue signals vary continuously.

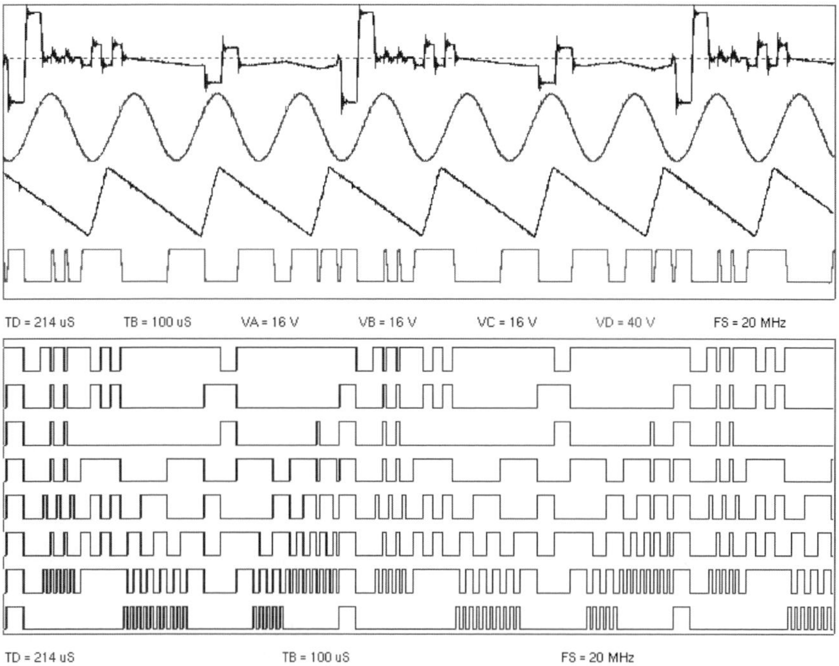


Figure 8: Analogue signals at top – digital signals below [www.bitscope.com]

Humans cannot process digital signals directly. The digital signals must be converted into analogue signals for us to see or hear. Computers can't process analogue signals directly. The analogue signals must be converted into digital numbers for the computer to process.

Complex analogue signals like speech or music are made up of the sum of many frequencies at varying levels. The purest analogue signal is a sine wave, which consists of a single tone at a constant level. This purity is why we often use sine waves as test signals. A square wave is still an analogue signal. It consists of a sine wave plus an infinite number of its odd harmonic frequencies all added together. That is the frequency of the square wave (f) plus 3f, 5f, 7fF, 9f, etc. It is impossible to transmit a completely square wave over a radio system because you would need an infinite radio bandwidth to include all of the harmonics. Systems that transmit digital data streams shape the square waves to limit the transmitted bandwidth.

CW transmitters shape the Morse code signal to reduce the transmitted bandwidth.

Most digital signals are binary which means that only two voltage levels are used. The signal is either a high level known as a 'one' or a low level known as a 'zero.' The actual one and zero voltages depend on the type of circuit and chips being used.

This is known as a 'logic family' and the high and low logic levels are called 'logic states.' Some digital systems use four or more voltage levels but that does not apply to the SDR systems discussed in this book.

Digital signals are used in two different ways. They are used in hardware circuits to perform logic decisions or they are grouped to represent numbers, forming the basis of digital computing.

Hardware digital circuits, using switches, latches, counters, registers, and other CMOS or TTL logic family devices, can be used to control things. You might design a circuit to turn on the lights and open the gate when you arrive home, or to open the gate but not turn on the lights when it is daytime. This type of digital circuit uses a set of logical rules. AND, OR, EXCLUSIVE OR, NAND, and NOR are the most common. Devices like the FPGA (field programmable gate array) are computer chips that can be programmed to connect themselves into complex arrays of these simple logic elements.

Computers also use digital signals and they can also be used to control things. But computers, microprocessors, microcontrollers, and so on, have the added advantage of programmability. You can make them perform operations on the digital signals in predefined sequences. This adaptability makes computers seem like they have intelligence. But so far, they don't. What you actually experience is a result of the intelligence of the hardware designers and the software developers. Fixed logic using non-programmable logic gates cannot be changed without altering the physical wiring or circuit. But, computers can be re-programmed, which means that you can change what the computer does with the digital information. Many digital logic circuits are now constructed using programmable devices to provide the ability to reconfigure the logic without having to redesign the circuit board.

Using only two voltage levels for the data signals makes the information easier for a computer to process. Computers are very stupid. They can only manage data streams and groups of binary bits. They can add binary numbers together, shift their bits left and right using shift registers, compare a number with another number, or store numbers in a memory device and retrieve them later. They can do all of those functions extremely quickly, performing millions of operations a second.

Computers can't use analogue signals directly. They have to turn them into digital numbers before they can be processed. This is done with an analogue to digital converter (ADC). After the computer processing has been performed, the data can be turned back into an analogue signal using a digital to analogue converter (DAC).

A PC soundcard is a good example of this process. It converts analogue audio signals from the line or microphone input to the digital signals used in your PC applications. On the output side, it converts the digital data stream into an analogue audio signal for you to listen to on your amplified speakers.

## CLOCK SIGNALS

Digital signals need to be read at the right time. Reading the data at a time when it is changing from the high state to the low state or vice versa is likely to cause an error in the data or logic system. The circuit must also avoid the possibility of missing a change of level or reading the same level twice. Reading the level of a data signal is called sampling. The best thing to do is to synchronize the system so that the sampling occurs when the data signal is stable at either the high or low logic state. This is achieved by using a clock signal to control the sampling times. The clock signal is a regular string of data pulses. It may be a square wave signal or it may be a string of more narrow pulses. The logic chips are designed so that they read the level of input signals at the time that the clock signal changes, (usually from a high to a low state). Outputs from the logic chips are synchronized as well. The output transition is often triggered by the rising edge of the clock pulse (from a low to a high logic state).

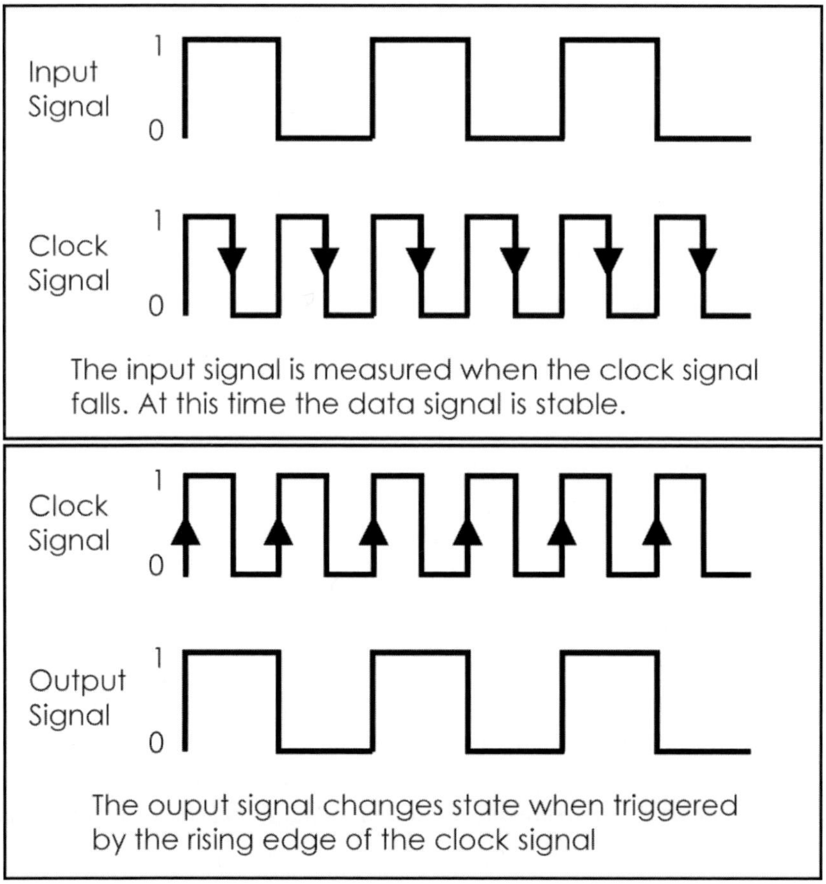


Figure 9: The clock signal controls the timing of a digital network

## USING DIGITAL LOGIC TO REPRESENT NUMBERS

At any particular sample time, the digital input or output of a digital logic hardware device can only be at a 'high' logic level or a 'low' logic level. You can use this information to represent numbers. Normally the 'high' logic level represents a binary number 1 and the low logic level represents a binary number 0. At the next sample time, the system looks at the digital level again. It might have stayed the same or it might have changed state.

Each time the data is sampled in this way the result is referred to as a binary 'bit.' A bit is a logic level at a particular instant in time. A single bit cannot convey very much information, only 0 or 1. But eight bits read together or in order can convey the 256 different combinations between 0000 0000 and 1111 1111. For example, an 8-bit binary sequence of 1000 0001 is equivalent to the decimal number 129.

*Technical TIP: Our normal number system is called the 'base 10' or decimal system. It uses ten numbers from 0 to 9. Binary is a 'base 2' number system. It only uses two numbers, 0 and 1. Binary numbers are easier for computers to handle because the microprocessor and other digital circuits only have to work with two voltage levels. To manage decimal numbers directly the circuitry would need to be able to detect and calculate with ten voltage levels. This is far too complex for computers, so in a computer, the decimal numbers are converted to their binary equivalent.*

Eight bits grouped together is known as a byte. A byte is much more useful than a bit because it can represent any integer between 0 and 255.

If we decide that in some situations the number 65 (0100 0001 in binary) is equivalent to an 'A' and the number 66 (0100 0010 in binary) is equivalent to a 'B' we can start to use the computer to manage text. It goes on from there but this is not a book on computers.

*Technical TIP: You may have noticed that I separated the binary numbers into groups of four. This makes them easier to read. But even grouped, long strings of binary digits are difficult to read and write down. We could convert the numbers to decimal but this is inconvenient and not particularly easy, so another numbering system is commonly used.*

*The problem is that the range from 0000 to 1111 represents the numbers 0 to 15. The binary number 1001 1100 1101 0010 is 40146 in the decimal system. It is not easy to convert a number like that from binary to decimal, or back to binary again. We could break the number down to 9, 12, 13, 2 but that is a bit clumsy as well. The answer is to use a base 16 numbering system called 'Hexadecimal.' This is often shortened to 'Hex.' The Hexadecimal system uses the digits 0 to 9 and A to F to represent the 16 binary numbers between 0000 and 1111. In the example above 1001 1100 1101 0010 becomes 9CD2 in Hex. This is much easier to convert back to binary, so it is the preferred way to write down binary numbers. Of course, Hex is just a notation system for humans. The computer can't work in Hexadecimal it always uses binary.*

Sometimes you need more detail or a bigger number and 8 bits is not enough. For example, an 8-bit analogue to digital converter can only sample the incoming analogue signal at 256 individual levels. If the analogue input signal ranges between 0 and 1 volt, each of the digital output numbers represents an input voltage step of 3.9 millivolts. At the instant that the ADC looks at the input signal, the incoming signal is almost never at exactly one of the sampling steps. So, the ADC outputs the closest number. This results in small sampling errors, which can't be fixed.

The best way to reduce these sampling level errors is to describe smaller voltage steps between each input level and to do that you need more bits. The sampling errors are called 'quantisation' errors because they are errors in measuring the quantity (level) of the signal.

Sixteen bits grouped together is called a word, it is two bytes long. A 16-bit binary word can store 65,536 distinct values. This is enough for a computer to do complex floating-point mathematics. Sometimes 16 bits is still not enough. Modern computers use 64-bit data words. This is done so that the computer can perform calculations on very large amounts of data very quickly.

A 16-bit analogue to digital converter can sample the incoming analogue signal at 65,536 levels. So now, if the analogue signal ranges between 0 and 1 volt, the digital output levels are only 0.015 millivolts apart. The sampling is much more accurate and the sampling error levels are much smaller.

Digital signals are simple in that they only have two voltage levels. But you need a large number of individual bits to accurately, represent the analogue signal levels. The other thing is that the analogue input signals are changing continuously but the digital signals coming out of the analogue to digital converter represent the analogue signal at distinct points in time. Not only is the level represented by each digital number not exactly the same as the original analogue level, but each sample only represents the input level at a particular time. We minimized the voltage sampling errors by using more bits. Now we need to keep the sampling rate fast enough to minimize the time errors. Faster sampling is better, but the faster we sample the analogue signal, the more 16-bit numbers we create each second. This creates more data for our computer or digital signal processing chip to manage.

Harry Nyquist (1889-1976) determined the fundamental rules for analogue to digital conversion. He found that as long as the sample rate was at least twice the rate of the highest frequency in the analogue signal, the original analogue signal could be recreated accurately from the digital data.

This means that to accurately, recreate a radio signal at 50 MHz, we would need to sample at a minimum of 100 Msps, (one hundred million samples per second). Multiply that by 16 bits and we have a data rate of 1,600,000,000 bits per second at the output of the ADC chip. The DSP stage in a direct sampling software defined radio has to work very hard to manage that much continuous data.

In most software defined radio models, the task is made easier by only using parts of the initially sampled data and also by parallel processing. Parallel processing involves splitting tasks so that different parts of the computer can work on them at the same time.

One way to route data from the 16-bit ADC to the computer is to have a wire carrying a continuous 'serial' data stream of 16-bit numbers. Another way is to use 16 'parallel' wires. That way a computer buffer can be presented with the entire 16-bit word at once. The required computer processing rate is much lower using 16 wires. This parallel structure is used inside the computer to speed up operation. For example, the data that is stored in computer memory is routed to the computer processor in parallel. Fast 16-bit ADCs like the one in the example above actually have 16 output lines. The parallel data lines on a computer motherboard that link internal parts of the computer together are called a digital 'bus.' You can think of the bus as a computer data highway with 16 lanes in each direction.

*The data bus in your PC is 32 or 64 bits wide. Some microprocessors and other computing devices like DSP chips may use an 8-bit or 16-bit data bus. Intel introduced the 4004, its first microprocessor in 1971. It was a 4-bit device, which used a 4-bit data bus.*

Conversely, most external connections between computers or between your software defined radio and the computer, are serial connections where the data signal is sent one bit at a time over one wire with an earth return. USB, Ethernet, RS232, and HDMI are examples of serial connections. The old outdated Centronics printer cable is an example of an external parallel connection.

Software defined radios and the DSP stages in conventional receivers convert analogue signals to digital data streams of numbers. The idea is to keep the signal in digital format through as much of the radio as possible. In the receiver, the signal is converted back to analogue just before the audio amplifier and speakers. In the transmitter, the signal stays digital until just before the RF power amplifier and low pass filters.

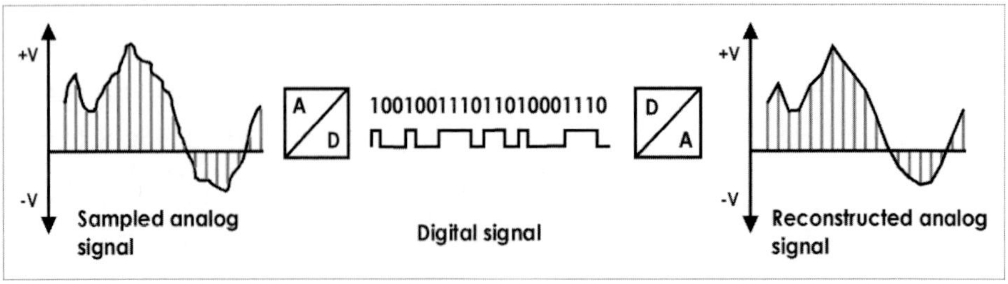


Figure 10: Analogue and digital signals

Understanding that the signals that you receive are converted into strings of numbers and then back into audio for you to listen to is initially an odd concept. But it has been happening in the DSP stage of your old radio for years.

# WHY IS DIGITAL BETTER?

The whole world seems to be going digital. Your cell phone has used digital modulation since the old AMPS networks were phased out in favour of GSM and CDMA networks in the 1990s. Microwave and satellite radio links went digital even earlier, in the 1980s. Those radios are digital but not necessarily software defined. Cassette tapes and vinyl records were made virtually obsolete by digital compact discs. The first commercial music CDs were ABBA's 'The Visitors' released in April 1982 and Billy Joel's '52$^{nd}$ Street', released in October 1982. Soon after that, CDs started to be used to store data. In most places, your television receiver uses digital technology and receives a digital signal as well. All of these things use digital rather than analogue technology. So it must be better. But why?

The first and most important reason is noise. As far as possible, we want to stop any noise being generated inside the radio. All active analogue circuits add noise. Digital signals can be amplified, filtered, shifted in frequency, modulated, or demodulated without adding any extra noise.

The second reason is interference. Conventional hardware mixers and oscillators can mix oscillator phase noise onto your wanted signal and the non-linear mixing process can introduce unwanted intermodulation signals and images. Software DSP, which is performed by manipulating numbers in data buffers, can perform the same functions as the hardware devices without introducing intermodulation distortion and unwanted signals.

The third reason is performance. Narrow hardware filters are difficult to make, and very careful circuit design is required to get good performance. Narrow filters are prone to ringing and they are usually fixed in bandwidth. Digital filters implemented in software can be variable and they don't exhibit ringing. Features like auto notch filters, high and low audio frequency cuts, I.F. width and I.F. shift controls, and band-scopes are common on radios, which use digital signal processing.

Another reason is cost vs performance. It is very difficult and quite expensive to create a top-class conventional receiver, but the DSP part of a software defined radio is just a few computer chips, which can be mass-produced relatively cheaply. Firmware loaded into a DSP chip or a software defined radio's FPGA can be easily updated, improved, and revised. With a hardware radio, the performance is fixed as a part of the initial design. Your 1970s transceiver may have a terrific receiver, but it cannot be improved further without adding more hardware such as INRAD filters.

Some early DSP filters did introduce a variety of weird-sounding artefacts, but the technology is more developed now. You might be able to adjust DSP filters to the point where the wanted audio becomes distorted, but you don't have to. It can be a trade-off. In some cases, a small amount of audio distortion is still far better than listening to the noise or interference that the filter has managed to eliminate.

With an analogue signal, virtually everything that you do to the signal adds noise, and once it is in there it is difficult to remove without adversely affecting the quality of the signal. In a typical radio system comprising a transmitter, a radio path, and a receiver, there are many places where unwanted noise becomes added to the signal that you want to hear at the far end of the path. The big problem is active devices like amplifiers, oscillators, and mixers, but even passive devices like resistors add thermal noise. With digital signals, so long as it is possible to determine whether each bit is a zero, or a one, even a noise-damaged signal can be re-created perfectly with no noise added.

Audiophiles point accusingly at the quantisation noise that is a byproduct of any analogue to digital conversion process, but this is very small compared to the noise created elsewhere in your stereo equipment. They say that the audio from digital signals sounds harsh, but I expect that this has more to do with the hugely improved dynamic range and frequency response of audio from a CD compared to the audio from an LP record or an analogue tape. Quantisation noise does add noise to the signal. It is a result of the errors that occur when you sample an analogue signal with an analogue to digital converter. The more bits that are used to sample a signal the smaller the steps in signal level. So sampling with more bits results in lower quantisation noise. There is more about this topic in the ADC chapter. After the signal has been converted to a digital stream of data bits, the DSP firmware code can modify the signal without adding any additional noise.

Getting back to radio systems, the optimal system would be a digital up-conversion transmitter using a digital modulation scheme. At the receiver, a digital down conversion radio would receive and decode the digital audio transmission. The only analogue parts would be the microphone and its preamplifier plus the audio amplifier and speakers in the receiver. Well actually, the RF radio signal is usually amplified with a standard analogue power amplifier followed by a low pass filter network and this does inevitably add some distortion. Also, there may be some filters and a low noise preamp before the analogue to digital conversion in the receiver. In a VHF or UHF receiver, there is often a tuner chip or a hardware mixer and oscillator as well.

Reducing the amount of noise and distortion being created inside the radio is a major advantage. The primary aim of software defined radio is to keep the signal in a digital format for as long as possible in the transmitter and sample it back into a digital signal as soon as possible in the receiver. This minimizes the number of stages and processes performed on analogue signals. Digital signal processing is at the core of software defined radio.

Of course, there is still noise being added as the signal traverses the radio path between the transmitter and your receiver. This can be overcome by using digital modulation, but there is a trade-off because digital transmissions require a higher received signal level.

Most amateur radio and a lot of shortwave signals still use analogue modulation modes like; SSB, SSTV, AM, or FM. But we also use digital modulation for digital voice modes like D-Star, C4M (Fusion), DMR, P25, and NXDN. Shortwave broadcast radio stations are migrating towards using digital modulation, often DAB or DRM. Point-to-point radio links such as satellite links, microwave radio links, cellular, and even WiFi all use digital modulation, usually variants of QAM, which is a modulation type that uses both phase and amplitude modulation to send digital data signals. Morse code is a binary digital mode. Yes, it really is! When the key is down the carrier is on. You can consider this as equivalent to sending a digital 'one' signal. When the key is up the carrier is off. This is equivalent to sending a digital 'zero' signal. It is a good example of what I am talking about. We all know that you can copy and decode (in your head) very weak and noisy CW signals. As long as you can hear a tone or no tone you can decode the message with no errors. It is the same with other digital transmissions. If the signal level or phase can be detected accurately, the transmitted signal can be recreated with no noise or distortion.

*Technical TIP: I have noted some confusion in online forums about what is digital and what is analogue. I have seen comments stating that various components like inductors in the low pass filter were "analogue." This is rubbish! Individual components are neither analogue nor digital. There is no such thing as a digital capacitor or an analogue inductor. If you don't believe me, I have some digital resistors I can sell you, cheap! Some circuits are designed to manage digital signals and others are designed to manage analogue signals. Some, like the analogue to digital converter, manage both. An RF carrier signal is analogue but the modulation imposed on it may be digital.*

In a conventional superheterodyne HF radio, the received radio signal is converted to an I.F. signal by mixing the incoming signal with a signal from an oscillator. In many receivers, this happens twice (double conversion) or three times (triple conversion). In most transmitters, the modulated signal is mixed up to the final transmitted frequency in the same way. Mixers are non-linear devices and they create all sorts of unwanted signals. These are often noted as 'birdies' or whistles at some frequencies. Noise and any frequency variation on the oscillators get mixed into the wanted signals and degrade the receiver's performance. In a direct digital sampling HF software defined radio, the frequency conversions are performed using mathematics on the digital representation of the incoming signal. You don't get any unwanted noise or mixing products. This is the main reason that cheap SDRs can produce audio quality that is as good as the audio produced by expensive conventional receivers.

# DEFINITION OF SDR

The definition of 'software defined radio' is a bit vague and is open to interpretation. These days it is usually used to describe radios, which use analogue to digital converters to sample the radio spectrum as close as possible to the antenna input. The signal remains in the 'digital domain' through all of the digital signal processing and is not converted back into an analogue signal until just before it is passed to the audio amplifier and speaker. In the transmitter, the reverse applies. The microphone or modulating signal is converted into a digital signal, processed with DSP, and not converted back to an analogue signal until it is sent to the RF power amplifier. The panadapter spectrum and waterfall displays are representations of the analogue signals being received by the radio, but they are just data displayed as dots on a display screen.

The ITU definition is not adequate. In fact, in my opinion, it describes any radio with any kind of computer control, including CAT (Icom CI-V).

"Software-defined radio (SDR): A radio transmitter and/or receiver employing a technology that allows the RF operating parameters including, but not limited to, frequency range, modulation type, or output power to be set or altered by software, excluding changes to operating parameters which occur during the normal pre-installed and predetermined operation of a radio according to a system specification or standard." ITU-R SM.2152 (09/2009).

The FCC manages the radio spectrum in the USA. It is primarily interested in devices that emit signals on radio frequencies. Their definition says software defined radios must always include a transmitter. There is no FCC definition for software defined radio receivers.

"Software defined radio. A radio that includes a transmitter in which the operating parameters of frequency range, modulation type, or maximum output power (either radiated or conducted), or the circumstances under which the transmitter operates in accordance with Commission rules, can be altered by making a change in software without making any changes to hardware components that affect the radio frequency emissions". Spectrum Use Employing Cognitive Radio Technologies, ET Docket No. 03-108, Report and Order, 20 FCC Rcd 5486 (2005).

I prefer a definition that I found on Wikipedia.

"Software-defined radio (SDR) is a radio communication system where components that have been typically implemented in hardware (e.g. mixers, filters, amplifiers, modulators/demodulators, detectors, etc.) are instead implemented by means of software on a personal computer or embedded system."

From, 'Software Defined Radio: Architectures, Systems, and Functions (Markus Dillinger, Kambiz Madani, Nancy Alonistioti).

See http://en.wikipedia.org/wiki/Software-defined_radio.

The definitions vary, but in general, terms, if the filtering and demodulation in a receiver, or the filtering and modulation in a transmitter, is done using software that is re-configurable without replacing chips in the radio, then the radio is 'Software Defined.' In other words, if the signal is converted to a digital signal and processed using reconfigurable software or firmware running on a PC, embedded computer processor, programmable DSP chip, or FPGA (field programmable gate array) it is a software defined radio. But if the signal is converted to a digital signal and processed in a dedicated (not user reconfigurable in the radio) DSP chip it is not a software defined radio.

Even that definition is not adequate anymore because the lines are blurred. Many recent conventional radios can have firmware updates, making them at least partially field upgradable. Often, the firmware running on DSP chips can be updated without removing the chips from the radio, so in some respects, those radios could be considered 'software defined.' But most folks don't refer to them as software defined radios. On the 'flip side', radios like the Elecraft KX3 and the new FlexRadio Signature 8000 series are usually considered to be SDRs even though they use dedicated DSP chips and a microcomputer and/or FPGA inside the radio to perform the modulation, demodulation, and other DSP functions.

These days what is, and what is not, an SDR is more usually defined by the hardware architecture. Conventional radios typically use a double or triple conversion superheterodyne design with hardware oscillators, mixers, and I.F. amplifiers, usually followed by a dedicated DSP chip. Software defined radios use direct digital sampling, or quadrature sampling detectors and exciters. Computer code replaces the hardware mixers, oscillators, filters, and amplifiers.

*Technical TIP: Some FPGA chips include a microcomputer, which runs on the FPGA chip hardware as a software emulation of a microprocessor. These are called 'soft processors.' Some FPGA chips are large enough to host several soft processors.*

*DSP chips are specialized computer processors that are based on the same type of structure as microprocessors. Some dedicated DSP chips include ARM or multicore processors. You could program these chips to manage the control and display aspects of the radio. That could include a bandscope and real knobs and switches.*

*By the way, the name of the ARM processor is an example of an acronym inside an acronym! ARM stands for 'Advanced RISC Machine.' RISC stands for 'Reduced Instruction Set Computer' or 'Reduced Instruction Set Computing.' An 'instruction set' is the list of commands or instructions that the microprocessor can perform. Computer programs ask the microprocessor to perform these instructions in a particular order. RISC computers are often used in embedded systems where the full complex instruction set that is required for your PC is not needed. Simpler instructions mean more instructions can be performed per clock cycle, so RISC computers are faster. The microprocessor inside your PC is a CISC (Complex Instruction Set Computing) device.*

To add a little more confusion to the story, some manufacturers claim that their radio is an SDR even though it uses a completely standard receiver and transmitter architecture. These radios have double conversion superheterodyne receivers, with an audio frequency IQ signal output from the DSP stage available on the back panel. You can use SDR programs to display a bandscope of the signals that are within the I.F. passband. But the radio itself is not a software defined radio. I think it is best to use the new term 'Superheterodyne + SDR' to describe this kind of receiver.

It's not even safe to assume that a radio with no front panel controls that are controlled using PC software is a software defined radio. Some WiNRADiO and Icom black box receivers are standard configuration superheterodyne receivers, which use PC control, instead of having knobs and buttons.

Finally, radios like your new handheld portable or mobile, may have a radio chip inside and be completely digital. They may use IQ streams and DSP internally. But unless the software code that performs the modulation, demodulation, and DSP filtering can be updated by the user without removing the chips from the radio, they are digital radios, but not software defined radios.

SDR is a logical extension of the DSP that has been inside our radios for years. I laugh at the online 'grizzlers' who say that they would never touch an SDR because they don't like the whole "digital thing." They are probably using a radio with I.F. or AF DSP and they probably have no idea that their radios already contain digital to analogue and analogue to digital converters. I suspect that they prefer to have their CDs 12 inches wide with the audio scratched on with a nail. I know that there are some 'die-hards' who use boat anchor radios from the '50s, '60s, '70s, or '80s and that is absolutely fine. It is great to collect and use antiques. Some of the early models have excellent receivers. There is no doubt about that. In the end, the definition of software defined radio is unimportant. What is important is that we have a new breed of radio with new features, including:

- Very wide frequency coverage compared to older technologies. Some SDRs cover from the LF to SHF bands.
- The ability to see signals as well as hear them and the ability to monitor several bands at the same time.
- Very good receiver performance, often at a fairly low cost.
- Very sharp, constantly variable, receiver passband filters which don't introduce ringing distortion.
- New techniques for noise reduction when receiving, and new techniques for improving transmitter intermodulation performance.

# GENERATIONS AND TYPES OF SDR

It is possible to buy software defined radios for frequencies ranging from 10 kHz up to 6 GHz or more. SDR receiver designs aim to take a range of frequencies, for example, the entire HF band, and convert the whole range into a digital data stream. After that, digital signal processing either on the receiver board or in computer software is used to provide the unique displays and functions that make operating a software defined radio exciting.

The latest SDR radio receivers for the HF bands use digital down conversion (DDC) where the RF band is sampled using an analogue to digital converter close to the antenna, normally immediately after a preamplifier and bandpass or low pass filter. In the case of UHF, or SHF radios, the analogue to digital converter will be placed after a hardware mixer and oscillator acting as a down-conversion stage. Other SDR radio receiver designs use a quadrature sampling detector (QSD) followed by an audio frequency analogue to digital converter.

Most SDR transmitters use the same kind of arrangement as the receiver. They use digital up-conversion (DUC) if the receiver uses digital down-conversion, or a quadrature sampling exciter (QSE) if the receiver uses a quadrature sampling detector. It is possible to use an SDR receiver in conjunction with a conventional transmitter. Some people use a black box or dongle SDR receiver to add a bandscope function to their conventional transceiver or receiver. This can be achieved by feeding the SDR receiver from an I.F. output. Some transceivers provide an I.F. output on the rear panel, especially for this purpose.

There is a strong trend towards moving the digital signal processing functions into dedicated DSP chips, or a dedicated computer inside the radio. This simplifies the SDR software running on an external PC and more importantly, it dramatically reduces the demands on the data connection between the radio and the PC. The FlexRadio Signature 6000 and 8000 series of radios already use this type of design, employing DSP chips and computer processing inside the radio. The Xiegu X6100 and similar QRP transceivers use a built-in ARM processor for the control and display functions and an STM32 microcontroller for the SDR and DSP functions. Apache Labs and the OpenHPSDR group, of volunteer SDR developers, have integrated a Raspberry Pi4 compute module into the latest ANAN G2 radio. The earlier Andromeda model had a built-in Windows PC. So does the Expert MB1.

Using computer power inside the radio to perform the DSP functions means that only the data for the audio from each receiver and the graphics required to display the panadapters have to be sent to the PC. The PC does not need to be as powerful, opening the way for SDR operation via Netbook computers, tablets, and mobile phones. The lower continuous data requirement also makes it easier to operate your SDR transceiver remotely over your local WiFi network or the Internet.

QSD based software defined radios are usually cheaper and they are fine for getting a taste of SDR or as a bandscope. But you only get part of the full SDR experience because you only get one panadapter. If you want to get the most out of using a software defined radio, there is no substitute for a direct digital or hybrid radio.

Some people are very happy using an SDR receiver in conjunction with a conventional transmitter. CAT control manages the tracking of frequencies. This arrangement gives you the advantages of an SDR panadapter and the comfort of using a familiar transmitter.

Shortwave listeners and anyone who only wants a receiver should look for a hybrid or direct sampling receiver, which covers the bands of interest. Conventional HF receivers often have a maximum frequency of around 32 MHz with some having VHF coverage as well. But many software defined radios have a much wider frequency coverage. The VHF and higher amateur radio bands are several MHz wide. To display wide sections of the VHF and UHF bands, you need a receiver capable of displaying a wide panadapter.

Some of the 'RTL dongle' receivers work up to around 2200 MHz, but they do not cover the HF bands. Be aware that using a receiver covering 60 MHz to 6 GHz may not be particularly interesting compared to an HF receiver. But they can be used for listening to aircraft traffic, beacons, and maritime AIS signals, looking at satellite weather maps, doing weak signal VHF or UHF work, and EME or satellite reception.

Some of the new software defined radios; for example, the Malahit DSP2, Icom IC-R8600, FUNcube Dongle, BladeRF, and HackRF radios are specifically designed for VHF to SHF (microwave) coverage. The RTL dongles vary, from a maximum frequency of 948 MHz up to a maximum of 2200 MHz, with most models working down to around 24 MHz. The FUNcube Dongle Pro Plus is a little different in that it covers bands from low frequency (LF) 150 kHz up to super high frequency (SHF) 2050 MHz, (with a gap between about 260 and 410 MHz). The SDRplay receivers cover a frequency range of 1 kHz to 2000 MHz.

Software defined radios are not ideal for monitoring local repeaters. Often the panadapter display is too narrow to cover all of the local repeaters at the same time. Receivers with a wide panadapter are better because they can display more channels across the panadapter. Software like SDR Console V3 supports several receivers across the displayed panadapter bandwidth.

The following table describes my interpretation of the definition of each SDR generation. Other interpretations probably exist, but these definitions make sense to me.

| TABLE 2: SDR GENERATIONS | |
|---|---|
| ***Generation*** | ***Definition*** |
| Generation 1 SDR | Software defined radio, usually employing a QSD (quadrature sampling detector) such as a Tayloe detector. It uses the soundcard in a PC to perform the analogue to digital conversion of the two audio streams from the QSD detector. Control, DSP, and demodulation are performed by PC software.<br>If the device is capable of transmitting, it will probably employ a QSE (quadrature sampling exciter) such as a Tayloe detector and will use the soundcard in a PC to perform the digital to analogue conversion.<br>These radios were originally single-frequency models with crystal oscillators. Later versions used SiLabs variable oscillators. |
| Generation 2 SDR | Software defined radio, usually employing a QSD (quadrature sampling detector) with onboard ADC to perform the analogue to digital conversion of the two audio streams from the QSD detector.<br>Control, display, DSP, and demodulation are performed by PC software.<br>If the device is capable of transmitting, it will probably employ a QSE (quadrature sampling exciter) such as a Tayloe detector and use an onboard DAC (digital to analogue converter) to perform the digital to analogue conversion. |
| Generation 3 SDR | As above, with the DSP and demodulation being performed by chips or a computer inside the radio.<br>Includes SDRs "with knobs" which can work as stand-alone radios without connection to a PC like the Elecraft KX3 and Xiegu X6100. |
| Generation 4A SDR | Direct digital sampling SDR, using DDC (digital down conversion) in the receiver and DUC (digital up-conversion) in the transmitter. The ADC and DAC are inside the radio and an FPGA may be used to perform decimation and filtering to limit the data bandwidth requirement to the DSP stage, which may be inside the radio, or running on SDR software on a PC or SBC. Includes FT-710, IC-7300, IC-7610, IC-705, Flex Radio, ANAN, Expert Systems. |
| Generation 4B SDR | A hybrid with one or more superheterodyne mixer, oscillator, and amplifier stages down-converting to a SDR receiver. DSP is usually on-board using a dedicated DSP chip or a microcomputer. May have knobs and a bandscope or use external controls and bandscope or panadapter. Often have a very wide frequency range. The transmitter has DSP on board, feeding a superheterodyne stage and then the RF amplifier.<br>Includes any non-QSD SDR that works above the VHF range, including RTL dongles, Funcube dongles, SDRplay RSP, IC-R8600, IC-905, and HF radios like the FTdx10 and FTdx101. |

## CONVENTIONAL RECEIVERS

Conventional receivers are usually double or triple-conversion superheterodyne designs, with hardware oscillators and mixers. They have I.F. amplifiers and some sort of AGC circuit. They often have DSP stages after the final mixer and I.F. amplifier. The I.F. signal is converted from analogue to digital and processed within the DSP chip. DSP functions usually include noise and notch filtering, I.F. width and shift control, and often a bandscope. Demodulation may be performed by the DSP, or the signal may be converted back to an analogue signal and passed through a hardware demodulator. This might be followed by an AF DSP stage, which shapes the audio response and provides noise filters. Even if the DSP firmware can be updated by downloading a file and sending it to the radio, we don't consider these radios to be 'software defined.'

## CONVENTIONAL TRANSMITTERS

Conventional transmitters often use the same local oscillator as the receiver. The microphone, other audio, or CW signal is converted into a digital signal, shaped and filtered using DSP processes, and then converted back to an analogue signal. It is amplified and mixed up to the desired RF frequency. Then it is amplified up to the final output power, passed through a low pass filter to remove harmonic frequencies, and sent out to the antenna port. An older transmitter does the same but without the DSP stage. The modulation may be performed by a hardware modulator stage or by the DSP stage.

## SDR RECEIVERS

HF software defined radio receivers are usually based on one of two common architectures. Generation one, two, and three SDR receivers are QSD designs that use a Tayloe detector as a direct conversion mixer to create I and Q audio streams that are passed on to two analogue to digital converters. Generation 4 SDR receivers use digital down conversion. The RF signal is sampled with a high-speed analogue to digital converter and then the digital signal is split into I and Q signals. This is often done in an FPGA, but it can be done in a dedicated receiver chip, by an embedded microprocessor, or a DSP chip.

Hybrid SDRs and receivers working at VHF and higher frequencies use a slightly different design. Typically a hardware oscillator and a mixer are used to mix the RF signal down to an I.F. channel and then a digital down conversion SDR performs the analogue to digital conversion and DSP functions. Some of these radios use a dedicated receiver chip rather than an FPGA.

The receiver tuner chips have internal hardware oscillators and mixers followed by an analogue to digital conversion. A variety of methods can be used to produce the IQ data output streams for the DSP stage. The RTL dongles use a Realtek 2832U tuner chip, and the Afedri receivers use a Burr-Brown (Texas Instruments) chip.

Other radios including the FCD pro+, and the SDRplay receivers use the Mirics MSi001 tuner chip. The HackRF board uses a Xilinx CPLD and an ARM Cortex processor. The BladeRF board uses an ARM9 general-purpose processor and an FPGA.

*Technical TIP: A CPLD is a Complex Programmable Logic Device. It is a field-programmable logic device like an FPGA. CPLD chips have fewer programmable gates than FPGA chips and their internal structure is different. The main practical difference apart from size is that a CPLD has non-volatile memory so it is ready to go as soon as power is applied. An FPGA does not usually have any non-volatile memory. It needs a program stored in a separate EPROM, which is loaded when the device is powered up. A big CPLD will have about the same number of programmable logic gates as a very small FPGA.*

## SDR TRANSMITTERS

There are not very many commercial generation one or generation two transmitters and almost all of them are QRP low-power radios. The SoftRock Ensemble, and the SDR Cube spring to mind. They use a Tayloe detector in reverse as a quadrature sampling exciter (QSE). The digital (IQ) signals carry the modulated audio frequency signal. The QSE mixes the audio up to the final RF frequency, which is amplified and passed through low pass harmonic filters to the antenna. Generation three SDR transmitters use a DSP followed by onboard digital to analogue conversion and a QSE to mixt the I & Q audio signals up to the RF frequency. Often these are software defined radios with knobs like the Elecraft KX3.

Then there are generation four, digital up-conversion (DUC) transmitters. The DSP stage creates a modulated digital IQ signal, which is up-converted to the final frequency as a digital signal and then converted to an analogue signal using a fast digital to analogue converter. This is followed by a power amplifier and a low pass filter. VHF transmitters might be straight digital up-conversion designs or hybrids.

UHF and SHF transmitters are usually similar to the equivalent receivers. A direct up-conversion SDR creates an I.F. frequency, which is mixed up to the final RF frequency by a hardware mixer and oscillator. The RF power amplifier and low pass filters are the same as in conventional radios.

## COMMERCIAL (NON-AMATEUR) SDRS

Software defined radio is well established in the commercial world. SDR technology is very common in cellular base stations and military radios. It is rapidly becoming common for mobile and hand-held radios as well. These types of radios are probably better described as 'digital' rather than software defined because the internal firmware is not usually user upgradeable. Many of the manufacturers of software defined radios for the amateur radio and hobby market also make radios for the commercial and military market.

## SUPERHETERODYNE + SDR RADIOS

There used to be a few manufacturers that used terms like 'Superheterodyne + SDR.' Most of them have faded away now or they are making genuine SDRs rather than trying to rebrand their radios as being a part of the trendy new technology. You did get some SDR functionality, even if it was only a bandscope or an IQ output from the DSP stage to use with SDR software on your PC. Some examples of + SDR radios are the Elecraft K3S, the Alinco DX SR9T/E, and the AOR AR-2300 / AR-5001D receivers. They all output IQ streams at audio frequencies that can be used for a bandscope display. But, they are superheterodyne radios with no internal SDR architecture.

## HYBRID RECEIVER

The Icom IC-R8600 is a good example of a modern SDR hybrid. A block diagram of the receiver is shown on the next page. Up to 30 MHz it uses direct digital sampling.

From 30 MHz to 500 MHz. A conventional oscillator and mixer stage converts the incoming frequencies up to a 778.700 MHz first I.F. Then a second oscillator and mixer converts the signals down to a 46.35 MHz second I.F. with a bandwidth of 10 MHz. That 10 MHz wide band is sent to the direct sampling HF SDR stage.

From 500 MHz to 1100 MHz the architecture is the same double conversion superheterodyne, except the first I.F. is at 278.700 MHz. The 2$^{nd}$ I.F. is at 46.35 MHz.

The radio uses triple conversion from 1100 to 1500 MHz. A 2000 MHz fixed oscillator converts the received band down to 900 to 500 MHz. Then that band is processed using the double conversion system described above. The 2$^{nd}$ I.F. is at 278.700 MHz, and the 3$^{rd}$ I.F. is at 46.35 MHz.

Finally, from 1500 to 3000 MHz, the same triple conversion method is used. Fixed oscillators at 2.5 GHz, 3 GHz, and 3.5 GHz convert each 500 MHz band segment down to 1000 to 500 MHz. Then the resulting 1000 to 500 MHz band is processed using the double conversion system described earlier. Again, the 2$^{nd}$ I.F. is at 278.700 MHz, and the 3$^{rd}$ I.F. is at 46.35 MHz.

To reduce image responses and SDR alias effects, 9 bandpass and 3 low pass filters are provided for HF bands, and 13 bandpass filters for the VHF and UHF bands.

The analogue to digital converter (ADC) samples at 122.88 Msps (million samples per second). A field programmable gate array (FPGA) converts the data samples to the frequency domain, for display on the spectrum and waterfall (band scope) display. It also performs digital down conversion to reduce the data rate to something more usable, and probably some DSP filtering such as noise blanking which is usually done early in the receiver chain. The radio has separate DACs for the audio, USB output, metering, and audio scope (panel). The FPGA also does the FFT for the spectrum and waterfall display.

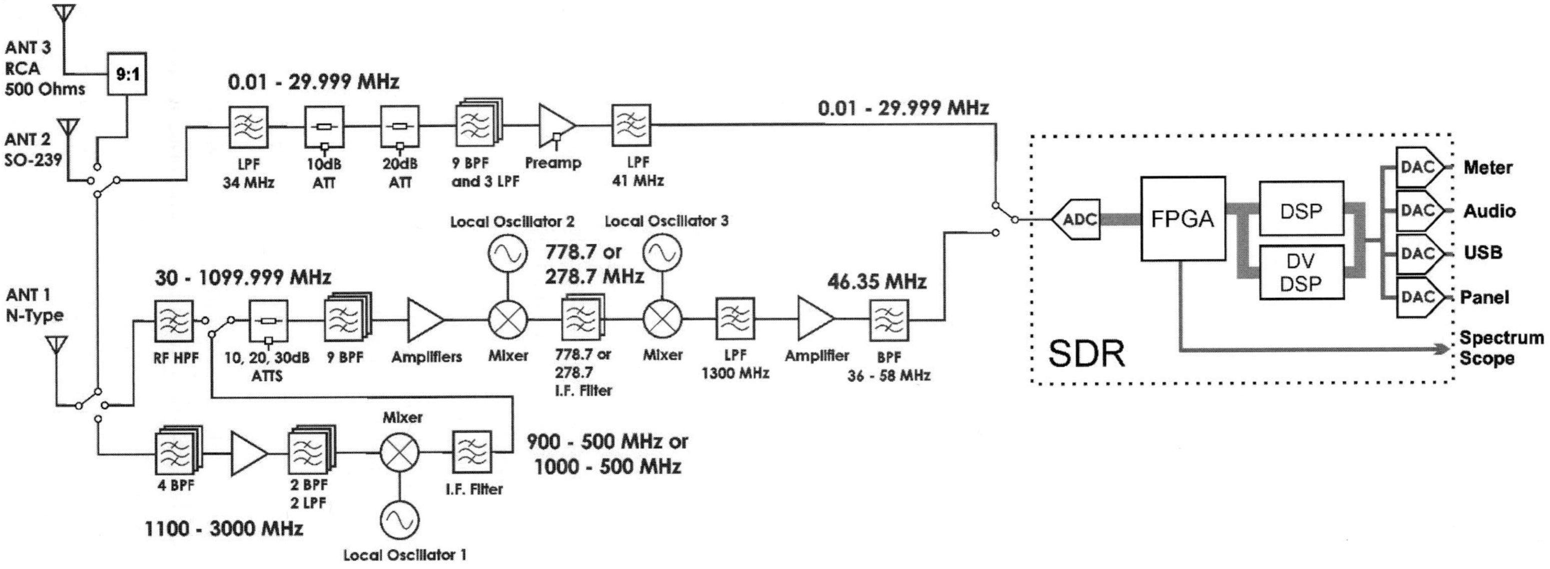


Figure 11: The Icom IC-R8600 receiver

# ARE SDRS BETTER?

Are SDRs better? The answer is "yes... and no." There is no doubt that the ergonomics, build quality, and 'feel' of using a top-end conventional transceiver or receiver have an undeniable attraction, and many hams love their beautifully crafted Collins or other "boat anchor" equipment. While SDR receivers and transceivers have amazing features, excellent audio quality, and superb performance, using them sometimes lacks the ambience of the older equipment. They work very well and can do amazing things but they don't warm the shack and "glow in the dark." Well actually, the computer monitor does glow in the dark, but you know what I mean. When it comes down to what you can do and actual on-air capability, I believe that SDRs have the edge. It is interesting that the last of the conventional radios still available new, have SDR features like a touchscreen and computer mouse control. The Yaesu FT-991A, Kenwood TS-890, and TS-990 are all superheterodyne radios with a bandscope.

As with most things, the more expensive the radio is, the more you can expect in the way of features and performance. Top-end conventional radios are better than cheap software defined radios. But a cheap SDR may be able to do some things that the expensive transceiver can't. Dollar for dollar, I believe an SDR will outperform a conventional architecture radio on some key tests. On the other hand, there are other tests where a conventional radio may be better. For example, a radio with a 50-volt power amplifier might have better transmitter intermodulation performance than a 13.8-volt SDR transmitter. One sure thing is that SDR technology is undergoing constant development. If you own an SDR, you can take advantage of the new developments simply by updating the software or firmware. A conventional radio gets left behind when each new model is released. I was interested to see just how poorly some of the older elite-class transceivers like the FT-1000MP compare with the new radios produced by the same manufacturer. It's not even in my top 20, and places 106$^{th}$ on the Sherwood list.

*Technical TIP: The new 'top end' SDR transceivers like the Elecraft K4, Icom IC-7610, and FTdx101 have significantly better lab-tested performance figures than 'elite' and 'contest' grade radios from the early 2000s. Recent mid-range radios like the FTdx10 and the IC-7300 are better as well.*

There are two ways to measure whether software defined radios are better than conventional ones. You can measure the technical performance, using test equipment and on-air evaluations or you can compare the way they operate, the things they can do, and whether you like the new features. In the end, it comes down to; **"do you like using this new radio?"**

Online reviews and videos add some value, but they are very subjective. People tend to like the brands they are familiar with and they almost always strongly approve of a radio in which they have just invested a lot of their hard-earned cash.

Some people will turn a blind eye to truly appalling equipment out of pure brand loyalty. However, cumulative online reviews do reveal equipment that has poor reliability or customer support, which can be valuable information when you are deciding on what radio to buy.

Software defined radios use a completely different technology from that used in conventional superheterodyne receivers and transceivers. You need some different tests to measure their performance. Measuring SDR performance using the old tests that were developed for conventional radio designs can produce unreliable results. Despite this, SDRs usually score very well in the lab tests published by ham radio magazines. The new methods used to test SDRs are interesting, so I have included a section on performance testing later in the book.

## SDR FEATURES

Software defined radios have some great features that are unavailable or less well implemented on conventional radios.

1. Some SDRs can monitor several bands at the same time. The latest black box HF SDR transceivers can place as many as four panadapters on one or sometimes multiple computer screens.
2. You can see the signals on the panadapter, which lets you precisely adjust the receiver bandwidth making sure you are not adding band noise to the audio, or limiting the fidelity of the transmitted signal.
3. SDR filters created in DSP software are exceptionally sharp and they don't 'ring' distorting the received audio. They are also fully variable.
4. Once the signal has been converted to digital, it does not suffer from noise addition within the receiver stages. A purely analogue receiver adds noise and intermodulation distortion at each mixer and amplifier stage.
5. SDR receivers are less affected by other signals on the band. The receiver's intermodulation performance may actually be improved when there are other signals on the band. The receiver two tone intermodulation distortion, blocking dynamic range, and RMDR (reciprocal mixing dynamic range) test results stay the same when measured close to the wanted signal. With superheterodyne receivers, the performance degrades when tested close to the wanted signal.
6. You don't need a heap of cables to connect up the radio for digital modes. Modern transceivers, both conventional and SDR, use a single USB cable to carry CAT (CI-V) control and audio signals. With a black box radio, your logging and digital mode software can run on the same PC as the SDR software.

7. Changing the SDR software or firmware can add new features keeping the radio up to date. While you can update the DSP firmware on conventional radios, they still tend to get out of date over time.

8. Using the latest technology is a cool part of being a radio hobbyist.

## CONVENTIONAL (NON-SDR) RADIO FEATURES

1. Knobs and buttons are tactile and nice to play with. You get to tune across the band using the big VFO knob. It feels nice. If this is important to you, buy an SDR with knobs and a bandscope.

2. Non-SDR radios are better for mobile operation because you don't need to look at a screen. SDRs are very visual devices. That is not useful, and potentially dangerous when you are operating mobile. Also, it is inconvenient if you need a PC as well as a radio just to trigger the local 2m repeater.

3. People used to say that non-SDR radios are better for CW. This used to be true, but not anymore. The latest SDR transceivers create CW by keying the data signal in the FPGA or microcontroller in the radio, rather than by generating a tone at the start of the transmit chain. This provides low latency and very fast QSK operation.

## THE TOP 20 RECEIVERS AND TRANSCEIVERS

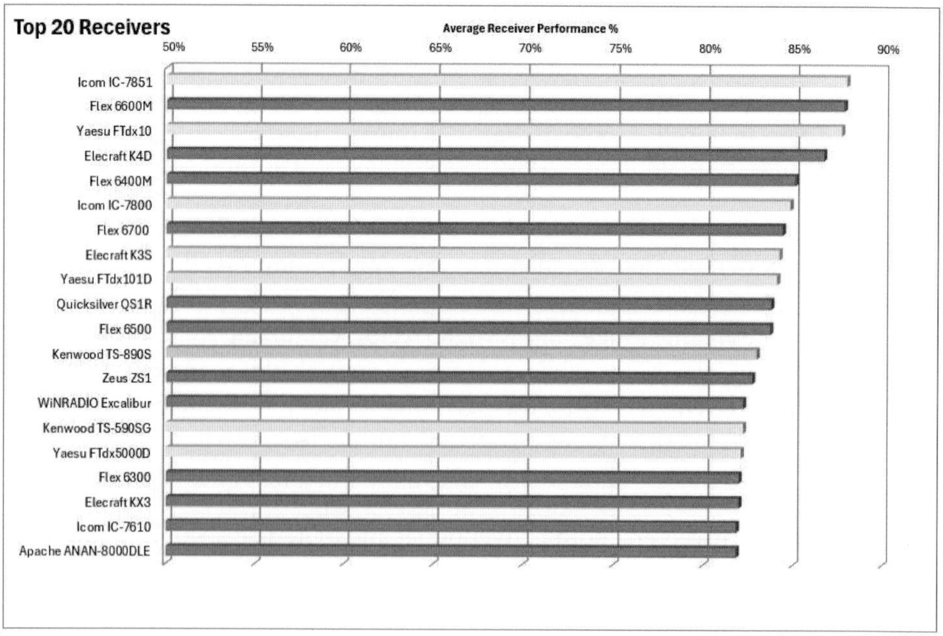


Figure 12: Andrew's list of the Top 20 receivers

The top 20 charts are based on a wide range of receiver and transceiver performance test data, mostly from ARRL QST reviews. In my opinion, they represent a balanced survey of performance, but it is only my personal ranking. As new radios are tested, more data will be recorded and the coveted top spots may change. These charts and a handy radio comparison tool are available at https://rtbooks.co.nz/Test/ and https://rtbooks.co.nz/Top20/.

Other excellent performance resources include the Sherwood Engineering tables by Rob Sherwood NC0B, reviews by Adam Farson VA7OJ/AB4OJ (SK), the comparison of CW keying spectra and phase noise published by Jim Brown K9YC, reviews in RSGB Radcom mostly by Peter Hart G3SJX, and reviews in ARRL QST magazine usually with testing by Bob Allison WB1GCM.

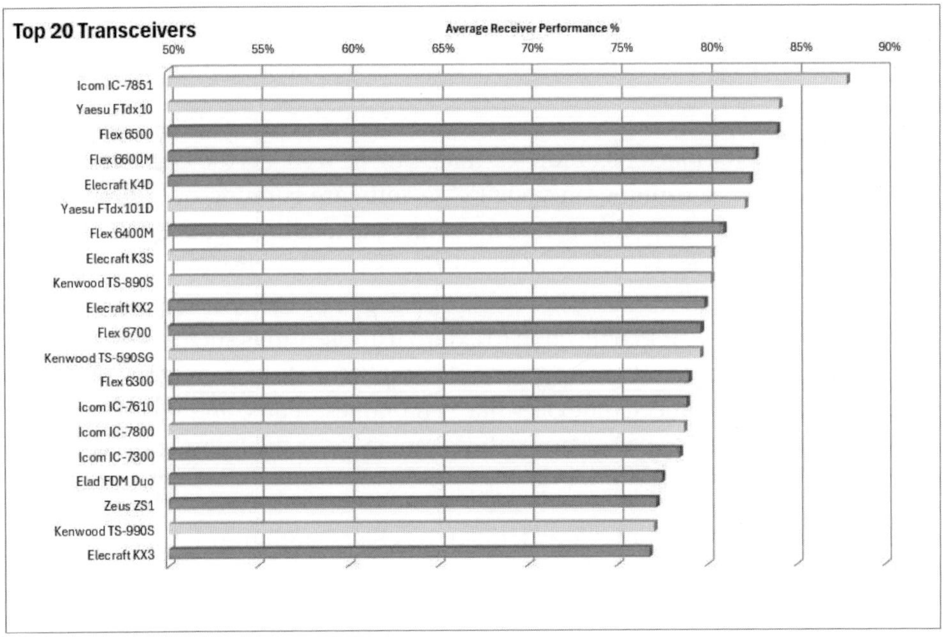


Figure 13: Andrew's list of the Top 20 transceivers

Twelve of the top twenty receivers and twelve of the top twenty transceivers are SDRs. The receiver tests include; sensitivity (MDS) 500 Hz BW preamp off, blocking dynamic range (BDR) 20m at 20 kHz and 2 kHz, noise figure, reciprocal mixing dynamic range (RMDR), S meter calibration, Receiver 3$^{rd}$ order IMD at 2 kHz, 5 kHz and 20 kHz, white noise (NPR), and image rejection. The transmitter tests include; Transmitter 3$^{rd}$ 5$^{th}$ 7$^{th}$ 9$^{th}$ and worst order IMD, CW keying spectrum (narrow 0.5-1 kHz and wide 2-5 kHz), worst harmonic on 20m and 6m, transmit composite noise at 100 Hz and 10 kHz offsets, SSB relay TR switching time, carrier suppression, and sideband suppression. I don't have results for all of these tests on all of the radios. Radios are not penalized for tests, which have not been measured. The FTdx10 and FTdx101 were not counted as SDRs due to their superheterodyne 'hybrid' design.

# NEW DEVELOPMENTS AND FUTURE TRENDS

The most obvious trend is the move away from simple hardware solutions with complex software running on an external computer, towards complex hardware solutions with more simple software running on a variety of computer-like devices. There seems to be an evolution happening as the capabilities of what can be achieved using software defined radio expands. SDR technology has been embraced by the big amateur radio manufacturers, particularly Icom which has more SDR models than any of the others.

SDR hardware has evolved from simple narrow-band, single-channel, SoftRock devices to complex HF transceivers and wideband receivers. We now have 100 watt software defined HF transceivers and radios that can cover frequencies from as low as 10 kHz up to 10 GHz. The bandwidth that can be passed on to the DSP processing stage has also increased. Allowing wider panadapter spans that are very useful at UHF frequencies and above.

The SDR software for black box radios is maturing as well. As the DSP functions are moved from software running on your PC to software running on a processor, or firmware running on a DSP chip inside the radio, the loading on the Ethernet or USB connection between the devices is reduced. The saved bandwidth can be used for advanced capabilities like; multiple panadapters, more receiver slices, and multiple narrowband IQ streams for connection to external applications such as CW Skimmer. You can run SDR client software on several computers at the same time, or control your radio via WiFi-connected devices like cell phones, iPads, and notebooks. Some radios include the ability to use your radio remotely over the Internet from virtually anywhere.

Modern direct sampling software defined receivers and transceivers may have several layers of software:

- FPGA software (usually Verilog) to perform mixing, decimation, and filtering. (Firmware is updateable, downloaded from a PC).
- RISC software running on an ARM microprocessor, soft-processor, or micro-controller, to manage control functions and switching. (Firmware is updatable, downloaded from a PC).
- DSP software running on a dedicated DSP processor chip in the radio. (Firmware is usually updatable by download from a PC).
- DSP 'server' software running on a built-in computer or single board computer inside the radio. (Software updates are directly downloadable from the Internet or downloaded via the host computer).
- SDR 'client' software running on a computer, tablet, or phone. (Software updates are directly downloadable from the Internet).

You might have different client software running on different devices for different modes of operation. One program might be good for listening to digital broadcast stations, while another might have a built-in FT8 decoder. In the future, your digital modes program could directly control the radio rather than through an intermediary SDR program. The QRP Labs QDX transceiver already does this.

Another trend is the adoption of software defined radio by the 'Maker' and 'Hacker' communities. These enthusiasts are developing software defined radios with very wide frequency ranges, well into the GHz region. Their activities include receiving signals from high-altitude balloons, aircraft, ships, and satellites. They might be interested in controlling household appliances or robots. Most importantly, this is a growing group of young, intelligent, people who are becoming interested in radio. Some of them are becoming licensed amateur radio operators so that they can legally transmit RF signals on the amateur radio bands. These enthusiasts are creating new software defined radios capable of working on the VHF, UHF, and microwave amateur radio bands. The new radios and software that they create can also be used for more traditional ham radio activities such as amateur TV, digital voice modes, repeater operation, weather stations, monitoring weather satellites, satellite operation, EME, and microwave DX.

## ADAPTIVE TRANSMITTER PRE-DISTORTION

Adaptive transmitter pre-distortion is a technique where a sample of your transmitted signal is taken back into the radio and compared with the modulating signal. By doing this it is possible to alter the phase and amplitude of the modulation in a way that reduces the effects of nonlinearity in the transmitter power amplifier. The result is a significantly improved transmission with a typical improvement of 20 dBc in transmitted intermodulation performance. The improvement is so significant that you can easily see the improved spectrum of a radio using adaptive pre-distortion on your panadapter display. Full duplex operation and wide transmitter bandwidth capability make this dynamic technology much easier to implement in a software defined radio than in a conventional radio.

The Pure Signal transmitter pre-distortion system in Thetis and PowerSDR mRX dramatically improves the transmitter intermodulation performance of the Apache Labs ANAN transmitters and it can be applied to the Hermes Lite 2.0. It relies on the software defined radio being full-duplex i.e. able to receive and transmit simultaneously, and having a wide (48 kHz) transmit baseband. There is more about the technical aspects of The Pure Signal transmitter pre-distortion system in the section about Thetis, on page 166.

The new FlexRadio FLEX-8400 and FLEX-8600 models will have internal couplers to allow for the introduction of adaptive transmitter pre-distortion in SmartSDR. Hopefully, this will be fully implemented when the radios start shipping in August 2024.

## NEW QRP RADIOS

QRP receivers and transceivers are becoming more sophisticated. Several new models include knobs, buttons, and a touchscreen bandscope or panadapter. The first of these was probably the Elecraft KX3 which was already available when I wrote the first edition in 2016. It is a stand-alone QSD SDR with an exceptionally good receiver. The KX3 does not include a panadapter or bandscope display, but you can buy an external panadapter for it.

Other low-power SDR transceivers and receivers with an integrated screen and controls include the Xiegu X6100, X6200, and other models, KN-990C, H-108, Q900, Zastone Wolf, HackRF Portapack, and the Recent RS-918 or McHF transceivers. Also the Malahit DSP2, Malachite DSP, ATS-25X2, and ATS25 receivers, and even an RTL dongle receiver with a screen. No doubt there are many more. The only ones I have any experience with are the Xiegu X6100 and the Malahit DSP2. They are both very good radios. The ELAD DUO transceiver and DUPr receiver include front panel controls and a display, but they don't have a bandscope or panadapter.

## SERVER/CLIENT RADIOS

The latest trend is to split the SDR software into a 'server' program working on a computer device inside the radio and a 'client' program operating on the same or a different computer. The Server performs the SDR and DSP functions. The 'client' handles the panadapter display and control functions, either as a mouse and keyboard-operated PC application or with a touchscreen display and real buttons and knobs. Sometimes both. Radios like the Flex 6000 and 8000 series can use the integrated M series front panel, or a remote client such as the Maestro, SmartSDR (Windows), SmartSDR (iOS), or Dog Park (macOS).

An API (application programming interface) describes the data requirements and control instructions for communication between the client and the server. If the API is made available, external developers can write specialist client software that communicates with the server application inside the radio.

The server program usually has no direct user access. It simply acts as a channel between the radio and the client software. The server typically does the DSP and signal routing. The SDR client software may be running on the same computer as the server or on another computer connected to the same computer network via an Ethernet cable or WiFi. It might even be on a remote computer connected via the Internet. The SDR client does all of the display and control 'GUI' functions.

The client software can work with different radios that use the same server application. It is also possible to have more than one client connected to the server at the same time. For example, you can use SmartSDR to monitor different bands on different computing devices. Two operators can share a single radio, to operate on two bands at the same time, during an amateur radio contest.

You could listen for a 6m band opening on your phone while talking to a station on 20m SSB in the 'shack.' Maybe you don't want to do all of that, but the client/server architecture does have several advantages.

Firstly, the server can do much of the DSP work. There will be a large continuous amount of data flowing between the radio hardware and the server program. However, the data between the server application and the client can be much reduced. That data connection only needs to carry the graphics for the panadapters being displayed, audio in both directions, information data like the S meter and frequency readouts, and the control commands (like volume, filter width, etc.). Less data allows you to use less powerful computers like phones, Netbooks, or tablets and WiFi-connected or Internet connected devices.

External developers can provide a range of client software designed to work with the server. This should mean faster development time, as they only need to create a GUI. The difficult DSP and FFT code will already be in the server program. You could have a choice of user interface styles based on the same radio and server. In the future, your favourite digital mode program could talk directly to the server, without having to work through a SDR program.

Although the SmartSDR is a proprietary application (not open source), FlexRadio has generously allowed for three levels of API access. This is interesting because it allows software developers to write open-source software applications that will work seamlessly with the FlexRadio SmartSDR software. The ability to add customisation will keep the radio fresh and stimulate innovation without pushing up the price. I am sure that Flex users will welcome the ability to add to what is already a first-class package.

The 'Works with Flex' section on the FlexRadio Systems website lists several clients and API addons for SmartSDR.

- Dog Park SDR – a MacOS client
- FRStack band stacking addon
- Slice Master easy digital mode integration
- N1MM logger integration
- SDRBridge integrates CW Skimmer
- K9UDR various utilities
- N4PY station integration software
- FRlogger is a logging program, and
- SwissLog is another logging program

The Apache Labs Andromeda and G2 transceivers also employ a client/server arrangement. The Andromeda had a Windows PC inside the radio. The G2 has a Raspberry Pi4 compute module running the server application for piHPSDR client software, or simply passing the raw data through to Thetis working on your PC.

## ADDING NEW MODES

Because software defined radios are upgradable by the user, it is relatively easy to extend the software to include additional modulation types and modes. Some software defined radios can decode PSK, RTTY, FT8, or CW. Some can receive stereo FM with an RDS (radio data system) display, or decode DRM digital signals.

There is an add-on for SmartSDR that uses the FlexRadio API (application programming interface) to add the FreeDV digital voice mode. Extendibility like this is a major advantage of software defined radio.

## DIRECT DIGITAL MODES

At present, there is a separation between the transceiver (SDR or conventional) and the digital mode software used for FT8, RTTY, PSK, SSTV, and so on. The digital mode software sends audio tones in preset patterns to the radio which usually modulates the RF signal, as a single sideband transmission. If the digital mode software was integrated with the SDR software, the digital text modes could be created as IQ signals and used directly. It should also be possible to invent new modes by manipulating the I and Q data rather than using tones.

The ability to send the digital mode program a wideband IQ signal rather than a 3 kHz wide audio signal promises the ability to decode digital signals from several bands at once. Imagine having PSK or FT8 from four bands on the same display. Just click on any signal to work the one you want. FlexRadio is working on enabling the ability to extend the transmission and reception of digital modes like PSK or FT8 from the current 3 kHz out to 40 kHz. FT8 users would be able to spread out easing the severe congestion that we see on the bands at the moment. Of course, they can spread out now, but if they transmit more than 3 kHz above the nominal frequency they run the risk of not being heard. With this improvement, you could transmit at 10 kHz above the nominated FT8 frequency and still hear the stations working in the traditional 3 kHz range. A DXpedition station would be able to send multiple FT8 signals, responding to the pile-up of calling stations more quickly.

In most SDR transceivers, CW modulation is created in the FPGA or microcontroller rather than switching an audio tone on and off. The same techniques could be applied to phase modulation schemes like BPSK. To create a PSK31 signal you just switch the phase of the signal by 180 degrees rather than turn it on and off. This would be very easy to implement in an FPGA. QPSK could be implemented in the same way, using 90-degree phase changes.

## A DIFFERENT MODULATION METHOD

The traditional way of transmitting the WSJT-X modes like FT8, WISP, and Q65, and other tone modes like AFSK RTTY, Olivia, ROS, Domino, etc. is to send tones from the digital mode software to the SSB modulator in your transceiver.

The new QDX and QMX+ QRP SDR transceivers from QRP Labs use a different technique. They determine the frequency of the audio tones coming from the digital modes software on your PC and offset the transmitted frequency by that amount. Rather than using an SSB modulator, these radios are transmitting FSK (frequency shift keying). This works for any mode that transmits a single tone at a time. It does not work for PSK which uses phase shifts, or for any mode that transmits more than one tone at a time, like Winlink.

Frequency shift keying is not new. We have been using it to transmit direct keyed RTTY for years. It is, or was, also used for commercial data traffic, particularly for maritime services. The difference here is that in normal FSK, the radio receives a digital 'keying' signal. In this radio, the radio receives tones and works out what FSK offset is required. As far as I know, this is a unique method.

The receiver in the QDX and QMX+ transceivers is a QSD (quadrature sampling detector) feeding a 24-bit stereo ADC at audio frequencies. The audio is sent to the WSJT-X or other digital mode software in the usual way. It has four bandpass filters on the input side of the balanced Tayloe QSD detector. The oscillator is a SiLabs SI5351A clocked by a 0.25ppm 25 MHz TCXO.

## CESSB

CESSB (controlled envelope SSB modulation) is an enhancement that is available on the latest HF SDR transceivers. CESSB is a way of maximizing the talk power of your SSB signal. It works like a compressor but it is impossible to overdrive the modulator. There is no clipping and overshoot is controlled. Your average transmitted power can be almost doubled. That is like adding another antenna on top of your existing antenna. The microphone audio is converted to a digital signal which is manipulated by the DSP software before being converted back to analogue, shifted up to the transmitter frequency, amplified, and transmitted. CESSB could be implemented in conventional radios with DSP, provided there was sufficient number-crunching capability in the radio. CESSB is always enabled as a limiter on FlexRadios, it becomes more active when the speech processor is set to DX or DX+.

## BETTER ADCS

There may be better ADC chips in the future, but the trend is more likely to be towards higher sample rates allowing for a wider Nyquist bandwidth rather than a larger number of bits. This is because greater bandwidth is more important than noise performance in the commercial world.

It is impractical to sample at super-fast rates, so radios working in the UHF and SHF regions will always need some sort of down conversion before the ADC. This is increasingly likely to be in the form of a tuner chip or a dedicated down conversion chip.

## MULTIPLE ADCS

Having two ADCs in radios with direct sampling can double the usable bandwidth of the radio. For example, you could have a receiver capable of receiving HF and VHF, or VHF and UHF at the same time. Phase coherent receivers with two ADCs can be used with multiple antennas to artificially steer their directivity, boost a signal, or cancel noise and interference.

If both ADCs have individual front-end bandpass filters, you can monitor multiple bands with a reduced risk of dynamic gain compression or ADC overload. This technique is used in the FLEX-6000 series, the FLEX-8000 series, and the ANAN G2.

*TIP: You can monitor multiple bands with a single ADC, but you usually have to bypass the pre-selecting bandpass filters leaving the receiver more susceptible to out-of-band interference signals. This was the approach taken with the ANAN-100 and I never experienced any problems with it.*

Another reason to use two ADCs in the receiver is to improve the dynamic range of the receiver, by lowering the noise floor. Averaging the digital output of two ADCs averages the quantisation errors, reducing the quantisation noise and therefore improving the dynamic range of the receiver and the panadapter or bandscope by up to 3 dB.

## STANDARDISATION

A lot of black box software defined radios can work with more than one SDR software program. Conversely, some applications can work with a variety of software defined radios. Standardisation of the interface requirements and data protocols extends this capability. Wouldn't it be nice if you could buy any black box radio and use any SDR software with it? A good example is the Hermes Lite 2 SDR which uses essentially the same software as the ANAN G2, although it does have some changes.

The server/client model extends the opportunity for standardisation. The DSP is done inside the radio and the control and display functions are performed by client software on your PC, tablet, etc. If a common API was agreed between the manufacturers, any compliant client software could be used with any radio using the same API.

To some extent, this is how extio.dll works as an interface between a software application and radios with differing communications protocols.

## REMOTE OPERATION OVER THE INTERNET

A lot of people are very interested in using the Internet to operate their radio from a remote location. There are some methods already being used to achieve this and they are not limited to software defined radios. But a software defined radio that already sends its data over an Ethernet connection is ideally suited to remote operation. The only requirement is to extend the connection through the firewall to the Internet and to add a measure of security.

There are many reasons for remote operation. You might have restrictions on the size of antennas that you can have at home, or you may have a lot of electrical noise to deal with. People who travel a lot may be interested in operating their home station from another city or another country. Club stations may allow members to operate the club station from home. Geography may block transmission in the direction that you want to use. Or, there could be situations like; not being able to see down to the horizon for satellite or EME operation, needing a high site for VHF DX, or a mountain range blocking the short path to Europe.

Many software defined radios use an Ethernet connection to the PC, which is essentially the same method that computers use to talk to each other. So there is no technical reason why the Ethernet connection can't be extended over the Internet. The main barriers to remote operation over the Internet are unpredictable delays in receiving data packets and the need for security. Data latency (packet delay) could cause slow control operation, intermittent panadapter display, and broken audio. The need for security is evident. You don't want to open your local network to unscrupulous people or allow your transceiver to be operated by unlicensed individuals. Radios that include a server are ideally suited for direct connection to the Internet. The thin client connection between the server software and the client software requires less continuous data, so the connection is less affected by latency problems.

## AI INTEGRATION

Artificial intelligence (AI) is one of the trendiest topics in recent years, and it is easy to integrate into software defined radios which already include computers and programmable logic devices. AI can enhance digital signal processing, especially in the area of adaptive noise or notch filters and adaptive transmitter modulation. Either to improve the linearity of the transmitted signal or to boost the average transmitted power by modifying the modulation. These techniques already exist in the form of adaptive transmitter pre-distortion, CESSB, automatic notch filters, and noise reduction algorithms, but AI can make them react faster and work better. Your radio may be able to analyse and precisely filter out an interference signal without impacting the received signal. Or it could evaluate received signals across several bands and decide which has the best HF propagation for a contact with a wanted country.

# COMMON QUESTIONS ABOUT SDRS

Times have moved on since the first edition in 2016. Many of the hams who "would never buy an SDR," now own the latest Yaesu or Icom and are very pleased with it. This chapter addresses the common misconceptions that people had about SDRs back in 2016. Hopefully, most people have now come around to the idea of using this, not so new anymore, technology.

There were many very vocal and uninformed knockers inhabiting the online forums. Not so many, these days. Most of what they said is outdated or incorrect. These are probably the same people who believed that eliminating the CW requirement for the amateur radio license would lead to the destruction of the hobby. They may also believe that the Earth is flat, DSP sounds bad, and that vinyl is better. Many of the criticisms levelled at SDRs are no longer valid because the technology and software are continually being improved. Some of the criticisms are just plain wrong, being based on an incomplete understanding of how software defined radios work. There is a solution for every 'real problem.'

## ARE SDRS ANY GOOD FOR CW?

Initially, many CW operators didn't like using black box SDRs because of the delay between pressing the Morse key and the sidetone being heard. This delay, known as latency, was due to the time taken for the CW tone to be processed in the SDR software and then sent to the radio. The problem can be avoided to some extent by using a keyer that generates its own side-tone. Another issue was the ability to work QSK at high CW speeds. QSK operation allows a proficient operator to listen to the receiver in between the Morse code 'dits' and 'dahs' that they are sending. To achieve this, the transceiver must switch from receive to transmit and back again, quietly and fast. Because the, transmit to receive (TR) switching is a signal that must be passed from the SDR radio to the PC software and back again, early SDRs had a poor transmit-to-receive switching time.

This problem has been eliminated in the new 100 watt SDR transceivers and most of the QRP transceivers. The switching and keying issues have been fixed by keying the transceiver using embedded firmware code in the FPGA rather than the older method of using a tone generated in the PC software. This has resulted in very fast CW generation with immediate side-tone and full QSK operation up to at least 60 wpm. These developments were introduced for the Apache Labs ANAN series of radios in July 2014 and a similar technique has been implemented in the FlexRadio Signature 6000 and now 8000 range of transceivers.

*Technical TIP: This advancement is a good example of learning that SDR is different. The old method of generating CW in an SDR was the same as the technique that is used in most modern SSB transceivers. The CW keyer turns on and off an audio frequency tone that is then sent to the SSB modulator and mixed up to RF.*

*In the new method, the FPGA simply generates a digital 'carrier' signal which is turned on and off by the keyer. The data is then sent to the DAC and converted to an RF signal in the usual way. In both cases, the waveform rise and fall times are shaped to avoid harsh-sounding key clicks.*

Sending CW is now the same with an SDR as it was with a conventional transceiver. The only difference is that your new radio will probably include a CW message keyer to automate sending CQ calls or signal reports. They are very good when you are contesting.

## DO SDRS HAVE A POOR DYNAMIC RANGE?

No! They have a very good dynamic range. Some online pundits claim that SDRs have poor dynamic range and that the owners are lying about how good they are. This claim is just plain wrong and it has led to the widespread but mistaken impression that 16-bit SDRs are inherently better than 14-bit SDRs.

In fact, the dynamic range of 14-bit radios is very similar to 16-bit radios. However, I am prepared to admit that 14 and 16-bit radios are generally better than 8-bit models. I cover this in more detail later in the section describing the ADC. But take it from me; SDRs have a much better dynamic range than conventional superheterodyne architecture radios, which need AGC to extend their dynamic range to an acceptable level.

A high-quality software defined receiver will typically have a dynamic range above 120 dB in a 2.4 kHz bandwidth.

The confusion arises from a lack of understanding about the formula for the maximum dynamic range of an ADC, *DR=20*log(2^Nbits-1) dB*. What is often overlooked is that the dynamic range as specified in the formula relates to the entire frequency response of the analogue to digital converter. As the bandwidth of the signal is reduced using frequency decimation and filtering down to the final receiver bandwidth of around 2.4 kHz for SSB, or 500 Hz for CW, the dynamic range increases. This effect is called 'process gain.' It is the ratio of the initial and final bandwidth expressed in decibels. It is not really a 'gain.' It is an increase in dynamic range resulting from the exclusion of noise. Think of it as having a sliding door open when your grandchildren are playing outside. If the door is open wide you get a lot of noise and it is hard to hear the TV. If it is only opened to a narrow slot, you get less noise and you are able to hear the TV.

Process gain is the reason that you use a narrow filter for CW. You get less noise, so you can hear weaker signals. You can try this at home. Tune your SDR to a part of a band that has no signals or perhaps a single CW signal. As you drag the receiver filter to reduce the bandwidth, the amount of noise you hear from the speaker reduces. Note that the spectrum display is not affected by changing the receiver bandwidth, because it is created separately.

*Technical TIP: Process gain is a fundamental part of the panadapter display. The data used to plot the panadapter spectrum display is derived from measuring the levels of thousands of narrow bandwidth frequency samples, known as 'bins.' The bins are obtained from the ADC data output using FFT, the 'fast Fourier transformation' process. Measurement of the amplitude of the signal in each of the bins is used to plot the points across the spectrum display. Because each bin represents a very narrow bandwidth, each bin has a low noise floor, so the panadapter can display a large dynamic range. In fact, the spectrum and waterfall display can display signals below the noise floor of the 2.4 kHz receiver slice. It is easily able to display any received signal that is strong enough to hear.*

## DO SDRS HAVE A POOR SIGNAL TO NOISE RATIO?

No! They have a very good signal to noise ratio as well. This is actually the same question as the one about dynamic range. The maximum receiver input level is limited by the clipping level of the ADC. The difference between that and the noise floor of the receiver is the dynamic range of the receiver and therefore the maximum possible signal to noise ratio. As the bandwidth is reduced, process gain reduces the noise level, lowering the noise floor. There is the same signal level but less noise, so the signal to noise ratio improves as the receiver bandwidth is reduced. Another factor affecting the signal to noise ratio is the IMD (intermodulation distortion) occurring in the receiver. IMD occurs when received signals mix with the local oscillator signal to create a variety of interference signals. In a superheterodyne radio and to a large extent a hybrid SDR with a superheterodyne mixer and oscillator before the SDR stage, the IMD products increase linearly with the increase in input signal level. In a direct sampling SDR the increase in IMD is not linear. It can change unpredictably with an increase in signal level and, in some cases, it improves with the addition of additional input signals.

Some mathematical wizardry produces the formula for the maximum possible signal to noise ratio for an ADC, *SNR = 6.02*Nbits+1.76 dB*. Like the dynamic range, the maximum possible signal to noise ratio is related to the number of bits used to describe each sampled input voltage level. This is the formula that causes all the aggravation in online forums because at first glance it indicates that the maximum achievable signal to noise ratio (SNR) is not very good and nowhere near the figures quoted in SDR product documentation and reviews.

The formula indicates significant differences between the performance of the 8-bit, 14-bit, and 16-bit ADCs used in various SDR models, which is much less apparent in real-world tests. The typical signal to noise ratio of most ADCs is a little worse than the maximum predicted by the formula. But, as the bandwidth is reduced from the ADC sampling bandwidth down to the receiver bandwidth, the dynamic range and the signal to noise ratio improves due to 'process gain.'

The number of ADC bits is not a very reliable indicator of the receiver's signal to noise performance. This is because an ADC can't sample very weak signals accurately. So even though a device may have a 16-bit output signal the effective number of bits 'ENOB' that carry valid information is less than 16 bits. It is the ENOB value that determines the real signal to noise ratio of an ADC.

## DO I HAVE TO USE A MOUSE? – I LIKE KNOBS AND BUTTONS

You don't need a mouse to operate the new SDRs 'with knobs.' Although most of them include a touchscreen display. In many cases, you can use a mouse, but it can be rather limited. For example, mouse operation on the IC-7610 is limited to the spectrum and waterfall display. You cannot use a mouse to control the radio.

You do have to use a mouse to operate a black box SDR. The software runs on your computer, and that's how you use a computer. But, you soon get used to using a mouse instead of traditional controls. Using the radio becomes the same as using any other program on your PC. You don't expect knobs and buttons on your Internet web browser or email client.

There are options available if you really need a tuning knob. FlexRadio offers a USB VFO knob with three buttons that you can dedicate to other transceiver functions. It is just a mouse with a knob instead of the mouse wheel. Some people use a 'Hercules' Midi music 'DJ console' to provide a suite of slide controls and buttons. I have tried some of these options, but I find mouse control easier and faster. FlexRadio offers a console called Maestro for the Signature range.

When you get used to the different ways of operating your radio, I think you will find that the mouse works very well, especially if it has a mouse wheel between the buttons for fine-tuning the receiver frequency. I am not the only person to believe that mouse control is better. The latest top-end conventional radios like the Icom IC-7850 and the Kenwood TS-890 have mouse-controlled bandscopes, to make operating them more like using a software defined radio.

## ARE SDRS TOO TECHNICAL AND EXPERIMENTAL?

Playing around with software defined radio can be as technical and experimental as you want it to be. You have the option of buying a software defined radio 'with knobs' and using it in the same way as any previous radio. Or you can get involved with high-specification wideband boards, building your own SDR, or writing your own software. In the case of most commercially available black box SDR receivers and transceivers, the experience is pretty similar to setting up and using any new radio. It might even be easier. Most of the small receivers and dongles use a USB connection. Simply plug the USB cable or dongle into your computer. Load an SDR program, connect an antenna and you are ready to play! You will probably have to determine the COM port number by looking at Windows Device Manager.

Other radios use an Ethernet connection. The easiest way is to buy a four-port Ethernet switch and install a CAT5 cable back to your Internet router. Connect the radio to that and the software on your computer (also connected to the network) will find the radio automatically. A wired Ethernet connection is usually better than a WiFi connection, especially if the DSP is being performed on the computer, rather than inside the radio.

Small box receivers and dongles using a USB connection to the radio are very often powered over the USB cable. Most black box SDR transceivers need an external 13.8V power supply, the same as regular transceivers.

Load the SDR program, connect an antenna and you are almost done. Often the sound from the receiver is played over your computer's speakers, but for some radios, you may need another set of amplified PC speakers and a microphone wired for the socket on the radio. Then you are ready to go.

I guess if you want to use cutting-edge technology with continuing development, you must expect things to be somewhat fluid and experimental. This is especially true with open source software defined radio because we are relying on the enthusiasm of amateur software and hardware developers who are collaborating. It does help if you have some computer skills, including the ability to install new programs and understand directory folders. To keep up to date, you might have to install a new software revision, or download a firmware file and use a utility program to send it to the radio.

If you want to run digital modes with a conventional radio, you need to install a computer program, connect a radio to a power supply and antenna, then connect up a CAT cable and audio leads to the PC. Doing that on an SDR is essentially the same. Install a couple of computer programs for the audio and COM port connections, connect the radio to a power supply and antenna, and plug in an Ethernet or USB cable. It only needs to be configured once. If you are in doubt or you do have problems ask a teenager, or raise a question on an online SDR forum.

Building SDR radio kits is fun and has much the same technical challenge as building a QRP radio or an electronics kit. In most cases, the kit is supplied with the 'tricky to solder' surface mount chips already soldered to the board. If you want to throw yourself into the deep end, building one of the QRP Labs kits is a great project. If you are interested in programming, you can give GNU Radio a try, or write an application that takes advantage of the FlexRadio API. There is huge scope for you to become involved in software or hardware development.

# PANADAPTERS AND BANDSCOPES

The spectrum and waterfall image, displayed on a bandscope or a panadapter, is probably the most popular feature of modern radios. SDR software running on a computer usually displays a panadapter. SDRs with knobs usually provide a choice of a 'FIX display' panadapter, or a 'CENTRE' bandscope.

A panadapter displays a range of frequencies. One or more receivers can be placed anywhere across the selected range.

A bandscope displays a 'span' of frequencies above and below the receiver frequency which is displayed in the centre of the spectrum and waterfall display.

Most of the Icom and Yaesu transceivers can display a maximum panadapter bandwidth of 1 MHz, or a bandscope showing 500 kHz on each side of the centre frequency. The IC-R8600 can display a 5 MHz panadapter or a ±2.5 MHz bandscope, and the IC-905 can display a maximum panadapter bandwidth of 50 MHz or a ±25 MHz bandscope span. Black box radios can support panadapters ranging from 48 kHz in the case of a FUNcube dongle to 14 MHz in the case of a FLEX-6600 or FLEX-8600.

Some software defined radios can display multiple panadapters at the same time. That means that you can monitor several bands or band segments at once. The ability to display signals on a panadapter display is definitely one of the major advancements offered by software defined radios. The panadapter display usually dominates the front of the radio or the computer screen.

A spectrum display shows the signal strength of received signals across a range of frequencies. Stronger signals show as higher peaks. The reference level should be adjusted so that the noise 'grass' at the bottom of the spectrum display is just visible, like a newly mown lawn. That provides the maximum dynamic range on the display, between the received noise level at the bottom and the level at the top of the spectrum scope. If you can see the noise floor 'grass' you will be able to see any signal that is strong enough to be heard above the background noise being received. A waterfall display shows the history of the spectrum display. Stronger signals are indicated as brighter colours. You can see signals that are not on the spectrum anymore such as a station that has just finished talking or sending.

Normally the radio is tuned to the frequency at the centre of the panadapter display. Signals above the tuned frequency are shown on the right side and signals that are below the tuned frequency are shown on the left. Note that on a panadapter the frequency that the radio is 'tuned' to is not the frequency of the 'receiver slice' you are listening to. One or more 'receiver slices' will be shown superimposed on the spectrum display. They represent the part of the band that you are listening to, as opposed to the part of the band you are looking at.

*TIP: You have to get used to the idea that the frequency of the receiver slice that you are listening to is not necessarily at the centre of the spectrum display. The radio has a wide bandwidth and the receiver slice has a narrow bandwidth.*

Receiver slices are often depicted as a vertical line indicating the 'carrier' frequency, next to a shaded area that represents the receiver's bandwidth. An upper sideband receiver will be shown as a vertical line with a 2.4 to 2.8 kHz wide shaded area on the right side. A lower sideband receiver shows the shaded area on the left side of the carrier. On AM or FM mode, the shaded area will be displayed on both sides of the centre line, indicating a double sideband signal. CW is the same as SSB but with a narrower bandwidth.

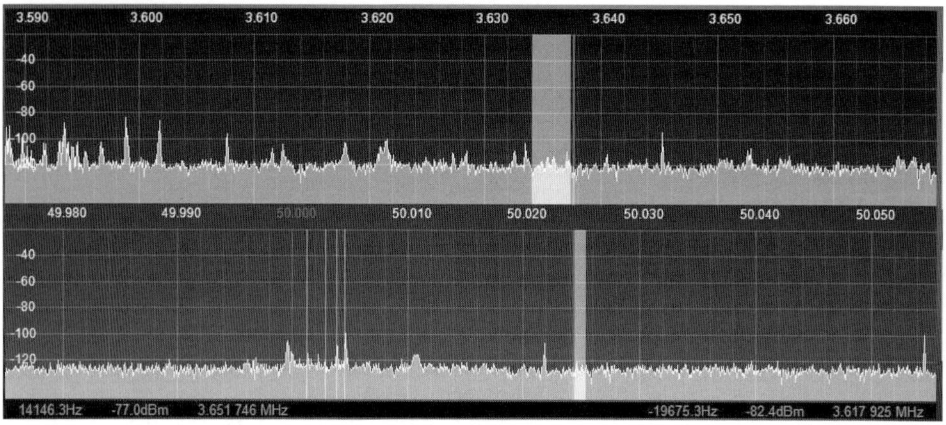


Figure 14: Panadapters on the 80m and 6m amateur bands

The upper image shows a lower sideband signal on the 80m amateur radio band. The panadapter is around 80 kHz wide, extending from about 3.588 to 3.668 MHz.

The lower image shows a slice receiver set to a 500 Hz CW bandwidth on the 6m amateur radio band. The panadapter is also about 80 kHz wide, extending from below the start of the 6m band to 50.056 MHz. The five vertical lines are tracking notch filters, a feature that is available on SmartSDR and Thetis software.

*TIP: The shading depicting the bandwidth of the receiver on the panadapters is not included on the spectrum and waterfall displays of the Icom and Yaesu receivers. I am disappointed that they decided against including it. Even though almost all PC-based SDR software does. I find it easier to adjust the receiver bandwidth if I can see it on the spectrum display.*

A 'receiver' or 'receiver slice' on a software defined radio is like a VFO on a conventional radio. Most HF amateur radio transceivers have two VFOs to facilitate 'split' operation or a quick jump to another frequency on the same band. Some SDR software supports two (or more) receivers on the same band, so you can run 'split' mode with the option of hearing both receivers at the same time, often one receiver on the left speaker and the other on the right speaker. This is great for listening to a DX station as well as the pile-up.

The more expensive SDR transceivers have two completely independent receivers, so you can listen on two bands at the same time, or two receivers on different modes on the same band if that's what you want to do.

One of the real joys of operating a software defined radio is the ability to adjust the receiver bandwidth very precisely, to eliminate unwanted noise or interference. When you are listening to an SSB signal there is no point in having the receiver bandwidth set any wider than the signal that is being transmitted by the other station. Setting the receiver bandwidth too wide just makes the signal sound noisy. Setting it too narrow cuts the audio frequency response. You can see exactly how wide the transmitted signal is by looking at the spectrum or waterfall display. You simply drag the shaded receiver bandwidth to be the same width as the waterfall image.

The panadapter on most SDR transceivers, is capable of displaying a representation of your transmitted signal while you are transmitting. This can be helpful to ensure that your transmitted signal is not distorted.

Next, we need to define some terms for the panadapter display.

## PANADAPTER OR BANDSCOPE BANDWIDTH

The panadapter or bandscope bandwidth is the maximum frequency range that the panadapter can display on the computer screen or the radio display. It is limited by the data 'sample' rate of the digital signal used to create the spectrum and waterfall display. The maximum panadapter bandwidth ranges from 48 kHz for some 'sound card' QSD models, to more than 60 MHz wide for a few hybrid designs.

With hybrid radios that sample at an I.F. (intermediate frequency), or dongle receivers with a tuner chip, the maximum panadapter span is usually the same as the sample rate of the analogue to digital converter (ADC). It could be around 2.5 MHz in the case of an RTL dongle or as much as 50 MHz in the case of the IC-905 transceiver.

In a QSD type of SDR, the maximum panadapter display bandwidth is equal to the sampling rate of the audio frequency analogue to digital converter. Most 'generation one' and 'generation two' QSD designs can only display a single 48 kHz or 96 kHz wide panadapter. QSD radios using a 24-bit USB soundcard may be able to sample at the 192 kHz rate and display a 192 kHz panadapter. These are the maximum bandwidths. The software can zoom in to show a smaller range of frequencies to suit your needs. For example, you could zoom in to view a CW sub-band or the FT8 signals.

The maximum bandwidth of a direct sampling receiver is governed by the Nyquist rule, to ½ the ADC sample rate. For example 61.44 MHz wide for an ANAN radio, or the Hermes Lite 2 with an ADC sampling rate of 122.88 Msps.

While the radio or SDR software might be able to display a wideband spectrum view of the entire HF spectrum, the required data rate places too much strain on the communications link between the radio and the SDR software performing the DSP on the computer. This is becoming less of a problem with the introduction of Gigabit Ethernet and USB4 (USB-C) connections. Normally you don't want to view the whole spectrum anyway. You want to look at a band or part of a band. I find a 192 kHz wide panadapter is good for SSB reception. Current direct sampling radios usually down-convert the data stream to a slower rate using a digital oscillator and mixer to shift the frequency and a process called 'filtering and decimation' to reduce the data rate. Because this introduces a mixer stage, a panadapter created using the new data has the same maximum bandwidth as the reduced data 'sample' rate.

### Nyquist bandwidth confusion

The Nyquist rule states that the maximum bandwidth you can use is ½ of the sample rate of the analogue to digital conversion. The formula for the 'Nyquist' bandwidth is BW=Fs/2. If you reduce the sample rate using filtering and decimation, the maximum bandwidth you can use becomes ½ of the reduced sample rate.

This leads to some confusion because SDR receivers usually display a panadapter that is the same width in kHz as the sample rate in ksps. This wider bandwidth is achievable because the received signals are presented to a mixer. Frequencies from above and below the oscillator frequency are mixed into the output data stream. By using both the I and the Q signals, we can eliminate the image signals and display signals from both sides of the centre frequency. The I+Q signal is used to create one side of the spectrum view and the I-Q signal is used to create the other side, relative to the oscillator frequency in the centre of the panadapter.

See more about quadrature sampling and image cancellation on page 182.

### Padapters on direct sampling radios

In direct sampling SDRs for the HF or low VHF bands, there is no mixer before the ADC, so the maximum bandwidth of the radio is Fs/2, i.e. half the sample rate of the ADC. The RX888 receiver has an ADC sample rate of 130 Msps. The DSP stage can process all of the ADC output, so the receiver has an amazing bandwidth of 1 kHz to 64 MHz. It can display the entire HF spectrum at the same time. I don't know of any software that allows a display wider than 32 MHz. The number of slice receivers depends on the software used. Theoretically, multiple slice receivers could be placed anywhere across the HF spectrum.

An ANAN radio or a FLEX-8400 has an ADC sample rate of 122.88 Msps (million samples per second). That means that these radios can cover from around 10 kHz to 61.44 MHz. However, due to the high data rate, the whole 61.44 MHZ bandwidth can't be processed by the DSP stage. The digital information is down-converted in frequency using a numerically controlled oscillator and a mixer in the FPGA. Note that this is done mathematically by manipulating the numbers from the ADC.

Then the data rate is reduced using 'filtering and decimation.' Because there is a mixer stage, a panadapter created using the new data has the same maximum bandwidth as the new, lower, data 'sample' rate.

The maximum panadapter width on Thetis is 1526 kHz. The panadapter is smaller than on the Flex radios because the ADC data stream has to be transferred over the Ethernet connection to the SDR software running on your PC. The FLEX-6600 and FLEX-8600 radios support a 7 MHz maximum panadapter. The data rate does not have to be reduced as much because the DSP is done inside the radio.

### QSD radios using a PC or USB soundcard

The panadapter display will be 48, 96, or 192 kHz.

### Direct sampling HF receivers or transceivers

Direct sampling SDRs can receive from low frequencies around 10 kHz up to ½ of the ADC sample rate. The maximum panadapter bandwidth depends on whether the DSP is performed in the radio, but it will usually be significantly less than the radio bandwidth. Radios like the IC-7300, IC-7610, FT-710, FTdx10, and FTdx101,support a 1 MHz panadapter (±500 kHz bandscope). The Elecraft K4 supports a 368 kHz panadapter or a ±184 kHz bandscope.

### Hybrid SDRs using a superheterodyne mixer or a tuner chip

Hybrid SDRs can cover huge frequency ranges by mixing the incoming radio frequencies down to a range that the ADC can sample. Sometimes down to a frequency in the HF or low VHF spectrum for a direct sampling SDR. Sometimes to an audio frequency I.F. such as 24 kHz which is delivered to a DSP chip with an integrated ADC. These radios can deliver quite large bandwidths of data to external software or DSP inside the radio.

## PANADAPTER SPAN

You don't usually want to display the whole panadapter bandwidth on your PC screen. The signals look too narrow and close together to easily select them. You will probably want to zoom in so that the panadapter displays the wanted shortwave band or a section of a band.

The part of the full panadapter bandwidth that you choose to show on the screen is known as the 'Span' of the display. It is usually fully adjustable from around 1 kHz displayed across the screen, up to having the full panadapter bandwidth displayed.

*TIP: If you change the panadapter zoom level, it may be possible to have an active receiver slice that is not visible on the span that you are displaying. It can be confusing to hear a signal from a receiver that is no longer being displayed on the panadapter.*

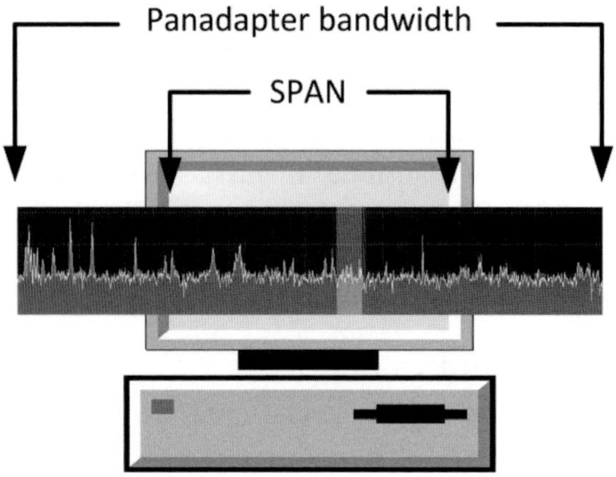


Figure 15: The span is the part of the panadapter that is displayed

## TIME AND FREQUENCY DOMAINS

The panadapter or bandscope spectrum display is created by displaying the level of the received signal at many frequencies across the screen. To do that the radio needs to measure the signal level at different frequencies. This is a difficult thing to do.

The digital data coming from the ADC represents the level of the sum of all the signals in the bandwidth being sampled by the ADC at each sampling time. This is known as a time domain signal because it represents the received signal level information at different times. While this data is great for receiving signals using digital signal processing, it is not good for creating a spectrum display.

Instead of knowing the signal level at different times, we need to know the signal level at different frequencies. The FFT (fast Fourier transformation) process converts the signal from the time domain to the frequency domain. It creates data containing the level of the signal in several thousand narrow frequency slices called bins. If we plot the level of each of the bins, we get a spectrum display.

*TIP: The FFT often operates on a slowed-down (decimated) version of the ADC data. It depends on the RF bandwidth of the radio. For example, the ANAN G2 receives up to 61.44 MHz, but the data presented to the FFT might be set to 384 ksps (384 thousand 16-bit samples per second), resulting in a panadapter bandwidth of 384 kHz. If you decide to set a 192 kHz panadapter the data rate at the input to the FFT will be reduced to 192 ksps.*

Each digital sample being presented to the FFT and each bin coming out of the FFT is represented by a 16-bit number, so the bit rate is sixteen times the sample rate. If the sample rate is 192 ksps the bit rate is 3.072 Mbits.

See the FFT magic chapter on page 218 for more on fast Fourier transformation.

## PANADAPTER RESOLUTION

The panadapter or bandscope resolution determines how close together signals can be before they can't be 'resolved' as different signals and they merge together on the spectrum and waterfall display. It is limited by the data 'sample' rate and the number of pixels the screen can display. For example, if the panadapter is 192 kHz wide and the screen can display 1024 pixels, each dot on the spectrum or waterfall display represents a frequency step of 187.5 Hz. If two signals are more than 188 Hz apart you should see both of them on the panadapter. If they are closer than 187 Hz they may merge on the display. If you reduce the sample rate to 96 ksps to create a 96 kHz panadapter or zoom in the display to a panadapter that is 96 kHz wide, each displayed dot represents a frequency change of 93.75 Hz. The panadapter resolution has doubled, but the panadapter span has halved.

The other factor to be considered is the FFT (fast Fourier transformation) process that converts the signal from the time domain to the frequency domain. The more bins that the FFT creates the smaller the bandwidth of each bin. Narrow bins mean a lower displayed noise floor without the antenna connected and more dynamic range on the spectrum display. The trade-off is that it takes longer to process a larger FFT and that slows down the rate that the spectrum can be displayed on the screen.

The width of each FFT output bin is dependent on the sample rate of the data being presented to the FFT stage and the number of bins that the FFT calculates.

A typical example would be a panadapter with a sample rate of 192,000 samples per second and an FFT of 4096 bins. This results in a bin width of 46.875 Hz. Since the display is 1024 pixels wide, each dot represents the average or peak value of 4 bins.

4 bins x 46.875 Hz = 187.5 Hz per pixel.

If we zoom in until the panadapter is displaying a 96 kHz span, as we did before, the panadapter resolution changes to 93.75 Hz, meaning that now only 2 bins are being averaged or measured per displayed pixel.

2 bins x 46.875 Hz = 93.75 Hz per pixel.

If we zoom in until the panadapter is displaying a 48 kHz span, the panadapter resolution changes to 46.875 Hz, meaning that now only 1 bin is being displayed per pixel.

Zooming in past that point results in more than one display pixel displaying the same bin value. The image starts to degrade. At high levels of zoom the spectrum noise looks rounded and the waterfall images are blurred. Decreasing the sample rate improves the image but halves the span of the display. Increasing the number of FFT bins reduces the bin width, increasing the resolution at high levels of zoom, but slowing down the refresh rate of the spectrum and waterfall display.

## PANADAPTER FREQUENCY

The panadapter frequency is the frequency that is at the centre of the currently displayed Span. It is the frequency that the radio is tuned to, i.e. the VFO frequency displayed on a 'radio with knobs.' If the radio is displaying a panadapter, the receiver you are listening to can be anywhere across the displayed frequency range. Some SDR programs indicate the panadapter and the slice receiver frequency, others only display the panadapter frequency or the slice receiver frequency.

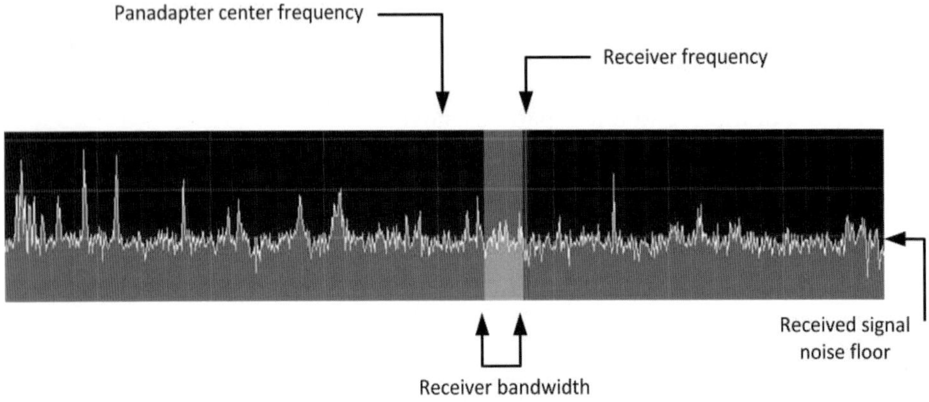


Figure 16: Panadapter features

If the radio is displaying a bandscope, the slice receiver frequency is at the VFO frequency and it is displayed at the centre of the display. Sometimes a sub-receiver may be shown offset from the centre of the display. Or, if 'split' operation is enabled, a marker may show the frequency that the radio will transmit.

## RECEIVER FREQUENCY

The panadapter shows many received signals across the band, but you only want to listen to one or maybe two of them at a time. The receiver graphic is an indication of the part of the spectrum that is being passed to the demodulator software and associated filters. It shows the frequency and sometimes the bandwidth of the signal that you are listening to. The receiver frequency is depicted as a vertical line on the panadapter spectrum display. The position of the line on the panadapter is equivalent to the frequency that would be displayed on the VFO frequency indicator of a conventional receiver. A shaded area may be displayed next to, or on both sides of, the receiver frequency to indicate the receiver's bandwidth.

You can click the mouse, or tap on the screen to move between the signals that are displayed on the panadapter. Sometimes this moves the bandscope so that the receiver stays in the centre of the display and sometimes the panadapter stays fixed and the receiver moves. With Thetis and the latest Yaesu and Icom radios, you have the choice of either option. You may be more comfortable with one mode or the other, or you might use different settings for different situations.

Some software supports two or more active slice receivers on the same panadapter. With headphones or stereo speakers, you can listen to the DX or contest station in one ear and the pileup in the other. This is surprisingly useful. Some software offers stereo effects such as CW signals that move from left to right as you tune up the band. This gives you a sense of spatial awareness about where the signals are in a CW pileup.

## RECEIVER BANDWIDTH

The shaded area shown next to the vertical receiver frequency line is an indication of the bandwidth of the receiver. It is usually adjustable in very fine steps so that it appears to be completely variable. Sadly this indication is not available on Yaesu and Icom radios.

Normally you can adjust the high-frequency cut-off frequency and also shift the receiver's passband. This is equivalent to adjusting the I.F. width and shift controls, on a conventional receiver.

*Technical TIP: You should set the receiver bandwidth to the same width as the other station's transmitted signal, as indicated by the width of the waterfall trace. For SSB it will be between 2.4 kHz and 3 kHz. For CW, you will probably set the receiver bandwidth between 200 Hz and 500 Hz. Setting the SSB receiver bandwidth wider than the transmitted signal unnecessarily increases the noise level on the signal you are listening to. Keep the bandwidth tight to maximize the signal to noise ratio. Reducing the bandwidth so that it is narrower than the transmitted signal reduces the fidelity (quality) of the received signal, generally the high-frequency audio response.*

*Reducing or shifting the receiver passband can be used as a technique to reduce splatter and interference from an overlapping QSO (signal). Often there is more than one way to adjust the receiver bandwidth settings. Typically you can drag the edges of the shaded image with your computer mouse, adjust slider controls, or simply press a 'bandwidth' button.*

*By the way, many hams don't realize that you can change the pre-set bandwidth buttons for every mode on Thetis. Just right-click any bandwidth button to change the preset high-cut and low-cut frequencies on any of the twelve buttons. You can also customize the button names to suit your needs. I configured my SSB bandwidth control buttons so that I have 2.4, 2.5, 2.6, 2.8, and 3 kHz pre-sets. I also set the low cut to 180 Hz on all of the SSB filters because it suits my sound system better than the default of 150 Hz. I renamed the DIGU filters to 'wide', 'medium', and 'narrow' since the actual bandwidth is not critical.*

## PANADAPTER NOISE FLOOR

The wiggly line at the bottom of the panadapter spectrum display represents the noise level that you are receiving with your particular antenna system. Its average level is commonly referred to as the 'received noise floor.'

Due to the magic of FFT (fast Fourier transformation), the noise floor displayed on the spectrum display should be well below the noise within the receiver pass-band, as indicated in dBm on the S meter (with no other signals present). This means that the panadapter will be able to display any signal that you can hear on the receiver.

If you disconnect the antenna, you will see the noise floor of the actual SDR receiver without any received noise. The displayed noise at the bottom of the panadapter should drop by around 10 to 20 dB. Ideally, with the antenna removed, the spectrum should be flat right across the panadapter with no spikes or stray signals generated in the receiver.

The noise level that is displayed on the panadapter when the antenna is disconnected is mostly dependent on:

- The signal to noise ratio of the analogue to digital converter. (This is related to the ENOB, not the number of bits that the ADC uses to describe the sampled voltage levels).
- The sampling rate of the analogue to digital converter. In a direct sampling radio, this also determines the maximum frequency of the receiver.
- The process gain arising from the decimation ratio from the ADC sampling rate down to the sample rate of the displayed panadapter.
- The FFT bin width and windowing filter shape factor.

When you connect the antenna, the noise level rises because you are receiving atmospheric and electrical noise on top of the noise being generated inside the radio.

Generally, the noise level will be higher in urban areas and lower in rural areas. If the level of the noise displayed at the bottom of the panadapter does not rise when you connect the antenna, the receiver is not good enough to display very weak signals that are just above the ambient noise level at your location. This can happen if you are using an 8-bit radio like an RTL dongle or a HackRF.

A typical noise level for urban areas is -103 dBm. The typical noise level for rural areas is -109 dBm. The actual noise level you see at your house, with your antenna connected, is dependent on the effectiveness of your antenna system and the level of received electrical noise at your location.

Exceptionally large incoming signals appearing at the input of the analogue to digital converter can cause 'ADC gain compression,' which results in a higher noise level being displayed right across the panadapter. The noise floor may jump erratically. This is what causes the spectrum image to jump when the radio receives lightning crashes. ADC gain compression can happen even if the offending signals are outside of the bandwidth currently being displayed by the panadapter. You might not be able to see the interference signals on the display.

Receivers that don't have band-pass filters to eliminate out-of-band signals, and radios using 8-bit ADCs which have much less dynamic range to begin with, are more likely to suffer from this effect. They have a higher noise floor and, therefore less room between the noise floor and the maximum 'clipping' input level to the receiver.

*The dynamic range of a receiver is the difference in dB between the clipping level, which is the maximum level that the receiver can handle without severe distortion, and the noise floor, which is the 'minimum discernible signal' (MDS) that you can hear on the receiver.*

## WIDEBAND DISPLAY

None of the Icom, Yaesu, or Elecraft radios offer a wideband display of the entire HF spectrum. Wideband panadapters seem to be limited to the "other kind" of SDR transceiver, made by FlexRadio, Expert Systems, Apache Labs, etc.

The wideband display is used to indicate band activity rather than to display individual narrowband signals. The displayed resolution is much lower than the standard panadapter display. Say for example the wideband display covers up to 61.44 MHz and the display window is 1024 pixels across. That would mean that each dot on the panadapter display is at a spacing of one dot per 60 kHz. You will be able to see which bands have a lot of activity and which bands your antenna is suited for, but you will not be able to see individual SSB signals.

As a comparison, a standard 192 kHz wide panadapter displayed on the same 1024 pixel display, shows bins that are spaced at 187.5 Hz apart. You can see individual FM, AM, CW, or SSB signals. You can see more detail as you zoom in, but there is a limit. The display may look blocky at extreme levels of zoom. Select a lower panadapter sample rate if you want to view a smaller span with high resolution. A 48 kHz wide panadapter displays bins that are spaced 46.8 Hz apart.

There are more FFT bins than pixels across the display, so each displayed dot represents the peak or average level of a group of bins. This gives the wideband spectrum display a sort of 'peak hold' effect.

## PANADAPTER DISPLAY

On most SDR receivers, the main panadapter is built from the same IQ data that is used for the DSP and demodulation. Either the I or the Q data stream can be sent to the FFT code block used to create the panadapter. Generally, the I stream is used.

The FlexRadio 6000 and 8000 series of radios do the FFT calculation inside the radio so the data that is sent to the PC is just graphics information used for the panadapter display. In the Thetis software, the FFT for the panadapter is selectable between 4096 and 262295 bins. The number of bin levels that are displayed as pixels on the screen depends on the zoom level. If you decrease the sample rate, the display resolution of the panadapter improves and the maximum panadapter bandwidth reduces.

## INFORMATION DISPLAYED ON THE PANADAPTER

The spectrum display shows frequencies across the screen and the levels of received signals on the vertical axis. The waterfall display shows frequencies across the screen, with the levels depicted as brighter colours. The vertical axis of the waterfall shows the signals at earlier times. It depicts a history of the received signals.

Have a look at the panadapter image below. At the top, numbers indicate the range of frequencies being displayed. You can zoom in to show a narrower range in more detail, or out to the maximum panadapter bandwidth to show a wider range of frequencies. The maximum bandwidth of the panadapter is determined by the sample rate of the IQ data stream, which is always an integer division of the ADC sampling rate. Some radios don't display the extreme edges of the panadapter.

The numbers at the left side of the spectrum display indicate the received signal level in dBm. You can adjust the maximum and minimum levels shown on the display by using slider controls, or by dragging the display axis with your computer mouse.

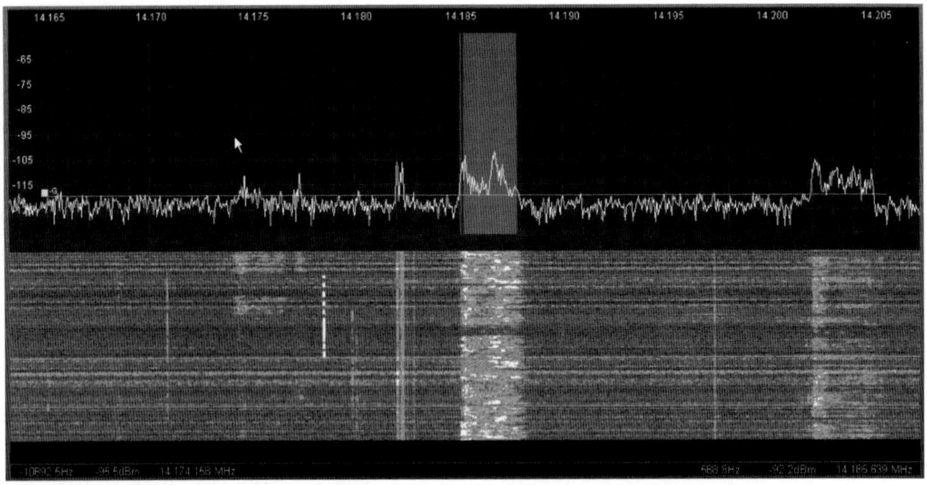


Figure 17: The spectrum & waterfall can show many interesting things

*Note that the dBm levels indicated on the spectrum display are the levels within the FFT bin bandwidth, not the levels inside the receiver bandwidth. The level within the receiver pass-band is shown on the S meter. It will be higher than the spectrum display noise level due to the difference between the FFT bin width and the slice receiver's bandwidth. If you can hear a received signal you will also be able to see it on the panadapter display.*

## MAKING THE MOST OF THE PANADAPTER AND WATERFALL

You can choose to display either the spectrum or the waterfall, but most people prefer to display both at the same time because they tell you different things. The spectrum display shows you the amplitude of signals across the band of interest.

You can see how wide each signal is, and on SSB or AM you get an indication of how flat the transmitted audio is. If you zoom in, so that the signal is a couple of centimetres wide on the screen, you can see any high or low-frequency roll-off in the modulated signal. You can see the effect of splatter caused by a transmitter or linear amplifier with poor linearity, as a spectrum image with wide skirts beside the wanted transmission. More importantly, you can see interference signals from other stations, carriers, Ionosondes, over-the-horizon radar, pirates, and in my case spikes from ADSL routers in the neighbourhood. Being able to see these interfering signals gives you a good start on adjusting your receiver bandwidth and filter settings to minimize the problem. You could precisely locate a manual notch filter, adjust the receiver bandwidth, or set a noise filter.

*Technical TIP: The spectrum display is very fast. I find it less irritating to leave the averaging function turned on. It slows down the flicker from the noise displayed on the spectrum display. If it still bugs you, adjust the display so that the noise floor is not showing at the bottom of the screen.*

The waterfall adds the dimension of time. It shows a history of the signals that have been on the spectrum display. Strong signals are shown as bright colours and weaker signals as weaker or darker colours. The background noise level should be very dark or black. One or two models allow you to scroll the waterfall image even further back in time, to show signals that were no longer being displayed. The Yaesu transceivers give you a 3DSS waterfall view. I don't believe it adds much value.

You can display the waterfall with a range of different colour palettes including a grey scale. Everyone has their favourite, so try them out. Changing the waterfall levels alters the colour range and the contrast between weak and strong signals. It can show very weak signals that are so far down in the noise you can't easily see them on the spectrum display, especially weak CW stations.

You can tell a lot from the waterfall display. Bursts of colour indicate short transmissions such as a CQ call. CW signals can almost be read off the screen and you can definitely see a 'clicky' signal as small horizontal streaks at the start and end of the 'dits' and 'dahs.' You can see a slow CW signal to the left of the SSB signal in the image above. Stronger signals are represented by brighter colours like orange or yellow and weaker levels are more subdued. With SSB signals, you can sometimes tell if the operator is a woman because their signal has more power (brighter colours) in the higher frequencies further from the carrier frequency and less in the lower audio range close to the carrier frequency. RTTY looks like railway lines. See the images of several digital text modes on page 98.

It is easy to differentiate between PSK and RTTY transmissions by looking at the shape and width of the tracks on the waterfall. Noise on a frequency will show as a fuzzy brighter region down the screen and stray carriers as a bright vertical line.

Lightning or other wideband noises will show as a bright line across the waterfall, often accompanied by a jump in the spectrum noise floor and a 'static crash' in the audio. Swept signals will zip across the spectrum screen leaving a diagonal line on the waterfall. Horizontal bands across the waterfall are caused by variations in the amount of the received wideband background noise. Ignition noise from nearby cars tends to look like repetitive bright dashes across the waterfall.

The spectrum is drawn like the trace on an oscilloscope from left to right across the display. A complete waterfall line is normally added to the waterfall image immediately after each spectrum line has been displayed. You can sometimes adjust the number of times per second that the spectrum and waterfall are redrawn on the panadapter. This is called the 'frame' rate.

The panadapter image above shows three upper sideband SSB signals with differing signal strengths. You can see from the shaded part of the panadapter that the operator is listening to the strongest signal. The waterfall indicates that there was a CW signal at about 14.178 MHz, and the lack of a corresponding signal on the panadapter spectrum indicates that it has stopped transmitting. It was probably someone setting up a linear amplifier before calling CQ. You can see two weak fixed carriers and an interference signal just to the left of the receiver slice probably from an ADSL router.

The receiver bandwidth as indicated by the shaded section on the spectrum display has been adjusted to almost the same bandwidth as the transmitted signal, which is clearly shown on the waterfall display. Note the small 150 Hz gap between the lower cut-off, of the receiver bandwidth and the receiver frequency indicated by the vertical line. At the top of the panadapter are the frequencies included in the displayed span. You can zoom in for more detail or out to show more of the band. The numbers on the left of the spectrum display indicate the received levels in dBm.

## USING A PANADAPTER TO SHOW INTERFERENCE

The spectrum display on a software defined radio is basically the same as the display on a spectrum analyser, so your SDR can double as a radio test set. The SDR panadapter displays the signal amplitude on the Y-axis and frequency on the X-axis, the same as a spectrum analyser does.

With software like Thetis, the panadapter vertical axis is marked in dBm and it is pretty accurate, so you can make the same sort of comparison measurements as you would with a spectrum analyser. Some useful features are missing such as the ability to add marker frequencies and have the test set measure the difference in frequency and amplitude between them. Nevertheless, it is still a very useful tool for looking at interference signals. You can also monitor the quality of your own transmitted signal. Unlike most spectrum analysers, you get a waterfall display, which can show interference in the time and frequency domain at the same time.

SDRs are great for helping to identify the type of modulation being used by intruders on the bands. Digital modes all have their own individual 'look' on the spectrum and waterfall displays.

The bandwidth and pulse duration can be read from the waterfall axis. A waterfall can indicate what type of digital mode is being transmitted. PSK looks like a railway track. A wider PSK trace indicates that a faster baud rate is being used (PSK63 or PSK128). RTTY also shows as two lines but the spacing is wider due to the 170 Hz tone offset. Modes like JT65 and Domino use tones sent in sequence so they look like little dots climbing stairs. Radio Fax, digital transmissions, SSTV, and phase shift modes all have distinctive images on the waterfall. With CW, you can see the individual dits and dahs as short and long lines. You may even be able to read the Morse code off the screen.

The spectrum display can show wideband noise from devices like LED lighting, plasma TVs, or other interference signals. You can see if a ham has a badly adjusted transmitter causing splatter across the band. You can see exactly how wide each SSB transmission is, or if a transmitter is off frequency. If you are in a contest, you can see fixed carriers, or stations transmitting on top of other stations. Stations that are 'tuning up' show clearly on the spectrum and the waterfall, as a fixed carrier or a CW signal.

Signals that are being swept across a band of frequencies show as a diagonal line on the waterfall and as a 'bunny' that hops across the spectrum display. You can also tell what the normal received noise level is at your location. This is handy to know if you ever experience wideband noise problems.

As a PSK operator, I often see over-modulated PSK signals on the waterfall display. They look like a wider than normal PSK signal with horizontal streaks flecking away from the sides. A clean PSK signal looks like a fuzzy railway track. You can see the same signal on the waterfall display of your digital mode software, but filtering in the software may hide the distortion caused by overmodulation. CW key clicks look interesting. They appear on the waterfall as horizontal streaks at the top or bottom of each dit or dah.

## A MISCONCEPTION ABOUT BANDSCOPES AND PANADAPTERS

I had an email from a person who had recently purchased an IC-7300 transceiver. He was very concerned that none of the fancy filters had any effect on the interference and unwanted signals he could see on the spectrum and waterfall display. I understood his confusion and had to point out that this is the way that the bandscope or panadapter is meant to work. One of the big benefits of an SDR, or any radio with a bandscope, is the ability to see interference, other signals on the band, or two stations transmitting on the same frequency. Noting the interference, you can avoid the frequency or use the DSP notch and noise filters to mitigate the effect of the problems, but the offending signals remain on the display.

The spectrum and waterfall shows received RF signals before the DSP stage. Some SDR software includes a second spectrum and waterfall display of the demodulated audio signals which will show the effect of the DSP filters. My reader was worried that large signals or interference showing on the display would affect the reception of signals on the wanted frequency. This is a valid concern.

In a traditional superheterodyne design or a hybrid SDR with a superheterodyne front end, multiple large signals at the input to the receiver cause intermodulation distortion (IMD) which could create noise on the wanted frequency. There is a linear relationship between the level of the large unwanted signals and the amount of intermodulation distortion that is created. There are two other problems. One is 'blocking' which is a desensitisation of the receiver caused by the large incoming signal(s), and the other is AGC 'pumping' where the superheterodyne receiver turns down the I.F. gain (level), to stop the receiver going into overload. Unfortunately, this also reduces the level of weak signals that you want to listen to or decode. The manufacturers tackle these issues with 'roofing filters' to limit the large 'out of band' signals being received. These filters are too wide to reduce the level of interfering or unwanted signals that are close to the wanted frequency. Yaesu uses narrow 'VC-Tune' variable filters that automatically track the VFO frequency, and older radios used manually tuned pre-selectors to further improve the receiver selectivity. The AGC pumping problem is mostly resolved by using DSP filters in the I.F. stages of the receiver.

SDR receivers are different. Large received signals do create IMD, but the effect is not linear. The amount of intermodulation distortion produced is not predictable. It depends on the levels and frequencies of the offending signals and sometimes it is improved when the band has many large signals. In most cases, the level of the IMD products remains below the local received noise floor until the offending signals are quite powerful, usually not far below the receiver's overload level.

SDRs do not suffer from blocking desensitisation. However multiple large signals can cause ADC gain compression, which has essentially the same effect. SDR receivers usually have very good IMD and blocking dynamic range (BDR) performance. They inherently have a larger dynamic range than superheterodyne receivers. The AGC is not usually used to reduce the receiver sensitivity. Instead, it is used to limit the increase in audio output level as the input level rises. This stops big signals "blowing you out of the room." As a result, SDRs don't suffer from AGC pumping effects. Strong signals do not affect the level of weak signals. That means that the reception of your wanted signal will probably not be affected by interference or large signals that are outside of the SSB or CW bandwidth of the receiver.

# AUDIO CONNECTIONS FOR DIGITAL MODES

It is very common to connect a USB cable or separate cables from the auxiliary audio input and output from your transceiver to your PC soundcard so that you can use digital mode software to send and receive RTTY, PSK, JT65, SSTV, Packet, CW and many other formats.

With black box software defined radios, the audio section of your receiver or transceiver is being implemented in the PC software. This presents a problem when we want to interface the radio for digital mode operation. You need a connection between the SDR software and the digital mode software that does not involve physical cables and soundcards. FlexRadio includes an integrated utility for managing the audio interconnection. But strangely, this functionality is not usually built directly into open-source SDR software. A separate utility program is used to provide interconnection of the audio signals.

## VIRTUAL AUDIO CABLES

**VAC** (virtual audio cable) is commercial software provided by Eugene Muzychenko. http://software.muzychenko.net/eng/vac.htm. VAC works like an additional soundcard on your PC. It does not use the physical soundcard, which remains available for the audio output to your speakers.

Using VAC is like having cables between the SDR software and the digital mode software. You have to set up an 'audio cable,' using the VAC setup software. This only has to be done once. After that, you can set the SDR software to output audio on VAC 1 and the digital mode software to receive the audio on VAC 1.

You are encouraged to download the free trial version before paying to register the program. There are three options at differing prices; no support, basic support, and full support. I went for the cheaper 'no support' option and have had no problems. Note that the trial version has an annoying embedded voice message, so you really do have to register the software before you can use VAC for transmitting digital mode signals.

Once the VAC software is installed it will be available for any SDR software you might install. VAC can also be used to connect other programs. For example, you can use it to record Skype conversations by sending the audio to a sound recording program.

**VB-Audio** offers an alternative to VAC. There are two options; 'VB-Cable' and 'HiFi-Cable' See http://vb-audio.pagesperso-orange.fr/Cable/index.htm. VB-Audio is 'donationware.' If you use the software, please donate to the developer so that he is encouraged to write more great programs. You get one 'cable' for free and additional 'cables' are available for a small donation.

HiFi Audio is similar except that it is more configurable and can support higher sampling rates. It is easy to get VB-Cable working. Apart from pointing the applications at the new virtual soundcard, there is no setup required.

## IQ OUTPUT FROM THETIS AND OTHER PROGRAMS

Some radios and some SDR software can send IQ data to other programs. Thetis can output two IQ signals to other software via VAC (or similar). FlexRadio SmartSDR can send up to four IQ data streams via the DAX interface. The ExpertSDR2 software can also send two channels of IQ data to another application. Other programs may have a similar ability. These IQ audio signals can be used by other SDR software applications, perhaps to decode specialist modes like DRM short-wave broadcasts or for CW Skimmer. The IQ output from receivers like the DSP2 or the IC-R8600 can be used with SDR Console v3, HDSDR, or SDR# to provide additional receiver audio or a large bandscope/panadapter display.

## DAX - DIGITAL AUDIO EXCHANGE

DAX is the audio interconnection manager provided with SmartSDR for the FlexRadio 6000 and 8000 series radios. It is much more powerful than the audio interconnection programs currently used with other radios.

DAX provides four virtual audio output cables for the FLEX-6600 and FLEX-8600 transceivers, to match the four available receiver slices. The FLEX-6400 and the FLEX-8400 have two virtual audio cables. DAX can route audio to and from programs on any computer device connected to your local network. This allows you to listen to audio from receiver slice one on one PC, while another device decodes RTTY at a different location. Depending on the radio, you could have four receiver slices connected to four different digital mode programs at the same time.

DAX also provides up to four audio IQ channel outputs. On the FLEX-6600 and FLEX-8600, they provide up to 384 ksps of data. This can be used as a single 384 kHz wide spectrum display, two 192 kHz spectrums, or four 96 kHz spectrums. The FLEX-6400 and FLEX-8400 only have two DAX IQ channels up to 96 kHz wide. The IQ outputs can be used to display large panadapters on different computers or to run CW Skimmer and other SDR software.

*Technical TIP: Windows defaults to an audio sample rate of 44.1 kHz, which is the standard for CD audio. The DAX connection works best when set to a multiple of the 24 kHz sample rate used by SmartSDR. It is suggested that you change the sample rate on all DAX audio RX, DAX audio TX, DAX reserved audio RX, and DAX reserved audio TX channels, to the 48 kHz sample rate.*

# ONLINE SDR

There are several online SDR platforms that you can use for free. Most of them allow you to control an SDR receiver including changing modes and frequencies. Unlike older platforms, the SDRs support multiple users at the same time. In most cases, you can contribute to the network by connecting your own SDR for others to use. These remote receivers are great for checking what bands are open in different countries, or for listening to stations when the bands are closed in your region. You can also listen to your transmissions, although there is often several seconds of latency.

### Map of SDR receivers worldwide (RX-TX info)

https://rx-tx.info/map-sdr-points

This website shows the location of web SDRs worldwide. You can zoom into a country and select an icon to get a link to the web receiver. You can select a frequency, band, or country, or search on title, QTH, or URL. The icons are colour-coded; green for WebSDR, blue for Open Web Rx, and purple for KiwiSDR. The site lists the available web SDRs in a preview mode, on a table, or a map. You can easily add your online SDR to the network. This is the best way to access the Web SDR networks listed below.

### Web SDR http://websdr.org/

Web SDR is a network of SDR receivers with differing frequency coverage and antennas. Most stations use omnidirectional antennas as different users may want to listen to stations from different directions. Each station link on the website includes a short description of the site's location and operator, its Maidenhead grid locator, antenna, the frequencies that the radio covers, and a hyperlink link to the receiver. There is a note about the number of currently connected users. Some WebSDR interfaces include links to switch to other receivers without going back to the main website. You may have to press the 'Start Audio' button on a popup window. WebSDR.org currently lists 152 servers with receivers covering a total of 731 MHz of radio spectrum.

*'A **WebSDR receiver** is a Software-Defined Radio receiver connected to the internet, allowing many listeners to listen and tune it simultaneously. SDR technology makes it possible for all listeners to tune independently, and thus listen to different signals; this is in contrast to the many classical receivers that are already available via the internet.' Websdr.org.'*

*A **WebSDR server** consists of a PC running Linux and the WebSDR server software, a fast internet connection (about a hundred kbit/s uplink bandwidth per listener), and some radio hardware to feed antenna signals into the PC. This radio hardware is typically a quadrature mixer connected to the PC's soundcard, like the popular SoftRock kits.' [Websdr.org].*

Several listeners using WebSDR can operate a receiver on the same wideband panadapter at the same time. Many of the sites have multiple bands available. This is different from websites where you can take over the operation of a conventional receiver or transceiver. In that case, only one person can take remote control at a time. Most of the WebSDR receivers operate in the HF spectrum but a few offer VHF and UHF coverage. There are even receivers operating on the 10 GHz band or dedicated to receiving signals from amateur satellites.

It can be interesting to tune around the band using a receiver located in another country, especially when propagation means the bands are dead at your location. You could use the technology to find out what regions have good propagation or to listen to a station, which is blocked by noise or interference at your place. Or you may be able to use a remote receiver to overcome antenna restrictions at your location.

The majority of HF band WebSDR sites have wire or vertical antennas because, with multiple users connected, it is not practical to use directional antennas. Everyone would want to point the beam in a different direction.

One of the cool things that you can do with WebSDR is to listen to your transmissions as they are being received in another country. Have you ever wondered what your signal sounds like? Or if you can be heard in Europe or Australia? With WebSDR, you can find out. Be aware though, that there is often significant latency in these systems. The signal you hear from the remote receiver may be several seconds delayed.

### OpenWebRX+

https://fms.komkon.org/OWRX/ and https://www.openwebrx.de/
"OpenWebRX+ is a multi-user SDR receiver that can be operated from any web browser without the need for any additional client software. It is the ideal solution to provide access to the HF spectrum at your location of choice to a wide audience. All you need is a computer, an SDR device, and network access."

The web-based software can decode FM, WFM, AM, SSB, CW, SAM, and many digital text modes. Optional additional software adds the DMR, D-Star, NXDN, YSF, M17, and FreeDV digital voice modes. Clicking on digital voice signals supplies the timeslot data, callsigns, etc. The ADSB decoder shows links that you can click to see the relevant aircraft's details. It also decodes ACARS aircraft information packets. If the digital information includes position information you can click on the map icon to display the aircraft, vehicle, or ship on the map. The positions on the map are updated when new position beacons are received.

For more information about OpenWebRX+, see the excellent 'TechMinds' video at, https://www.youtube.com/watch?v=YGd9Ym96pFY.

You can get a taste of Open Web SDR RX+ using the links on the RX-TX info site mentioned above, or at this website where 'Frank' has made several receivers available at his location in Germany.
http://hgdyn.spdns.org:8099/#freq=145600000,mod=nfm,sql=-56

To use OpenWebRX+ with your radio, you need a USB-connected SDR radio, and a Linux computer to run the software. It works well on a Raspberry Pi 4 or 5. You can get the program at the Ubuntu or Debian repositories, or there are Rpi SD card images.

Common radios include; Airspy (HF+, Discovery, R2, and mini), HackRF, Perseus, RTL, SDRplay, Hermes Lite 2, and HPSDR.

There is more about OpenWebRX+ in the SDR Servers chapter on page, 85.

### KiwiSDR http://kiwisdr.com/public/

KiwiSDR is another web-based SDR network. It is based around the KiwiSDR receivers designed and produced by John Seamons, a ham radio operator (ZL4VO/KF6VO) who lives in New Zealand.

You can use any web browser to connect and listen to one of the 800 KiwiSDR radios all over the world that have been made available by their owners for public access. See https://rx-tx.info/map-sdr-points, http://rx.kiwisdr.com/, or http://rx.linkfanel.net/.

They tune from 0-30 MHz and can receive and decode many signals including:

- Amateur radio AM, FM, USB, and LSB voice signals around the world.
- Decode amateur radio FT4, FT8, and WSPR digital signals and CW (Morse).
- Decode and display weather FAX signals.
- Medium, Long Wave, and Very Low Frequency (VLF) stations.
- Shortwave international broadcast stations.
- DRM (Digital Radio Mondial) digital broadcast decoding.
- International airline traffic directly from the cockpit during flight and the control tower (HFDL).

You can purchase a KiwiSDR and add it to the worldwide network, or just use it over your local LAN. The radios are available from WiMo.com, Martin Lynch and Sons (hamradio.co.uk), and https://kiwisdr.nz/.

You do not have to connect the radio to your computer. The KiwiSDR is always accessed over a network connection using a web browser running on a computer or mobile device. This can be from your local network or by anyone on the Internet if you have chosen to make your KiwiSDR publicly available to Internet users by listing it on http://rx.kiwisdr.com/.

## QO-100 geostationary satellite

The QO-100 Es'hail-2 geostationary satellite has a narrowband transponder for traditional amateur radio narrowband modes such as SSB, CW, and SSTV. No FM and no digital voice modes are allowed. There is a web SDR for the narrowband transponder, hosted at the Goonhilly Earth Station in Cornwall, England. https://eshail.batc.org.uk/nb/. It allows you to listen to the satellite's 10 GHz downlink on USB, LSB, or CW. You can record the audio and use it to decode other modes using digital mode software.

The wideband transponder is used for amateur radio digital TV. The web SDR receiver is at https://eshail.batc.org.uk/wb/ It is designed for people to check their uplink power. You can't decode the images or download the audio stream.

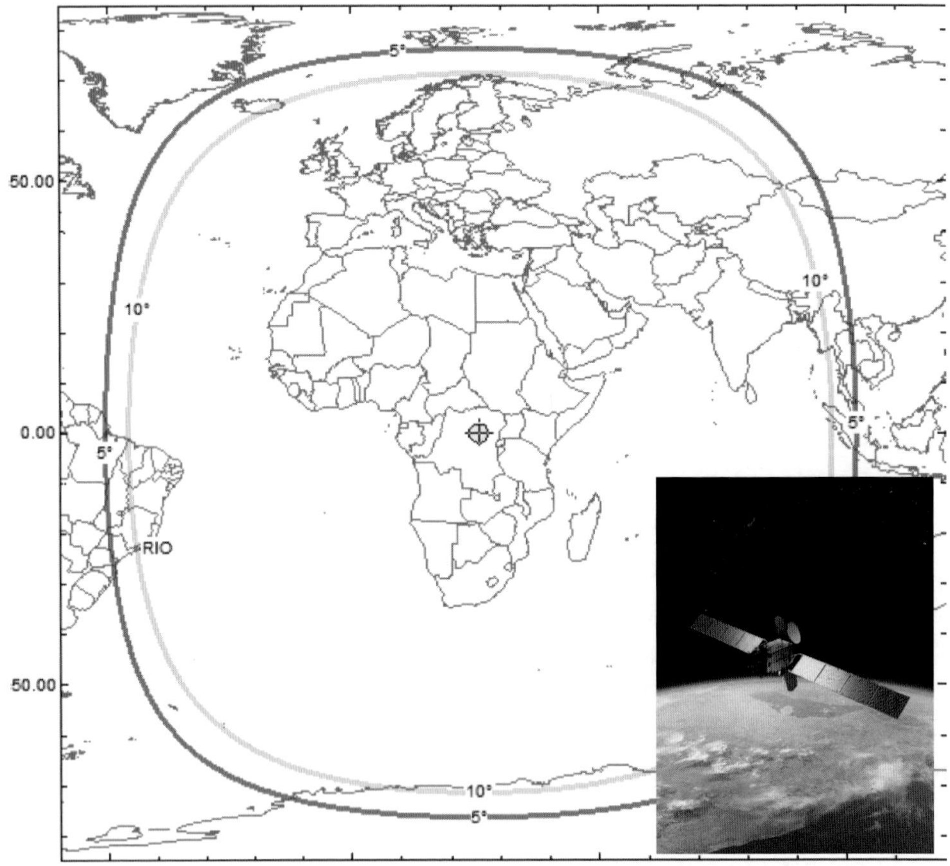


Figure 18: QO-100 Es'hail-2 satellite coverage

## SDR-Radio.com

SDR Console v3 allows you to operate remote SDR receivers as if the radio was attached to your computer. It will work with a receiver that you have at home, or with a receiver connected to one of the remote servers around the world. You can try out the software by connecting to a radio over the Internet and then continue to use it after you have purchased a software defined radio.

You can download SDR Console V3 at https://www.sdr-radio.com/console.

The program was written by Simon Brown, HB9DRV who was the author of the very popular 'Ham Radio Deluxe' software. It supports a wide range of SDR receivers including the RFSpace radios, AirSpy, most dongles, and Softrock receivers. Version 3 supports transmitting using the Apache labs ANAN transceivers, Hermes Lite 2.0, and probably other software defined radio transceivers. It also has a 3d waterfall feature a little like the Yaesu transceivers, although it looks a lot nicer than the Yaesu version.

Figure 19: SDR Console 3D waterfall

### Using SDR Console with a remote radio

When you have installed the software, you can connect to one of the online servers. Click Home > Select Radio > Definitions > Search > V3 Server (at the bottom) > Browser Online Servers. Click a server that has a green tick and then Select. If the server accepts a guest the username and password will automatically be added and you can click OK. If 'username' stays blank you need to enter a password from the server's website.

There is a list of the servers that have guest passwords on the selection screen (scroll right). Save the selection as a new radio definition so you can find it listed as a radio next time. Select the server on the Select Radio window, and click Connect. You should see some information arrive. Click Start.

If you want to, you can configure your computer as a server and add your receiver to the worldwide list of radios that others can use. The server software installs itself at the same time as it installs SDR Console V3. You may have to open ports through your firewall to allow others to access the radio.

### Benefits of Internet-based SDR

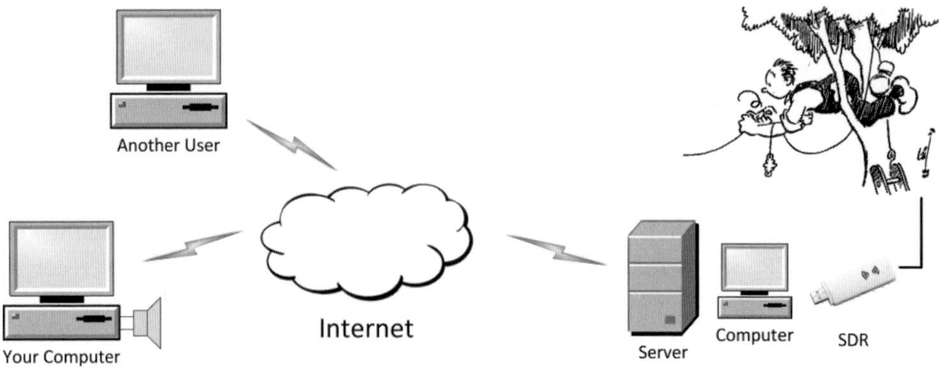


Figure 20: Internet based SDR allows several people to share the same radio

Internet-based software defined radio servers allow several users to operate receivers on the same SDR panadapter at the same time. You need a PC with an Internet connection but you don't need a radio. All of the online users can tune in to any signal that is displayed on the panadapter. Often, each user can select a mode and receive filter bandwidth. The SDR-Radio.com, SDR Console system even lets you configure the remote radio's panadapter bandwidth.

Internet SDR allows you to monitor your transmitted signal and it can allow you to work stations when you can't receive them at your location. It is up to you whether you consider this kind of remote operating to be 'cheating.' It is illegal in most contests and I don't think that contacts made in this way are valid for certificates like DXCC. Maybe they are since the transmitter is still at your house. You are only extending the reach of the receiver. This type of remote operation is good for shortwave listeners since they can continue to tune the bands when the propagation is poor at home. But, again... is it cheating? It's up to you.

You might be able to combine the signals from a remote server with those that you receive at home. But I suspect that the latency from the remote radio would make this type of operation impractical.

Using a remote radio, either a receiver or a full transceiver might be the only way that you can continue with your hobby. Many residential developments have strict restrictive covenants that ban any kind of antenna system. You can also use a remote radio to operate when you are away from home, while on holiday, or at work.

If you set up a server at your home, you can operate your radio from anywhere that has a fast Internet connection.

# SDR ON PHONES

There are SDR applications for Android and iOS (Apple) phones. Some are Internet browser-based applications that work with networked SDR radios, as discussed in the last section. Some use client software for server/client radios like the FlexRadio 6000 and 8000 transceivers, and some are SDR applications that will run on the phone hardware.

## ONLINE WEB SERVERS

The web servers linked at https://rx-tx.info/map-sdr-points (discussed in the previous section) will work on a phone, but the software is not optimized for small screens. The receivers work but they can be a challenge to control. They should work better on Internet connected devices with a bigger screen such as an iPad, Android or Windows tablet, or Linux single-board computer.

## IPHONE APPS

### Apple iOS SDR apps

An online search revealed; SmartSDR (FlexRadio), SDR control for Icom, rtl_tcp SDR, MagicSDR, WebSDR, SDR radio utilities, SDR-Control Mobile, SDR Receiver, DRM+, and SDRangel. They all require payment for use. I have no experience with any of these apps.

**SDR-Control Mobile** by Marcus Roskosch is available for iPhones, with a different version for the iPad's bigger screen. It will control Icom IC-705, IC-7610, IC-9700 transceivers, and the IC-R8600 receiver. You get remote control including the panadapter display and integrated FT8 operation. It also has a built-in station log. The application is excellent for remote operation while you are away for work or on holiday.

**Rtl_tcp** can control an RTL dongle, HF+, SDRplay RSP, or a Hermes Lite 2 via UDP control.

**Magic SDR** supports connections to KiwiSDR, RTL-TCP, Airspy, Hermes Lite, and HiQSDR.

**DRM+** decodes DRM broadcast from audio input to the phone.

### Using an SDR with an Apple iOS device

As far as I know, it is not possible to connect an SDR receiver directly to an iPhone, in the same way as you can with an Android phone. But you can control an SDR over the Internet if you have the right client software on the phone. There is a **SmartSDR** client by Marcus Roskosch for the iPhone which can also be used on an iPad.

## ANDROID SDR APPS

An online search revealed; SDR Touch, SDR++, SDR driver, MagicSDR, Dump1090, SDRangel, ADS-B Radar RTL-SDR, HAM FM radio, Aerial TV, ADSB Receiver, and MWK-SDR. I have no experience with any of these apps because I don't have an Android phone. SDR++ has support for more SDR receivers than most of the other phone apps.

### Connecting an RTL dongle receiver to an Android phone

You will need a USB Type C male to Type A female OTG (on the go) adapter cable to connect an RTL dongle to an Android phone. They are very cheap from online sources. Download and run the SDR driver software by Martin Maranov.

The video shows 'RF Analyser' by Dennis Mantz as the SDR software, but other options should work as well.

See the video at https://www.youtube.com/watch?v=COUb3Twd3B4.

**MagicSDR** processes signals that are streamed over the local network from the rtl_tcp server, which is running on the host computer. The smartphone which is running MagicSDR, can also act as the host computer.

To start playing with MagicSDR, you need to set up a server on a host computer to which SDR peripherals such as an RTL dongle will be connected, or you can connect an SDR directly to an Android smartphone via a USB OTG cable.

**ADS-B Radar** RTL-SDR https://ads-b-radar-rtl-sdr.en.softonic.com/android. There are two ways to use ADS-B Radar RTL-SDR. The first method involves using an RTL-SDR dongle and an SDR driver, which can be downloaded from Google Play. By selecting the Driver option in the application menu, users can easily access and install the necessary driver. The second method involves connecting to an instance of rtl_tcp running on a remote computer, potentially over WiFi. It is important to have a strong network connection for this option to work reliably.

**ADSB Receiver** is another ADSB app for receiving beacons from aircraft using an RTL dongle with an OTG cable.

**Dump1090** is intended to receive the wideband ADSB data from aircraft beacons on 1090 MHz. It requires an RTL dongle with an OTG cable.

**Ham FM Radio (RTL-SDR)** *"This application is about having a bit of fun with those amazing little RTL-SDR dongles, whilst listening to HAM radio conversations. It's nice and easy to use so why not give it a try?"* Bands: 144 - 148 MHz (2m) and 150 - 174 MHz.

**MWK-SDR** is an SDR-based news feed. Not a radio control program.

**Aerial TV** is a program for using the RTL dongle in its intended mode as a digital TV receiver.

# SDR SERVERS

SDR servers are becoming very popular. You can attach a small box SDR or an RTL SDR dongle to a small computer such as a Raspberry Pi and leave it running all the time. When you want to use the radio, you simply open an app or web browser on your phone, tablet, or PC and operate the radio. Most of these systems are for receivers only. There are two types which both use Internet and/or local LAN connectivity. One type uses a web browser interface. Typically these radios can be accessed over your local LAN, or via the Internet if the required ports have been opened on the Internet router. The second type requires a copy of the client software to be running on the client computer or phone. Again, the radio can be accessed over your local LAN, or via the Internet if the required ports have been opened on the Internet router. In most cases, WiFi is supported but it may not have the continuous data capacity required for a good result.

### SmartSDR

SmartSDR for the FlexRadios has always been a client/server application. The radio contains the server and it can be accessed via a client running SmartSDR. One or two users can connect to the radio over the local LAN, WiFi, or the Internet. The client could be a Maestro control unit, a computer, tablet, iPad, or iPhone. See the SmartSDR chapter on page 140 for more information.

### SDR++

SDR++ is a multi-platform (Windows, Linux, macOS, and Android) SDR receiver program. It has a built-in server which is started with a command line on Linux or Windows. I'm not sure about the other platforms. This lets you access the radio from any connected device running SDR++. You would probably have to open a port on your router if you want to access the radio from outside of your local LAN. You can plug a small box or RTL dongle into a USB port on your Raspberry Pi4 or 5 and listen to it on a desktop PC, tablet, or Android phone.

After the usual messing around, I managed to get SDR++ to work on the Raspberry Pi5. The server function is started from the command line, `sdrpp -server`. You don't have to have the main SDR++ application running on the Rpi or other host computer.

*TIP: If it is installed, you need to stop OpenWebRX+, before attempting to use SDR++.* `sudo systemctl stop openwebrx` *stops the OpenWeb server.*

I found that I had to slow the data rate on the Windows client down from 2048 ksps to 1024 ksps to stop the audio from breaking up when I was using the WiFi on the Raspberry Pi. It can be used at full speed over a wired Ethernet connection. The SDR++ server did not seem to work as well as the OpenWebRX server, which worked fine at 2.048 Msps and also at 3.200 Msps.

### OpenWebRX+

I mentioned OpenWebRX+ earlier. It is an SDR receiver server that can be operated from any web browser. You can view radios that are enabled for public use from the map of SDR receivers at https://rx-tx.info/map-sdr-points.

I like OpenWebRX+ a lot, it decodes modes that previously required separate digital modes programs. The correct window pops up when you select a mode. If the data contains position information you can open a map to view it. Press the Map icon twice to toggle the map type. Clicking on an aircraft or ship identifier opens a website with information about the craft. The digital voice decoders work well and you get the same sort of information about the stations as you would using a 'native' radio with a display.

To use your own radio, either on your local LAN or to make it available worldwide, you need to have a connected SDR radio and a Linux computer to run the software. The server software works well on a Raspberry Pi 4 or 5. You can get the program at the Ubuntu or Debian repositories, or there are Rpi SD card images. I hate Linux and found the program tricky to get working. Download and burn the Rpi image file to an SD card, if you are not experienced with Linux. I used Balena Etcher to write the SD card. That was the most foolproof way, although there is no Linux desktop in that distribution. I did eventually make the Debian Bookworm version load. All versions are at https://luarvique.github.io/ppa/.

*TIP: For Raspberry Pi SBCs running the latest operating system* ***DO NOT add the Debian Bullseye repository****. Ignore the first two instruction sections and* ***only*** *run the installation lines in the Debian Bookworm instructions. The Bullseye instruction will not work and if you install it, you will have to manually remove it from the directories before the Bookworm version will work! I found this out the hard way. I hate Linux!*

Set the Sudo shell with `su -` or prefix the lines with `sudo`. Run the three lines of installation code. I used SSH from my PC because I find it easier to cut and paste the code. Then `run apt install openwebrx`. Reboot the Pi and the server should run. If not, then run `openwebrx.`

Once loaded and running on the RPi. It runs automatically when you boot the Rpi. I could access it by setting the web browser URL on my PC to, the IP address of the Raspberry Pi5. http://192.168.1.12:8073/ The address will be different on your network. You do need the `:8073/` port number. Using http://localhost:8073/ did not work on my computer. The program runs on your web browser so you don't need to install anything on your computer. Select 'settings' to set your name and location details. **Settings** lets you add 'profiles' for different bands. The RTL dongle only has a 2 MHz panadapter bandwidth, so you need a profile for each 2 MHz segment.

The distribution does not include the DMR, YSF, NXDN, and D-STAR digital voice modes. They have to be added later. The suggested method of `sudo install-softmbe.sh` did not work for me.

I followed the instructions listed under 'Installation' on https://github.com/knatterfunker/codecserver-softmbe. Also, see the YouTube video by K9SWX at https://www.youtube.com/watch?v=VEayBGl-xLA&t=1420s.

`sudo systemctl stop openwebrx` stops the server.
`sudo systemctl start openwebrx` starts the server.

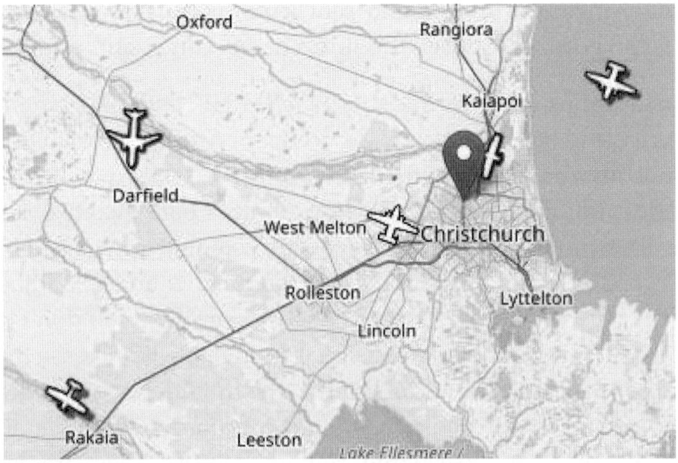


| Flight | Aircraft | Squawk | Dist | Alt (ft) | | Speed (kt) | | Signal |
|---|---|---|---|---|---|---|---|---|
| ZKTZZ | C8173A | 5502 | 5 km | 768↑ | 3450 | NE | 105 | -11.7 dB |
| SKL8 | C822DF | 5255 | 51 km | 1536↓ | 22600 | S | 313 | -38.2 dB |
| ANZ746M | C827F1 | 5522 | 13 km | 832↓ | 1975 | NNE | 152 | -38.4 dB |
| | C82986 | | 32 km | 1600↓ | 12550 | SW | 277 | -40.5 dB |
| ZKEQE | C80493 | | 42 km | 192↓ | 4300 | S | 109 | -39.1 dB |

Figure 21: ADSB information on 1090 MHz

The image shows my RTL dongle receiving ADSB on 1090 MHz. The aeroplane images move as the flight information is updated. This is live data received at my location so it can only show aircraft that are within range of the receiver.

### RTL-SDR Scanner (Borys Dyczko – 'shajen')

The scanner can record as many as 40 frequencies within the panadapter bandwidth of the receiver. That might be 2 MHz for an RTL dongle but as much as 10 MHz for some other receivers. You set the frequency and mode for each recorded channel. The software would be great if you wanted to record several trunk channels, or repeaters, for example, multiple air-band or marine-band channels. I have not tried out this software because the Linux installation looks difficult. The web page is at; https://github.com/shajen/rtl-sdr-scanner-cpp

There is a Tech Minds video describing the RTL-SDR scanner at;
https://www.youtube.com/watch?v=YzQ2N0VkKvE

# SDR FOR LISTENING

A lot of software defined radios are used for listening to the shortwave bands and other interesting signals like the aircraft or marine band. You can also listen to commercial traffic and unencrypted military communications. Aircraft ADS-B positioning information can be presented on maps showing commercial flights. ACARS text data includes information sent from aircrew to the ground stations. Maritime VHF AIS beacons indicate the ship's name, cargo, and position.

Software defined radios are great for shortwave, VHF, and UHF listening. You get great performance and often a much wider frequency range. Most SDR software has good noise suppression filters, which are important when you want to listen to broadcast-quality audio like music and news. Another big advantage is that you can set the bandwidth to match the signal that you are listening to. The AM filters in older receivers that were designed for SSB may be too narrow and not adjustable. Being able to adjust the receiver's bandwidth precisely means that you don't have to listen to noise from outside of the transmitted signal. Being able to increase the bandwidth greater than 3 kHz means that you can listen to the full fidelity of the AM signal. The audio sounds much nicer. Also, software defined radios have a better ability to drag out weak signals that are near large signals because they are less prone to intermodulation distortion.

## AM, SAM, DRM, DAB

There are a dwindling number of commercial **AM** broadcast stations on the HF shortwave bands. Many of the remaining ones are religious stations. Some countries still broadcast general music, news, weather, and current affairs. The BBC and Radio New Zealand Pacific are still broadcasting 24 hours a day on several frequencies. The frequencies are changed at night to maximize the reach into various regions as daily propagation changes.

The main reference book for SWLs (shortwave listeners) is, 'The World Radio and TV Handbook.'

You can check out https://ccrane.com/shortwave-frequency-list/ for shortwave frequencies. Also;
https://www.bbc.co.uk/programmes/articles/2x9tqt6mc05vB2S37j8MWMJ/global-short-wave-frequencies

https://www.rnz.co.nz/international/listen

**SAM** is an acronym for synchronous amplitude modulation. The SAM mode is useful when the received signal is fading because the resulting audio doesn't suffer from the loss of intelligibility that the audio from a normal AM detector does. The SAM detector in your SDR software uses a different algorithm to demodulate the AM broadcast signal.

Rather than using a simple 'envelope detector' the SAM detector phase-locks to the transmitted AM carrier and then uses a beat frequency oscillator with product detection. It is basically the same as an SSB receiver but it receives both sidebands. The advantage is that noise on the AM carrier frequency is not translated into noise on the received audio because the carrier is not used for demodulation. A nice clean oscillator is used instead.

**DRM** stands for Digital Radio Mondiale (French and Italian for World). It is a method of sending MPEG-4 coded digital audio signals over HF or VHF radios. It is widely used in Europe and it has replaced many of the old AM shortwave broadcast stations. DRM can deliver FM-comparable sound quality on frequencies below 30 MHz and the DRM+ variant is used for VHF frequencies. It is rapidly becoming the standard for digital broadcast radio, competing with DAB and satellite broadcast radio. DRM is more spectrally efficient than AM or FM, which allows more stations to share the band. Most DRM stations are in 'band I' (47 to 68 MHz), 'band II' (87.5 to 108 MHz), and 'band III' (174 to 240 MHz). Most people use the Dream program to decode DRM with their software defined radio. Winrad, SoDiRa, and Fraunhofer DAB/DRM are other options.

Most SDR software does not support the DRM mode directly because the DRM coding is quite complex. Even when the DRM coding has been demodulated, the audio is still in an MPEG format. However, you can connect the audio from your SDR to a specialist DRM decoder in the same way as connecting to a digital mode program. The Elad FDM-S1 and FDM-S2 models have DRM integrated into the FDM-SW2 software.

**DAB+** is Digital Audio Broadcast. It is another digital format for radio broadcasting. It is not used below 30 MHz and is typically allocated to channels between 174–240 MHz and between 1.452 – 1.492 MHz. SDR-J seems to be the only open-source SDR software for DAB. https://www.sdr-j.tk/index.html

## FAX AND WEATHER FAX

You can decode unencrypted commercial radio fax and weather fax pictures. Some of these are on the HF bands and some use satellite downlinks. Software defined radios are excellent for receiving the 137 MHz NOAA weather satellites because you need a wideband FM receiver. Unlike software defined radios, most VHF communications receivers and amateur band transceivers capable of receiving 137 MHz only have narrow band FM.

There are weather fax transmissions on many HF frequencies. Radiofax is also known as HF FAX, or WEFAX, although WEFAX is also used to describe satellite downlink based systems. Reception of WEFAX requires an SSB receiver. Note that the published frequencies for weather fax refer to the frequencies marked on dedicated weather receivers. An SSB radio should be tuned for the upper sideband at 1.9 kHz below the nominated frequency.

## APRS

APRS (Automatic Packet Reporting System) is the amateur radio positioning system. It includes position and other data relating to; ham radio operators, experimental balloon and rocket flights, vessels using inland waterways and seaports, and even satellites. APRS is also used by amateur weather station operators, so you can use it to check the real-time weather conditions at various locations around your city or elsewhere.

The data package sent to the APRS server is called a 'Beacon.' Using a repeater or reflector to send a beacon every few seconds is very annoying to other users. So please do not send automatic beacons at less than 10 minute intervals. Turn off the automatic beacon mode if you are talking on the radio fairly often because your position will be sent every time you transmit, as well as the regular automatic beacon.

APRS data is typically transmitted on a single shared FM simplex frequency (depending on the country) to be repeated locally by relay stations (digipeaters). In most countries, APRS is transmitted on the 2m amateur radio band. Your local APRS repeater receives the APRS beacon from your mobile or handheld radio and then repeats it to the local area. This store and repeat process is called digipeating. Any other APRS repeaters in range will receive the repeated beacon, modify the data to show that it has already been repeated, and then repeat it to its local area. Usually, the transmission should not be repeated more than three times. Most but not all APRS digipeaters have a connection to the Internet. They relay the APRS beacon to the https://aprs.fi/ website where it is plotted on a map that anyone can view.

## D-PRS

D-PRS is Icom's digital implementation of APRS. Most D-Star repeaters have D-PRS implemented on their Internet gateway. If you turn on D-PRS and transmit through an Icom D-Star repeater, the D-PRS packet should eventually be displayed on the https://aprs.fi/ website. Go to APRS.fi, enter your callsign in the search box, and see your location plotted on the map.

## MARINE AIS

The Marine AIS (automatic identification system) is used on vessels of all types to track individual craft and provide users with important information such as course, speed, and position. Even quite small boats use it in busy harbours and waterways. AIS 1 on 161.975 MHz simplex channel 87B, transmits information intended for other boats. AIS 2 on 162.075 MHz duplex channel 88B, transmits information intended for shore stations. Regular short beacons contain; status, 'at anchor' or 'under way,' speed, rate of turn, heading, location, and MMSI number. Full beacons are sent at least every six minutes. They send all of the above; plus callsign, name, ship type or cargo type, ship size, draught, destination, and possibly the ETA.

# SDR FOR CW

There used to be a lot of criticism of the CW performance of software defined radios. This is partly due to the way that CW was created by the SDR software and partly due to a reluctance to adapt to new technology. The early problems have been eliminated. These days CW operators can expect excellent performance from SDR transceivers.

The historical problems were caused by the delay between the time the Morse key was pressed and the time that the computer took to process and send the CW signal.

The issues were:

- The delay between CW 'key down' and the time that the 'side-tone' is heard. This really upsets your ability to send CW while listening to the 'side tone.'
- The delay between CW 'key down' and the time that the signal is transmitted. This is annoying and it could cost you contacts in a pressure situation like trying to work a rare DX station in a pileup.
- Slow change over from transmit to receive and receive to transmit. This limits the speed of QSK operation and it could potentially cause stations to 'double.'

It is a pity that these historical problems have limited the acceptance of software defined radios by CW operators because they are fixed now and there are significant advantages in using an SDR for CW operation. Using the waterfall and panadapter spectrum displays you can see the strength and quality of each CW signal on the band. This allows you to easily, skip from station to station across the band. Tuning each signal is a breeze. Just click on each station on the waterfall. The very narrow filters allow you to listen to only a single CW signal without hearing other QSOs. At the same time, you can run CW Skimmer or other software like MRP40 and see all of the activity on the band, or on several bands.

*Technical TIP: The old problem was that the CW signal was generated by SSB modulating an audio tone created by the SDR software running on the PC. This meant that the radio had to detect the Morse key operation, run the raw CW signal through the keyer software on the radio to trim the dit or dah length and spacing, and then send it over the Ethernet or USB interface to the PC. The PC software had to detect the information and create bursts of tone, suitably tailored to eliminate key clicks. This was then sent back to the radio to be transmitted. All of this takes time and the speed is limited by the capacity of the data path between the radio and the PC and by the processing power of the PC. The solution was to eliminate the traditional method of generating CW as an audio tone and instead generate it directly at the transmit frequency within the radio. Because the CW is now generated within the radio, the side-tone is not delayed and the radio can work at QSK up to 60 wpm.*

## CW SKIMMER

CW Skimmer by Afreet Software (Alex Shovkoplyas, VE3NEA), has been developed with wideband receivers like software defined radios in mind. When used with a conventional radio, Skimmer is restricted to decoding CW signals within the radio's 3 kHz passband. But when using the IQ signal from a software defined radio, Skimmer can decode up to 700 signals in parallel. You can download CW Skimmer for free for a 30-day trial but after that, the program must be registered.

CW Skimmer is widely used by CW enthusiasts, DXpeditions, and contest stations. It decodes many CW signals simultaneously allowing the operator to quickly identify stations without having to listen to them. You can easily move up the band working each station in turn, without duplicating contacts or wasting contest time listening for call signs.

## SDC (SOFTWARE DEFINED CONNECTORS)

SDC (Software Defined Connectors) is a useful utility program written by Yuri Pavlov UT4LW. It provides methods of joining COM ports and virtual soundcard audio between multiple applications. One of the items that is included in the download is SKM Server which works much the same as CW Skimmer except that it can decode RTTY and BPSK as well as CW. When configured with a TCI or Telnet interface it can place the identified callsigns onto the panadapter display. The program works with SmartSDR and other software including Thetis. I have seen the Tech Minds video on configuring Thetis with SDC, but I have not been successful in making it work.

This software could be extremely useful and it is definitely worth checking out at https://www.lw-sdc.com/.

## BUILT-IN CW

Most new 100 watt transceivers include CW message keyers. Some include a CW decoder as well. The Elecraft KX3, K3S, and K4 radios can decode and display CW, RTTY, and PSK. If you are proficient with a bug, the KX3 can also convert the CW that you send via the 'key' into RTTY or PSK transmissions. The Malahit DSP2 receiver includes decoders for CW, RTTY, and FT8.

*TIP: The Kenwood TS-890 and TS-990 transceiver also include a CW decoder, although they are not SDR transceivers.*

# SDR FOR DIGITAL TEXT MODES

The main advantages of using an SDR for digital text modes are the ability to adjust the filter bandwidth so that you can see and decode the whole of each digital mode band segment, the ability to operate on several bands at the same time, and the ability to operate several digital modes at the same time. You also get the same advantages as you get when you use your SDR for SSB operation such as a range of noise reduction filters and blankers.

Using software defined radios for digital modes is not much different from using a conventional radio. You do get the advantage that you don't need to connect audio cables between the radio and the PC for digital mode operation. The audio signal between the SDR software and the digital mode software is carried by a single USB cable in the case of radios with knobs, or a 'virtual audio cable,' in the case of black box radios.

Many people use Virtual Audio Cable 'VAC' for the audio connection. See the section about 'audio connections for digital modes.' Rig control is performed using CAT commands sent to the SDR software over a virtual com port. The radio may have a built-in Com port, or you can use com0com or Virtual Serial Port 'VSP' to connect the digital mode software Com port to the SDR software Com port.

If you have a FlexRadio transceiver, you will use the built-in DAX (digital audio exchange) to send and receive audio signals from the digital mode software. SmartSDR is able to send audio or IQ audio signals from individual receiver slices to different digital mode programs. This means that you can decode PSK, RTTY, and CW from the same or different bands all at the same time. In a contest situation, you could watch RTTY signals on three bands and then 'search and pounce' across the three bands. The same would apply if you were looking for unusual DX. Thetis, PowerSDR, ExpertSDR2, and probably others can send the output of receiver 1 to one audio cable and the output of receiver 2 to a second audio cable. They also support separate CAT control for each receiver.

One difference when you are using an SDR is that you can change the bandwidth of the signals sent to the digital mode software. For example, you can view much more than the normal 3 kHz of PSK signals.

## DOLLY FILTERS

Thetis has two twin peak 'Dolly' filters, especially for receiving RTTY signals. They are similar to the APF (audio peak filter) used for CW except they emphasize two audio frequencies that are 170 Hz apart. If you use Thetis for RTTY, you really should try them out. The Dolly filter allows you to copy with 100% accuracy signals that are just random strings of letters without the filter. This type of filter is not available on most non-SDR transceivers. Thetis, Setup > DSP > Audio.

## FT8 OPERATION

The WSJT digital mode software is open-source software initially developed by Joe Taylor, K1JT for very weak signal work. WSJT includes digital modes designed for ionospheric scatter, meteor scatter, EME, and QRP operation. The FT8 mode has taken over the digital modes world, especially during the low point in the cycle 24 sunspot cycle. It can successfully transfer short messages when the received signal is at or even below the ambient noise floor.

## SMARTSDR API

The SmartSDR API allows the developers of digital mode software to write applications that interface directly with FlexRadios. There is scope for specialized, mode-specific, DSP filters and transmitter management. Transmit filters and pre-distortion could be modified to maximize transmit power on digital modes. It should be possible to decode and send common digital modes from within SmartSDR. Perhaps a spectrum plus waterfall plus decoded text or pictures.

## FUTURE POSSIBILITIES

There is significant scope for improvement in the way that SDR software interfaces with digital mode software. FlexRadio Systems are leading the way and I am confident that others will follow. Thetis could be modified to interface with an external .dll file in the same way that many programs use extio.dll to interface differing hardware with programs like SDR# and HDSDR. A defined API could allow digital mode software to become an extension of the SDR software, effectively adding the digital modes as new modes to the SDR software. This integration would remove the need for virtual COM ports and audio cables because the audio and control signals would be directly connected. Different programs could add specific individual modes using the same .dll interface. In addition to the current and future digital modes, we could have, digital voice modes, DRM, DAB, or other decoders for shortwave, and modes for sending data files, pictures, or video.

Alternatively, the QRP Labs QDX and QMX+ transceivers interface directly with WSJT-X or other digital modes programs without using any SDR software on your computer. WSJT-X can change the band and set the appropriate frequency for FT8 or other modes. The radio is controlled using CAT commands over the USB interface. The audio to and from the radio is via a virtual soundcard also over the USB interface. Rather than 'plugging' the digital mode software into an API on the SDR software, as discussed above, this approach simply bypasses the SDR software and the radio becomes an SDR modem attached to the digital mode software. It works very well. You don't need to have two programs running or set up virtual COM ports and audio cables. Just turn on the radio and start WSJT-X, JS8Call, or similar, and start operating.

# SDR FOR CONTESTING

I love using my software defined radio in contests. It is one of the areas of the hobby where using an SDR helps a lot. Most SSB contest stations are spaced 1 kHz apart. If you set a 1 kHz or 500 Hz tuning step, you can turn the mouse wheel to step to the exact frequency of each contest station on the band. It is great for 'search and pounce' contest operation. You will quickly know where the 'big gun' stations are operating because you can see them on the panadapter. You get to recognize the "look" of each station, so you won't waste time carefully tuning them in over and over again, through the night. Some radios helpfully add 'spotted' station callsigns to the panadapter display, so you will know which have already been worked.

If you are in 'run mode,' you can see if your frequency is being squashed by big stations nearby. Or check the status of other bands by opening a second panadapter. No more wasting time and possibly losing your run frequency while you tune around a dead band.

In a busy contest, there will inevitably be stations overlapping. You will be able to see this on the waterfall and panadapter displays. With careful filtering, you can usually eliminate enough of the interference to work both stations. Conventional receivers without I.F. DSP don't have that advantage. Being able to see exactly where the two signals are, really helps you to set the filters in a way that accentuates the wanted signal and masks the unwanted one. You can watch the relative levels of each station as you turn the beam as well. If you are lucky enough to own a radio with two ADCs and receiver diversity, you can alter the phase between signals received on two antennas to maximize a wanted signal and reduce the level of an unwanted signal on or near the same frequency. This is known as 'electronic beam steering.' Two antenna diversity reception can also be used to reduce received noise.

The receiver filters are sharper than non-digital filters and they can be adjusted to exactly the right bandwidth. There is no point in having a receiver passband that is any wider than the transmitted signal. All that you get outside that bandwidth is unwanted noise. The waterfall display will show you exactly how wide each transmitter is operating and you can adjust the receive filter width for the best signal to noise ratio.

On a radio using SmartSDR or Thetis, you can see exactly where to place a tracking notch filter. The tracking notch filters stay in place, so you won't have to set them every time you tune across the band.

The AGC action in conventional receivers reduces the receiver gain if there are strong signals within the I.F. passband. This desensitizing makes it hard to hear weak stations that are near strong signals. Another really big advantage of using a software defined radio for contesting is that this desensitisation does not occur in software defined radios. Making it much easier to work weak stations.

*Technical TIP: Usually the blocking dynamic range of a software defined radio is equal to the clipping level of the radio. SDRs do not exhibit any loss of sensitivity when there are big signals within the receiver passband or displayed on the panadapter.*

During a contest, the band is packed with many signals operating close together. Receiving a lot of signals in the receiver pass-band at the same time is a problem for conventional radios. The signals mix together and with the local oscillators inside the receiver causing intermodulation products, which can be heard. This distortion adds to the cacophony of noise that is so typical in a contest situation. A lot of old-timers who complain that they "can't hear anything" during contests are experiencing this kind of intermodulation distortion. They believe that it is poor transmitters causing splatter, when in fact it is intermodulation distortion occurring within their own receiver. This intermodulation problem is why we prize receivers with very good 'close in' (2 kHz spacing) IMD performance.

You will note that the Sherwood Engineering list which ranks the best amateur radios, is sorted on the results of the 'narrow spaced two tone IMD' performance test. Software defined receivers have very good narrow-spaced two-tone IMD. Unlike conventional receivers, the receive IMD performance is usually the same at any spacing from the nominal receiver frequency. Not only that, software defined radios can actually get better when the band is busy! It is a peculiarity of SDR designs that when the ADC samples a mixture of signals at varying levels, the IMD performance improves.

Most SDR software has a selection of different noise reduction filters and noise blankers. You should be able to find one to combat any kind of noise interference. In some SDR software, you can change the mathematical parameters of these filters, but you wouldn't want to do that while competing in a contest.

Another advantage of using a software defined radio for contesting is that you will probably be using a contest logging program and you might be using digital mode software or an application like CW Skimmer. It is convenient to be operating the radio on the same computer. The best way is to have the SDR displayed on a second computer monitor. Thetis has a function that will return active control to the N1MM logging program or other nominated software as soon as you have finished making any adjustment on the radio. This is specifically intended for use during contest operation. It gives the logging program priority over the radio to stop you from inadvertently entering keystrokes into the SDR program when you want them in the logging program.

With the latency and delayed side tone problems now fixed, SDRs are great for CW contesting. The advantages mentioned above also apply to CW contesting.

You can click on each CW signal and use the **0-beat** button to automatically net the tuning to the preset sidetone. You can see and possibly read the Morse code on the waterfall display.

The software for the 100 watt HF transceivers includes configurable keyers with keyboard macros. With an SDR, you can use super narrow 'brick wall' filters, which can isolate a single CW signal even when the band is frantic. With careful filtering, you can make a received signal sound as clean as a code practice oscillator in the shack.

Thetis and SmartSDR can both use CESSB (controlled envelope SSB) which can compress your transmit audio without any risk of clipping or distortion. This means that you get a higher average 'talk power.' In a busy contest, more talk power usually equals more contacts and more points.

Finally, although it may not be a significant advantage to you, using Pure Signal or another form of adaptive transmitter pre-distortion, makes your transmitted signal cleaner. You will be able to operate closer to another station without causing interference on their spot. Having a transmitted signal with better transmitter intermodulation performance may give you some advantages. If a contest station is listening slightly off frequency the lack of IMD noise on the edges of your signal will make it sound much nicer. It might make it easier for them to copy your callsign and signal report. If you are a 'run' station, your signal will be cleaner and easier to tune in to. Plus, if you are not transmitting intermodulation products, the additional power adds to your wanted transmission. This extra amount is probably insignificant, but hey, everything counts in a contest!

So, in summary;

- You can see where the signals are and click to tune to them.
- You can see if other bands are open without actually switching over to them.
- The receiver filters are sharper, and they can be adjusted to exactly the right frequency and width.
- CW filters are very narrow and very sharp. You can isolate the single CW signal that you want to listen to.
- You can adjust your receiver bandwidth to exactly the width of the received signal, avoiding unwanted noise.
- The SDR has excellent noise reduction, blockers, and notch filters.
- SDRs have better receiver intermodulation performance on busy bands and no desensitizing due to strong signals near the weak signal that you want to receive.
- SDR software offers integration of the radio with logging and digital mode software.
- CESSB gives your signal more 'punch' without the risk of distortion.

# SDR FOR INTERFERENCE MONITORING

Some people like to monitor the bands for interesting signals and intruders on the amateur radio bands. Software defined radios offer many advantages for this type of activity. For a start, a software defined radio is effectively a spectrum analyser, so fairly accurate measurements of the received signal level and frequency can be made. You can monitor a wide range of frequencies on the panadapter and waterfall display, so you will see bursts of illegal transmission that might be missed if you were just tuning around with a conventional receiver.

## IONOSONDES AND SIGNALS THAT CHANGE FREQUENCY

'Whistlers,' Ionosondes, and other signals that are swept across a frequency range may only show up as a blip on a conventional radio, but they are displayed as a diagonal line on a waterfall display. Since you know the rate the waterfall is updating in frames per second and its width in kHz, you can calculate how fast the signal is changing frequency. Fast sweeps cause a flat line on the waterfall and slow sweeps cause a steeper diagonal line.

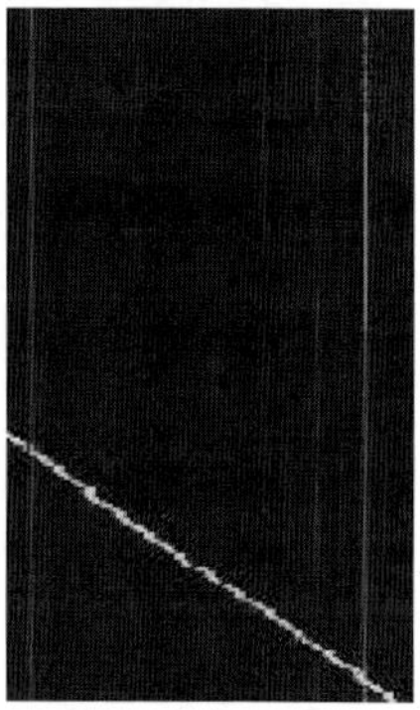

*An Ionosonde is a Radar that examines the Ionosphere by sweeping the HF band and receiving the echoes. Ionosondes sweep from 1-40Mhz. They send a line of very short pulses up the band. It looks like a slanted line when observed on a waterfall display. http://www.sigidwiki.com/wiki/Ionosonde.*

## RECORDING

Many SDR programs include the ability to record signals or even the whole panadapter. If you have recorded the IQ stream for a whole panadapter, you can listen to signals that you were not tuned to when the recording was made. This means that you can leave the receiver recording a band overnight, and then observe and listen to any of the signals on the captured spectrum at a later time. This is a really neat feature if you are watching a band for intruders or waiting for a signal. You make a recording to find out when a band is open to a DXpedition, to watch the downlink from a satellite that is passing over at an inconvenient time, or to catch a pirate who is operating intermittently.

## IDENTIFICATION

The waterfall and spectrum displays can help you to identify what mode is being transmitted. Digital modes and signals like fax and packet have distinctive patterns, which can be observed on the waterfall. You can see and hear what some modes sound like at: http://www.astrosurf.com/luxorion/qsl-ham-history15.htm (Picture below).

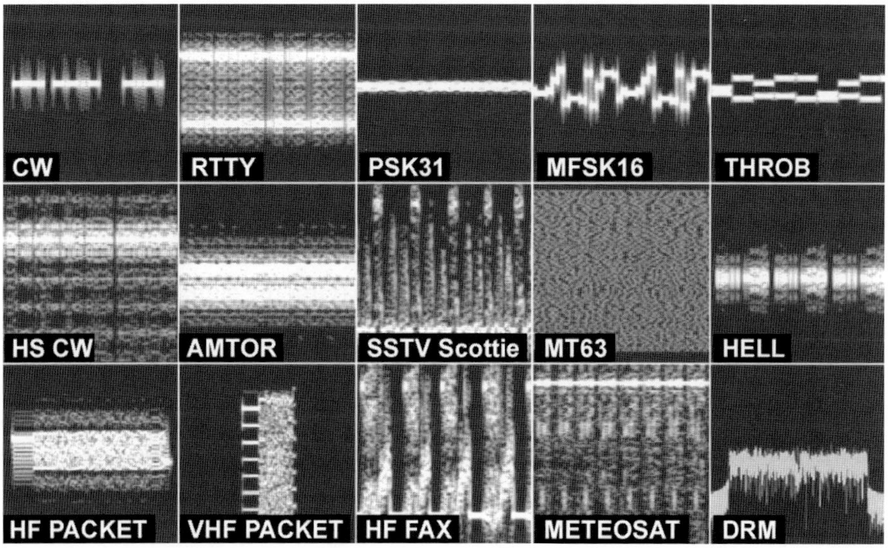

Figure 22: Digital modes have distinctive patterns

## INTERFERENCE

You can use the panadapter to see what kind of interference you are experiencing. Wideband noise, wobbly frequency variations, and multiple interfering signals show up very well. Combs of signals and swept signals are also easy to spot. Some signals carry an identifier that is readable on a waterfall display. Sometimes you can see interference that is generated inside your building. Likely culprits are plasma TVs, switch-mode power supplies (especially cheap ones powering computers and other portable devices), and low-voltage lighting systems. I have found that some computer monitors cause very localized wideband noise. I had one that wiped out the 40m amateur band when I used it in the shack, but it was OK when relocated to another room in the house. ADSL routers used for Internet connections are another source of local interference. They typically radiate a collection of closely spaced carriers at intervals right across the HF spectrum.

Figure 23: Embedded text at the end of a digital mode transmission

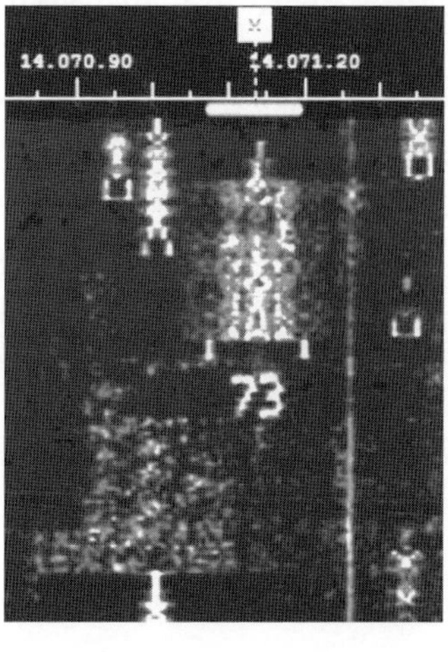

## DIVERSITY RECEPTION

Software defined radios which have two-phase coherent receivers can be used in some very interesting ways. 'Phase coherent' just means that both receivers are clocked from the same source. This means that if both receivers are connected to the same antenna and are tuned to the same frequency the receiver outputs will be at the same level and exactly in phase. If you use two antennas, the phase difference between the received signals can be used to enhance a wanted signal or reject an unwanted one. The SDR software does this by delaying the signal from one receiver in relation to the other and changing the relative levels of the two signals.

You can use the diversity mode to make the signals from two antennas add together. For example, on the low bands, you might have a Beverage receive antenna and a long wire or V beam transmit antenna. If you are receiving two stations and you only want to hear one of them, you can change the phase relationship to null out the unwanted signal. Say for example you want to hear a signal from New Zealand, but there is a European station on or near the same frequency. The phase relationship of the ZL station will be different from the EU station. So you can manipulate the diversity mode phase and level to pull the ZL out. Then if you want to work the other person, just change the phasing to emphasize the European station. This technique is called 'beam steering' because you are emphasizing signals from a particular direction. Beam steering can also be used to find the direction that a signal is coming from without having steerable antennas. This is useful when you are trying to identify and track down the location of intruders and pirates on the HF bands.

If you are suffering from received noise or interference, you can use diversity reception as a noise filter. It works best if the second antenna is picking up signals at around the same level as your wanted signal. But a wire dipole or a long wire 'noise antenna' can be surprisingly effective. Nige Coleman G7CNF has posted a very good video demonstrating diversity noise suppression at www.youtube.com/watch?v=qdQm17udCSo.

The idea is to use the phase cancellation mode to null out the high noise level. You tune the diversity phase and level control on the SDR software for a null in the noise level rather than for a peak in the wanted signal. Diversity noise suppression can be very effective and make the difference between barely hearing a signal and an easily heard station. Up to 40dB of noise suppression should be achievable. For this application, your secondary noise antenna should not be placed close to your transmit antenna. You don't want to couple too much energy from your transmitter into your sensitive receiver. Avoid purposely placing the noise antenna near a noise source as that would just make the situation worse by coupling more noise into the system. The idea is for the noise antenna to receive approximately the same noise level but with a different phase relationship to the wanted signal.

# SDR FOR EME, MICROWAVE, AND SATELLITES

Software defined radios are becoming popular for VHF, UHF, and microwave operation. Many software defined radios can be locked to a GPS disciplined or another very stable clock source, which makes them good platforms for use with receive converters or transverters. Many of the new SDR receivers and transceivers work well up into the microwave region so you can work microwave frequencies directly, without a transverter. Previously an LNB (low noise block) would be connected at the feed point of a microwave dish to convert the signal down to a lower frequency. Now you can use an LNA (low noise amplifier) and an SDR directly at the wanted receive frequency.

The Icom IC-9700 covers the 2m, 70cm, and 23cm (1.2 GHz) bands, and the IC-905 adds the 2.4, 5.6, and 10 GHz bands. The Icom IC-R8600 receiver works up to 3000 MHz. The SDRplay radios and the Malahit DSP2 cover up to 2000 MHz. Receivers like the RTL and FUNcube dongles and transceivers like the HackRF and the Ettus USRP are much cheaper than conventional radios covering the same range of frequency bands.

*Technical TIP: Very stable oscillators are important when you are using a converter or transverter. Frequency errors become multiplied. For example, a good HF radio may have an oscillator with a 0.5 ppm (parts per million) frequency stability. At a frequency of 30 MHz, this is a respectable frequency error of plus or minus 15 Hz. At 6 GHz the same oscillator has a frequency error of plus or minus 3000 Hz. Frequency stability is normally measured after the radio has been operating for at least an hour because oscillators are often sensitive to temperature variations.*

The panadapter is great for weak signal work such as EME (Earth – Moon – Earth) operation or VHF DX. You can see signals that are outside of the passband of a conventional receiver. For example, if you were working 2m EME with a conventional receiver you might easily miss a station transmitting a few kilohertz up or down the band. The waterfall can show signals that are very weak or on a different frequency than what you expect.

The SpectraVue program is popular for weak signal work. With careful adjustment, it can display signals, which are very difficult or impossible to see on the waterfall displays of other SDR programs.

The same is true when working amateur radio satellites with linear transponders. You can display the entire transponder width and see any downlink signals. You can see the relative strength of the downlink signals and compare them to the level from the beacon if there is one. Sometimes working out if a signal is actually from the satellite is confusing. You can get spoofed by local signals. The waterfall display from a software defined radio will show a slanted signal from a satellite because the frequency of the signal will be moving due to Doppler shift.

A signal from a terrestrial source will not show any slant. You will even be able to tell if the satellite is coming towards your location or heading away. *(If you are correcting the receiver frequency for Doppler shift, it would be the terrestrial signals that show as slanted).*

Cross-polarisation loss is a big problem, particularly for EME operations. Serious EME stations often have both horizontal and vertical polarized antenna arrays, but they don't usually use them at the same time. Combining two antennas or antenna arrays into a circularly polarized array causes a 3 dB penalty in received signal strength from a linearly polarized transmitter antenna, which EME operators can't afford.

Software defined radios, which have two front ends can use diversity reception to combine signals from vertical and horizontal polarized antenna arrays and adjust the phase relationship to enhance the received signal. This will make use of both antennas or antenna arrays, at the same time providing gain and eliminating the deep nulls that often result from cross-polarisation.

Your antenna's height above the ground can cause nulls in the received signal strength as the Moon rises. This is called ground gain (or loss). It happens when the signal coming directly from the Moon is enhanced or partially cancelled by a signal reflected from the foreground surface. If you have antennas at different heights, a software defined radio with two receiver front ends can be used to eliminate the nulls and maximize the wanted EME signal. The advantage of diversity reception over having a fixed antenna 'stack' connected with coax cable matching sections, is that the phase relationship between the signals is adjustable, not fixed.

## SDR FOR RADIO ASTRONOMY

Some people are using software defined radio receivers for other weak signal work including radio astronomy and propagation studies. One project is 'Sudden Ionospheric Disturbances' (SID monitoring). There is special software for astronomy use at www.radio-astronomy.org/rasdr.

If you are interested in using an SDR for radio astronomy there is an interesting paper by Dr. David Morgan www.dmrdas.co.uk on the topic, at; http://www.britastro.org/radio/projects/An_SDR_Radio_Telescope.pdf.

'A 21cm Radio Telescope for the Cost-Conscious' by Marcus Leech, of Science Radio Laboratories, Inc. is another interesting article on using software defined radio for radio astronomy. The pdf is available online at; http://www.sbrac.org/files/budget_radio_telescope.pdf.

There is an article by Pieter-Tjerk de Boer, PA3FWM on using the WebSDR at Twente University to create 'Ionograms' from data obtained by tracking the ionospheric reflections of signals generated by 'Chirp' transmitters. http://websdr.ewi.utwente.nl:8901/chirps/article/.

# SOFTROCK RADIOS

The SoftRock SDR receivers and transceivers are simple circuit boards available either as a kit or built-up. They were massively popular in the early years of SDR development because they provided experimenters with a cheap introduction to software defined radio. SoftRock boards are generation one SDRs which are now overshadowed by newer designs with better performance.

*"The SoftRock SDR kits came into being in 2005 based on Tony Parks KB9YIG's experience with an SDR 1000 Software Defined Radio and the series of seminal articles in QEX by Gerald Youngblood K5SDR. The kits were introduced as a low cost technology sampler for Software Defined Radio and Tony continues to offer the kits as such. Many people have contributed to the design and testing of the various kits to lower the cost of the kits and to improve performance and functionality. It is estimated that, since the SoftRock's inception in 2005, over 15,000 kits have been shipped to builders in about 60 different countries. http://www.wb5rvz.com/sdr/."*

All SoftRock models use the Tayloe QSD detector patented by Dan Tayloe N7VE in 2001. The transmitter section of the TXRX model uses the same circuit in reverse as a QSE (quadrature sampling exciter). See the Tayloe detector, on page 182.

*"Tony Parks (KB9YIG) and Bill Tracey (KD5TFD) designed the original SoftRock-40 kit." http://softrocksdr.wikispaces.com*. There have been many models since that first release.

The SoftRock boards are very important from a historical point of view. It was this design that started amateur radio enthusiasts on the SDR journey that has led to unprecedented cooperation in developing open-source SDR hardware and software, and ultimately the shift towards all amateur radio transceivers and receivers using SDR technology. These little boards were the first introduction to software defined radio for thousands of hams and experimenters, including me. My first SDR was a fixed-tuned SoftRock on 9 MHz connected to the IF output of my old Yaesu FT-301. They are fantastic value for money and they are still a great low cost option, especially for use as a bandscope on the I.F of a conventional receiver.

A few of the SoftRock kits are still available from SoftRock kits are available from https://www.softrock40.com/ but most of them have been retired now. It is fun to build these kits, but there are many other options, such as the kits by SDR Labs, RTL dongles, and radios like the SDRplay receivers.

## SOFTROCK PERFORMANCE

It is amazing how well these tiny, super cheap radios work. The audio quality is often excellent compared to conventional radios. The reason is mostly due to the SoftRock receiver not having several oscillators and mixers contributing phase noise and intermodulation. The main limitations are imperfect image signal suppression and only being able to support a single narrow panadapter.

The panadapter bandwidth is limited by the performance of the analogue to digital converter in the PC soundcard. Most soundcards can sample at up to 96 ksps (96,000 samples per second), which means that the radio will be able to display the centre frequency, plus and minus 48 kHz. This results in a 96 kHz wide spectrum and waterfall display on the panadapter. Some PC and USB soundcards and QSD radios with built-in audio frequency analogue to digital converters can manage a sample rate of 192 ksps, which provides a 192 kHz panadapter bandwidth.

PC soundcards do not work well at very low frequencies and the direct conversion process contributes noise near the local oscillator frequency. This combines to produce a spike of noise at the centre of the panadapter. Very good soundcards will have a reduced noise peak and many SDR programs can filter it out. Isolating the audio lines between the radio and the PC with audio transformers or isolators can help to kill mains hum, which is often a big part of the problem. If you see a noise peak in the middle of the panadapter you can be sure that the radio is a QSD-based SDR. The noise spike is not a problem when you are listening to signals away from the centre frequency, but it is rather annoying if you are using the SDR as a bandscope.

The RTL dongles are the main competitor to the SoftRock series. They are sometimes cheaper and they are hybrid (tuner chip) receivers. The dongles come fully assembled in a package like a flash drive 'memory stick.' They have a much wider frequency range, but most of them don't cover the HF bands below 24 MHz, or sometimes 64 MHz unless you add an HF up-converter before the dongle.

## TECHNICAL DESCRIPTION

SoftRocks are 'generation 1' SDRs. There is minimal hardware with the analogue to digital conversion done using the PC's stereo soundcard and all of the DSP processing being performed by the PC software. They work with a wide selection of SDR programs.

Rocky written by Alex Shovkoplyas VE3NEA, was one of the very first SoftRock compatible programs. Alex invented the adaptive image-cancelling algorithm, used in Rocky and subsequently used in almost all other SDR software for QSD-based receivers. At the time, those few lines of software code were a major advancement. SDR guru Phil Harmon VK6APH reverently referred to these lines as, "The four lines of code that changed the SDR world."

QSD receivers are single conversion receivers, a process that results in signals from above and below the local oscillator becoming mixed in the audio output. On the right side of the SDR panadapter, we want to see signals from above the local oscillator frequency without image signals from below. On the left side of the SDR panadapter, we want to see signals from below the local oscillator frequency without image signals from above. The I and Q signals are used to cancel out the image signals, but the cancellation is never perfect across the receiver pass band.

Alex's algorithm learns the characteristics of the receiver by comparing on-air signals and continuously adapting the phase and amplitude compensation to reduce the corresponding image signals. His technique vastly improves image cancellation and it revolutionized SDR software. This is another example of the way that open-source software benefits everybody. Without the open-source movement, his clever algorithm might not have been available to other programmers.

The heart of the SoftRock is the Tayloe detector, which converts the incoming RF from the antenna into two audio signals. One of the audio signals is delayed with a 90-degree phase lag to create the I and Q streams that are fundamental to software defined radio. The PC soundcard converts the I and Q audio to two digital streams, and the PC software does the DSP and the graphical user interface. Newer SoftRock models use an ATTiny 8bit microcontroller to process control information from the PC via a USB cable. You can control the frequency of the Si570 oscillator chip so that a much wider band of frequencies can be covered. One SoftRock model covers the entire HF band from 1.8 MHz to 30 MHz. On some models, the microcontroller also switches bandpass filters, and on the TXRX version, and manages the PTT.

There is an excellent resource for available and no longer available SoftRock boards at https://www.wb5rvz.org/. The pages include block diagrams, component lists, board layouts, and schematics.

Figure 24: SoftRock RX/TX Ensemble SDR

The Tayloe quadrature sampling detector (QSD) in the 'Ensemble' is slightly different from the one shown on page 182. It uses four inputs and two outputs from the FST3253 instead of the other way around.

# RTL DONGLES

Dongle SDR receivers have a similar physical format to USB 'memory sticks.' They are designed to be plugged directly into a USB 'Type A' port on a computer or USB extender hub. You will need to load a free SDR program such as SDR# (SDR sharp), HDSDR, or SDR Console, and you may have to load a USB device driver called Zadig and/or a data format interface utility called extio.dll. Most SDR dongle receivers use the R820T/T2 as the tuner chip, and the Realtek RTL2832U chip as the analogue to digital converter, digital down converter, and USB port interface. They typically cover frequencies from 24 to 1766 MHz or up to 2200 MHz in some models. Some are bundled with cheap up-converters to provide coverage of the HF bands. The maximum panadapter bandwidth is typically about 2.5 MHz.

The RTL dongles were designed as a cheap way to receive DVB-T digital television on your notebook PC. Online sources credit; Eric Fry, Antti Palosaari, and a company called Osmocom, for working out how these devices work and how they can be used as general-purpose SDR receivers.

I am not a big fan of these radios. The performance is limited and they can be very tricky to get working. They have a fixed RF filter in front of the preamplifier but they do not have separate bandpass filters for individual bands, so they are prone to overload and ADC gain compression. The 8-bit ADC means that they don't have a particularly large dynamic range to begin with. However, they are very sensitive and they cover a huge frequency range. Most importantly they are very cheap. They provide a great introduction to software defined radio and they have been extremely popular.

You must make sure to buy one with the right chipset. Not all models are compatible with the 'Zadig' driver. The ones that are not compatible won't work as a general SDR receiver. The dongles that use the Realtek RTL2382U chip will work and most dongles using the Elonics E4000 tuner should work. I once bought a dongle with the wrong chip and it did not work with the Zadig driver, although it did work with the bundled TV software. I also bought one that failed to work at all.

*"DVB-T dongles based on the Realtek RTL2832U can be used as a cheap SDR since the chip allows transferring the raw I/Q samples to the host, which is officially used for DAB/DAB+/FM demodulation."* *http://sdr.osmocom.org/trac/wiki/rtl-sdr*.

## TECHNICAL INFORMATION

The RTL2832U has an 8-bit analogue to digital converter and outputs an 8-bit IQ stream over the USB port to the PC. The Zadig driver converts the data into a format that can be used by the SDR software. The maximum sampling rate is 3.2 Msps but USB 2.0 ports can't transfer continuous data at that rate so the actual sample rate is more often 2.048 Msps. It is possible to display a bandwidth of around 2.5 MHz.

That's a lot more than the FUNcube dongle pro+, which only has a panadapter display bandwidth of 192 kHz.

The frequency range varies depending on which manufacturer made the dongle. Most models cover 24 to 1766 MHz. Check http://sdr.osmocom.org/trac/wiki/rtl-sdr.

They are generally very sensitive with typical MDS (minimum discernable signal) figures around -139 dBm. Clipping occurs at an input level of around -60 dBm.

The RTL dongles work well at VHF and above, but many don't cover the MF and HF bands. NooElec makes a USB powered up-converter with a local oscillator on 125 MHz which allows you to receive the HF spectrum. The up-converter costs about twice as much as the RTL dongle, but the whole package is still very cheap.

RTL dongles have a restricted dynamic range due to the 8-bit ADC. They are prone to overload and ADC gain compression which is a condition where large signals reduce the receiver's dynamic range. The large signals do not have to be visible on the panadapter. Any signal present at the input of the analogue to digital converter can contribute to the problem. ADC gain compression will cause the noise floor on the panadapter to rise or jump erratically in time with a large signal. You see this a lot when using RTL dongles. The dongle might be affected by big signals from TV or FM broadcast stations. Or high-power AM broadcast stations if you are using an HF up-converter with no front-end filtering. Overloading typically causes a high noise level and many large spikes across the panadapter display. You may hear distortion and noise that wipes out any wanted signals.

Walter Schellenberg HB9AJG published a very comprehensive performance report on two of the most commonly used RTL dongles. It is available at http://www.rtl-sdr.com/wp-content/uploads/2013/08/Some-Measurements-on-E4000-and-R820-Tuners.pdf. He points out that some of the dongles are direct conversion receivers, which are prone to image signals and are likely to show a noise spike or hump in the centre of the panadapter. Other RTL dongles use an I.F. frequency so they are less prone to these issues. The data shows excellent sensitivity but the direct conversion model had a fairly poor image response. The Intermodulation performance is generally poor and there are several unwanted signals visible on the panadapter, mostly harmonics of the clock signal. Given the very small size of these receivers, it is extremely difficult to shield the sensitive input to the pre-amplifier and ADC from the digital signals in the receiver. The usable dynamic range is only about 60 dB.

*"Of course, these neat, little toys are no match for 'real' software defined receivers. They do have some limitations, but you get a LOT of radio for almost no money! They are fun to experiment with, and to learn more about SDR technologies." Walter Schellenberg HB9AJG.*

## BUYING AND RTL DONGLE RECEIVER

Check out the excellent information about the features of good RTL dongles at, https://www.rtl-sdr.com/buy-rtl-sdr-dvb-t-dongles/

Figure 25: NooElec.com RTL dongle receiver

The NooElec SDRs are significantly better than most of the other dongles. There is a range of models. Some have metal cases to improve their noise performance, and some models include high-accuracy 0.5ppm oscillators and a bias-T to feed voltage to a masthead preamplifier. The range includes a downconverter for receiving frequencies up to 3 GHz and an upconverter for receiving HF and frequencies down to 300 Hz. NooElec also makes a range of excellent SAW (surface acoustic wave) filters for receiving NOAA, Inmarsat, Iridium, and GEOS satellites, or the 1420 MHz Hydrogen line for astronomy. See https://www.nooelec.com/store/.

I have heard that the RTL-SDR RTL models are good as well. They cover 100 kHz to 1.7 GHz. https://www.rtl-sdr.com/.

Figure 26: RTL-SDR dongle receiver

# FUNCUBE DONGLE

The FUNcube Dongle Pro+ (FCD) has a similar USB stick layout and case as the RTL dongles, but it was designed to be a radio receiver, not a cheap digital TV receiver. They are a bit more expensive but they have significant technical advantages. The FCD is very easy to set up because the computer reads the dongle as a USB HID (human interface device), essentially the same as a computer mouse. Just plug the FCD into a USB port and it will work. You don't have to install any driver software, just the SDR program. The FCD will work with any computer operating system that can recognize a USB device.

The FUNcube Dongle was designed in 2010 by Howard Long, G6LVB, as the ground station companion to the FUNcube satellite being built by AMSAT-UK. The satellite was finally launched on November 21st, 2013. It was awarded an OSCAR (Orbital Satellite Carrying Amateur Radio) designation of AO-73. Five more FUNcube satellites have been launched, with FUNcube-6 (JO-97) completing the series in 2018. I don't think any of them are still active.

The idea behind the FUNcube dongle receiver was to create a cheap, high-performance receiver that schools could use as a part of the educational goals for the FUNcube satellite program. Initially there was to be a version specifically for the 2m and 70cm bands that the satellite used, and a wide coverage version at a slightly higher cost. But in the end, only the wide coverage version was produced. Around June 2012, Howard experienced problems sourcing the Elconics tuner chip and the manufacturer eventually went into liquidation. So he designed an improved FCD version called the FUNcube Dongle Pro+ and released it in September 2012. The Pro Plus is far superior to the original version.

The FCD Pro+ has a typical frequency range of 150 kHz to 260 MHz and 410 MHz to 2.05 GHz. This means that it covers every ham band from the 1800m band to the 23cm band. In New Zealand, that is L.F to L Band. Seventeen amateur radio bands from a receiver you can carry in your pocket!

*TIP: Don't try to use a random piece of wire as an antenna. Initially, I didn't have the right adapter and tried using a piece of hook-up wire poked (carefully) into the RF input of the FCD receiver. If you do this, you will be disappointed to see spikes all across the panadapter display. I believe that the radio picks up its own clock signal. This does not happen if no antenna is connected, or if a good antenna is used. The FCD works best with a proper antenna located at a distance from the USB dongle.*

## FCD PRO+ TECHNICAL INFORMATION

Front-end filters are important in any SDR to avoid ADC gain compression and overloading of the low noise pre-amplifier. The FCD pro+ has 11 automatically selected filters before the tuner chip, including very narrow SAW filters on the amateur 2m and 70cm bands.

The receiver was initially designed for receiving a satellite that used those two amateur bands. The filters are important to stop out-of-band signals particularly FM broadcast stations from overloading the receiver. There is a low noise amplifier to provide additional sensitivity when receiving on the VHF and UHF bands. The preamplifier and I.F. stage gain are both adjustable.

*TIP: A SAW filter is a surface acoustic wave filter. It is a ceramic or crystal filter in a small sealed package. They are commonly used in cell phones and can be ordered on common frequencies from electronic suppliers. A typical size for a UHF one is 3mm x 3mm. The cellular band filters can be as small as 1.1mm x 0.9mm.*

## TECHNICAL DATA

- Typical frequency coverage: 150kHz to 260MHz and 410MHz to 2.05GHz
- TCXO specified at 0.5ppm (in practice about 1.5ppm due to thermal issues)
- 192kHz sampling rate
- Standard SMA female antenna port
- USB type A interface to the computer
- Eleven discrete hardware front-end filters including:
- 6 MHz 3dB bandwidth (10MHz at -40dB) SAW filter for the 2m band.
- 20 MHz 3dB bandwidth (42MHz at -40dB) SAW filter for the 70cm band
- Third- and fifth-order LC bandpass filters for other bands
- Front end LNA OIP3 30dB
- Typical noise figures:
    - 50MHz 2.5dB
    - 145MHz 3.5dB
    - 435MHz 3.5dB
    - 1296MHz 5.5dB
- Typical NFM 12dB SINAD measurements:
    - 145MHz 0.15uV
    - 435MHz 0.15uV
- No additional drivers are required for Linux, OSX, or Windows
- Integrated 5V bias T switchable from software

*Data sourced from the FUNcube website at http://funcube.org.uk/.*

# USB CONNECTED RECEIVERS

Small 'black box' SDR receivers with a USB connection to your PC are very popular. There are many models available. This type of radio is the most difficult to evaluate because there is so much choice. Some are old QSD designs, some are hybrids, and a few use direct digital sampling. Price is not necessarily a good indicator of quality or technical performance. To make things worse some of them work with a wide range of SDR software and others will only work with specific software supplied by the radio manufacturer. It is a bit of a minefield! The table on page 129 shows a list of SDR software and compatible radios. Almost all of the black box and dongle radios use a tuner chip that employs direct sampling up to low VHF frequencies and a hybrid architecture for the higher bands.

*USB Technical Information: There are four classes of USB interface, plus a few revisions and subclasses. The initial USB 1.0 standard only allowed a data rate of 1.5 Mbits (1,500,000 bits per second) but later versions managed up to 12 Mbits.*

*The USB 2.0 interface has a four-wire connector. It uses three wires for data transfer, and outputs +5V at up to around 300mA on the fourth connector pin. The USB 2.0 standard works up to a maximum of 480 Mbits. It is half-duplex meaning that it can send or receive data but not both at the same time. USB 2.0 connectors have black or grey plastic inside.*

*The USB 3.0 interface uses nine pins and can support asynchronous full duplex (i.e. it can receive and transmit data at the same time). Its +5V output can support loads up to 900mA. The plugs and sockets can be identified by the 9-pin connector which has a blue plastic inner. The USB driver software included with your SDR software may not be compatible with the USB 3.0 port(s) on your computer. If you have problems, try a USB 2.0 port. USB 3.0 ranges from 2.5 Gbits to 4.8 Gbits per second and USB 3.1 can work up to 10 Gbits.*

*USB 4 uses the USB-C connector. USB 4 v1.0 supports bit rates to 80 Gbits, and USB 4 v2.0 supports bit rates up to 120 Gbits.*

The high data transfer rate of USB 4 over a USB-C cable, is ideal for software defined radio. It can easily handle ADC data for the entire HF spectrum without having to reduce the data rate in an FPGA.

If the radio supports the USB 3.0 standard, you should use a USB 3.0 port on your computer. Using a USB 2 port might work but the data bandwidth is reduced and that would probably limit the width of the panadapters and/or the number of panadapters that can be displayed. An example is the RX888 MKII receiver which uses a USB 3 interface to send a 62 MHz bandwidth to the host computer. It has a 16-bit ADC sampling at 130 Msps so the raw data rate is 2.080 Gbits plus a frame structure for the data and the Ethernet packet structure. I guess the real overall data rate is close to 3 Gbits.

# USRP, NOCTAR, HACKRF, AND BLADE RF BOARDS

These boards are designed for experimenters, recently known as 'hackers.' They are not primarily aimed at radio amateurs, but of course, radio hams are 'hackers' too. We have been experimenting with new radio technologies for many years.

The Ettus Research USRP (universal software radio peripheral) boards are bare board versions of their more expensive commercial products. The other boards and boxed radios are for industrial research and integration into commercial applications. The B200 and B210 boards are the most accessible to hams and hackers on a budget.

The Ettus, Noctar, Blade RF, and HackRF radios all have amazing frequency coverage from HF or VHF right up into the GHz microwave frequencies and they all have, very low power, transmitting capability.

These boards and radios may be of interest to radio amateurs building microwave frequency DX or EME stations. You can operate directly on the bands of interest without transverters or converters. Of course, you would still need a low-noise amplifier mounted at the receive antenna and a UHF or microwave power amplifier. Ultimately, these radios may also be of interest to amateur radio satellite operators and maybe satellite makers. There are other options such as the Pluto +, the DX Patrol systems for QO-100 satellite, and the Icom IC-905.

The ability to observe received signals on a wide panadapter is a real plus of this kind of radio compared with older technologies. Small frequency errors or drift in the oscillators of frequency transverters, exciters, or receivers can lead to large errors when mixed or multiplied up to microwave frequencies. This can be very frustrating. Most of these boards can be locked to accurate frequency standards and the panadapter can show received signals that are off frequency and possibly outside of the passband of conventional converters.

Some of the boards have dual coherent receivers and even transmitters, which allows the use of techniques such as beam steering of signals, diversity reception, and diversity noise reduction.

Some of the radios are full duplex which is great for satellite operation where being able to hear your downlink signal while transmitting on the uplink frequency is virtually mandatory. The wide panadapter can help with Doppler control, as you will be able to see the received signal frequency changing.

There are online communities of interested people who can assist you with your projects. Most of these boards already have established GNU Radio software and a range of open source, software code sources.

# SDRS WITH KNOBS

Back in 2016, this group was the exception. Now SDRs with knobs are the predominant 'mainstream' radios. These radios usually use dedicated Texas or similar DSP chips, with the control functions managed by a microprocessor or microcomputer inside the radio. Some, like the ANAN G2 and Expert MB1, don't use a DSP chip. They use a microprocessor or microcontroller to perform the DSP code. For example, the Xiegu X6100 uses an STM-32 microprocessor to take care of the SDR and DSP functions. The radio's control and display functions are performed on a separate Allwinner R16 ARM (advanced RISC machine) quad-core processor running Linux.

Here are some examples of the various types of software defined radios 'with knobs.' A brief summary of each radio's technical specifications such as frequency coverage, ADC sampling rate and transmit power is included in the 'Catalogue of software defined radios,' on page 241 near the end of the book.

## ELECRAFT

*Tip: note that the Elecraft K3, K3S, and KX2 transceivers use conventional superheterodyne architecture. They are classed as + SDR radios.*

### KX3

The Elecraft KX3 is a QRP transceiver with a maximum transmit power of around 10 watts and exceptionally good receiver performance. The radio was designed for field and portable use, but many people are using it as their primary base station because it has such a good receiver. It is a generation three SDR design. The receiver uses a QSD (quadrature sampling detector) to generate audio frequency I and Q streams, which are sampled with an onboard audio frequency ADC chip. A dedicated DSP chip processes the audio. While an MCU (microcontroller) manages the inputs from knobs and buttons, the display, memories, and CAT commands. The audio frequency IQ streams coming out of the QSD stage are available on a jack at the side of the radio for use with optional soundcard-type SDR software.

The KX3 does not include a spectrum and waterfall display, but Elecraft does offer the PX3 companion bandscope with a maximum span of 200 kHz. Unlike the IQ output, it doesn't display a noise spike. I understand that the radio was designed for portable operation and not as a base station, but it is a pity that the IQ and control signals were not integrated into a mini USB port instead of separate serial and audio signals. Note that it is probably cheaper to buy a radio that has a built-in panadapter than to buy the KX3 and PX3 combination.

The radio is excellent but the external SDR functionality provided by the audio IQ output connector is disappointing.

It suffers from the limitations inherent in any soundcard-based software-defined radio. Depending on the sampling rate of your soundcard, you usually only get a 48 kHz or 96 kHz bandscope span and you are likely to experience ground loop hum and a DC noise spike in the centre of the spectrum display. The IQ output has a roll off at high audio frequencies so the resultant panadapter shows significant droop at both edges, especially when displaying a 96 kHz panadapter. The KX3 supports CAT commands over a USB to serial adapter, so your SDR software should be able to display and change the frequency. Unless you have a very good PC soundcard, the effectiveness of using the KX3's IQ output as a bandscope is limited by the noise spike, which appears on the bandscope right at the frequency that you are listening to. Some SDR software can eliminate the 'DC' noise spike and having a top-quality soundcard really helps. But overall, it is probably not worth the bother.

### K4, K4D, K4HD

The K4 transceivers are direct sampling 100 watt HF transceivers, with impressive connectivity and good specifications. Although the K4 is placed below the K3, K3S, and KX3 on the Sherwood list. http://www.sherweng.com/table.html. All models support main and sub-receivers operating on different bands. The K4D has two ADCs and supports two completely independent receivers each having its own set of bandpass filters. It is capable of diversity reception with two antennas. In my opinion, this is the model to buy. The K4HD is not yet available. It has all the features of the K4D but also includes a 'dual superhet receive section' making the radio an optional hybrid with an 8 MHz I.F. before the SDR stage. I can't see much advantage in this idea other than perhaps improving the receiver specification a little.

### EXPERT ELECTRONICS MB1

The MB1 is a direct digital sampling HF/VHF SDR transceiver with an Intel Core i7 based PC inside the radio box. The computer runs on either Linux or Windows. It covers the 2m VHF band or the HF band but not both at the same time. The radio supports two panadapters on the same or different bands, with two receivers on each panadapter. Four receivers in total. Output power is 100 watts on HF and 50 watts on the 6m and 2m bands. The radio has an impressive list of specifications, with a quoted RMDR of 110 dB, and BDR on HF of 130 dB. The main oscillator has a 0.5 ppm accuracy. The ExpertSDR3 software runs on the built-in touchscreen, or an external Windows, macOS, or Linux computer.

See the videos at https://eesdr.com/en/products-en/transceivers-en/mb1-en#video

### MIDNIGHT DESIGN SOLUTIONS SDR CUBE

The SDR Cube is a QRP radio with onboard DSP and a microcomputer to manage the controls and display. It was designed to be an add-on for a SoftRock transceiver that you build separately, but now there is an RF board available, which will cover the complete HF band. It is available assembled and tested, or in a 'quick kit' form.

## ELAD FDM-DUO

I rather like the look of the ELAD products, they are very stylish. The two DUO models have knobs and a small display, but they don't include a bandscope or a panadapter display. You can connect them to a PC and use the supplied FDM-SW2 Software. Four receivers can be configured. Each with its own CAT control and audio path. The FDM-SW2 panadapter bandwidth is 192 kHz.

The FDM-Duo is a direct sampling QRP transceiver. It has a fast LTC2165 16-bit ADC with a 122.88 Msps sample rate, so it should have good performance figures. The receiver covers 10kHz - 54MHz and the radio can transmit on the 160-6m amateur bands. The transmitter power is 5 watts. The ELAD range also includes the FDM-DUO, FDM-S1, FDM-S2, and FDM-S3 receivers.

## ICOM

Icom has shown exceptional innovation and a willingness to extend the boundaries of amateur radio over the past few years. It has released six new SDR models since 2016. The IC-7300 brought SDR technology and a heap of features previously only available on very expensive transceivers to the general amateur radio community. The IC-705 is a portable all-mode HF, 2m, and 70cm rig for those who like to go hiking, activate SOTA peaks, or POTA parks. The IC-9700 is great for working satellites, DX, and D-Star, and it includes the 1200 MHz band. Now the IC-905 has been added to the range. It is the first amateur radio transceiver to cover the 2m, 70cm, 1.2 GHz, 2.4 GHz, 5.6 GHz, and optionally the 10 GHz, amateur radio bands. Plans for a 24 GHz transverter have been announced.

### IC-7300 (2016)

The IC-7300 was the first software defined radio to be released by one of the 'big three' amateur radio manufacturers. It is a 100 watt HF direct digital sampling SDR with onboard DSP. The IC-7300 has proved to be a 'game changer' for amateur radio worldwide. It has many features that were previously not available, or only available on much more expensive radios. These include the touchscreen bandscope, VOX, antenna tuner, a CW keyer, an RTTY keyer and decoder, off-air recording, plain English or Japanese menu rather than cryptic codes and numbers, and a single USB cable interface to your computer. I found the IC-7300 easy to use and well laid out, more intuitive than the IC-7610. The IC-7300 is an excellent first HF transceiver.

### IC-7610 (2017)

The IC-7610 was the second software defined radio transceiver to be released by one of the 'big three' amateur radio manufacturers. It is a 100 watt HF direct digital sampling SDR, with onboard DSP and two completely independent receivers, each with its own front-end filters and panadapter or bandscope display. The IC-7610 shares the same menu structure and features as the IC-7300.

It has a PSK decoder in addition to the RTTY decoder that is in the IC-7300. The addition of the second receiver makes this radio a great choice for contesters and people seeking DX contacts.

### IC-R8600 (2017)

The IC-R8600 is a top-quality SDR and hybrid receiver capable of receiving signals from 10 kHz up to 3000 MHz. There are several regional variants. Most have small gaps in the coverage to meet the requirements of the jurisdictions where they are sold. The frequencies were usually regulated to protect the old analogue cellular bands from casual listeners. They are not very relevant today since cell sites now use digital transmissions and different bands. The IC-R8600 is very capable but quite expensive.

I was very impressed with the IC-R8600. It is still rare for 'communications' receivers to include a built-in bandscope or panadapter display. The bandscope has a 5 MHz span which is much wider than the Icom or Yaesu HF transceivers. The radio includes an RTTY decoder, but not CW or PSK decoders. I found this surprising considering that the IC-7610 has a PSK decoder.

The IC-R8600 has the usual AM, SAM, SSB, FM, and wideband FM modes, and it also supports internal decoding of the P25 (phase 1), NXDN, dPMR, DCR, and D-Star digital voice modes. It does not decode DMR (digital mobile radio) which is not surprising since Icom sponsors D-Star. It does not support FM stereo or decode the FM RDS data, and there is no support for digital broadcast modes like DRM or DAB.

The radio has 2000 regular memory channels split into up to 100 groups. High-speed scanning is supported with 50 pairs of programmed scan edges which store scan start and stop frequencies. There are also 200 auto memory write channels which can be filled automatically when the radio encounters active channels while it is scanning.

### IC-9700 (2019)

The IC-9700 is in the same family as the Icom IC-7300 sharing an almost identical front panel layout, weight, and size. It is a perfect companion to the IC-7300, adding the 2m, 70cm, and 23cm bands. It is a transceiver for FM and D-Star digital repeaters, VHF and UHF DX operation, working satellites, EME, and digital text modes.

If you already own an IC-7300 or an IC-7610 you will be familiar with the touchscreen display and the positions of many controls. But you will quickly find that the IC-9700 is very much more complicated than either of those radios. The radio is crammed with exciting new features. For a start, it has two receivers, so you can receive signals on two bands at the same time. It is also capable of full-duplex crossband operation. There is provision for the connection of a 3rd party GPS receiver to provide location information for use with D-Star. You can also connect an external 10 MHz frequency reference for superior oscillator stability. The 100 watt power output on the 2 m and 70 cm bands is great for long-distance work and meteor scatter operation.

There are an astounding number of memory slots and scanning functions. 99 channels per band, plus 99 satellite memories, 300 GPS memories, and 2500 D-Star repeater memories.

The IC-9700 is the first full-power, multi-mode, VHF/UHF amateur radio transceiver to be based on SDR technology. The receiver uses direct digital sampling for the 2 m and 70 cm bands and a single down-conversion stage followed by direct digital sampling for the 23 cm receiver. The transmitter uses direct digital up-conversion for the 2 m and 70 cm bands and a heterodyne up-conversion stage for the 23 cm band. It is also the first VHF and UHF transceiver to feature a spectrum and waterfall display.

### IC-705 (2020)

The IC-705 is a QRP radio with coverage of the HF bands, the 6m and 2m VHF bands, and the 70cm UHF band. This radio takes ham radio back into the wilderness, urging you to "be active in the field." I am sure it will become very popular for those of us who enjoy the great outdoors. You may be activating mountain tops for SOTA, camping near a stream, hiking in a forest, operating as a cycle or pedestrian mobile, or just walking the dog. Whatever your passion, if you want to operate your radio outside, this radio is an excellent choice. Built-in Bluetooth allows you to use a wireless headset while you are on the move.

The transceiver is a little bigger and heavier than the Elecraft KX3 and much heavier than the KX2, although the stated weight does include the battery. It is the same weight as the Yaesu FT-817 or FT-818, but a fraction larger. However, it outclasses all the competition with its amazing set of features. For a start, it has a 4.3" colour touchscreen spectrum display which is ideal for displaying weak signals across the band of interest. I imagine that this will be useful when you are operating on a windy peak and looking for those four elusive contacts. Calls can be recorded, saving you from having to log contacts while you are outside. You can just play them back when you get home. You also get a voice keyer to save you from having to make repetitive CQ calls and CW and RTTY message keyers.

The radio covers all of the ham bands from 160 m to 70cm and it has receiver coverage of the FM broadcast band, the Aircraft band, and the VHF marine band. Great for receiving weather reports. The transmitter output is 5 watts on battery power and 10 watts on an external 12 - 13.8 volt DC supply.

The built-in GPS receiver indicates your current location and Maidenhead grid locator which is handy whether you are activating a peak or just operating from a camping spot. It can even be used as a compass to find your way home.

The D-Star digital mode can be used in simplex mode, through a local D-Star capable repeater to digital radio 'gateways' worldwide, or as an Internet connected 'hotspot'.

### IC-905 (2023)

The IC-905 is different from most if not all, commercially made amateur radios. It has a remote-mounted RF unit. This makes operating on the microwave bands possible as the loss in traditional feeder cables would be excessive at frequencies above 2 GHz. The radio covers the 2m, 70cm, 1.2 GHz, 2.4 GHz, and 5.6 GHz amateur radio bands. The 10 GHz band is covered with the optional CX-10G transverter module. Plans for a 24 GHz transverter have been announced. The radio works on the amateur bands only. It provides full coverage of the entire band, unlike transverters that can only work on a single frequency. The menu system and features are similar to the IC-705, with changes to accommodate the different bands. There are no RF components in the head unit. All of the RF takes place in the remote-mounted radio unit.

The IC-905 has voice, RTTY, and CW message keyers and an RTTY decoder. The QSO Recorder can record received and transmitted audio, which is handy as a log when you are operating portable. Most microwave setups don't have built-in recording. The radio can transmit and receive the usual AM, FM, SSB, and RTTY modes, plus D-Star. There is also an analogue video mode so you can send video and sound to another IC-905.

The transmitter power varies across the bands. It generates 10 watts on 2m and 70cm which is the same as the IC-705, and 10 watts on 1.2 GHz which is the same as the IC-9700. It will probably radiate more power than an IC-9700 on this band as there is almost no feeder cable loss. The transmitter power on 2.4 GHz, and 5.6 GHz is 2 watts, and the amateur CX-10G transverter can output 0.5 watts on the 10 GHz band.

## YAESU

The FTdx10, FTdx101D, and FTdx101MP models follow on from the previous models such as the FT-5000. They are similar inside, but the second I.F. stage has been replaced with an ADC and FPGA. As far as I have been able to find out, this stage does not include any DSP functions. It simply provides data for the bandscope and down converts from 9 MHz to 24 kHz for the Texas DSP stage. On the transmit side, the FPGA appears to mix the modulated signal from the DSP straight up to the RF output frequency. I'm not sure if this is enough to qualify their radios as SDRs. But, anyway, the technical performance numbers are excellent. Yaesu has opted for the hybrid technology to ensure top-of-the-range IMD (intermodulation distortion) performance. The radio does include a 'direct sampling SDR' receiver but it is only used for the Band Scope display.

### FTdx10 (2020)

The FTDX10 has a similar styling to the FTDX101, and it is easy to see that it is from the same Yaesu family. The front panel is dominated by a bright full-colour 5" touchscreen display and the VFO tuning knob which has an outer multi-function tuning ring surrounded by the buttons that enable the functions it controls.

There are "proper" buttons and knobs for often-used functions such as Noise Blanker, Split, and VFO switching. Other functions such as the band and mode switches, direct entry of a frequency, and adjusting the transmitter output power have been moved onto touchscreen controls.

Under the skin, the radio employs the same hybrid SDR receiver architecture as the FTDX101. The second I.F. stage has been replaced with an ADC and FPGA.

The FTDX10 is a truly exceptional radio for contesting or working DX stations, with a fantastic receiver. The spectrum scope can display a band of frequencies up to 1 MHz wide. More than enough to show contest stations across any of the lower bands, or the pileup waiting to work a DXpedition. I love the SSB voice message keyer. It makes calling "CQ Contest" or just saying my callsign repeatedly, so much easier. You can operate CW, PSK, or RTTY without relying on a connection to digital mode software running on a PC. There are built-in decoders for all three modes.

### FTdx101D (2017) and FTdx101MP (2019)

The FTdx101D has superb technical specifications. It currently holds the top position on the highly regarded Sherwood Engineering transceiver performance table. These two radios share the same architecture as the FTdx10 except they are much larger, with two independent receivers. Following the previous models, the D model has 100 watts of transmitter power. The MP variant has 200 watts of output power with an external mains voltage power supply and speaker. The D model has a 'VC Tune' variable preselector on the main receiver, the MP has one on each receiver. It also has a 300 Hz CW filter which is available as an option on the FTdx101D. They have a 7" screen. Most of the menu functions and features are the same as the FTdx10.

These radios are intended for serious amateur radio operators interested in contesting and DX.

### FT-710 (2022)

The FT-710 is Yaesu's first direct sampling 100 watt HF SDR transceiver. It has excellent receiver performance and is listed at number four on the Sherwood list. One behind the FTdx10. In fact, it has the same narrow-spaced dynamic range performance as the FTdx10, proving that direct sampling SDR receivers can be just as good as the best conventional (hybrid) receivers. If the FT-710 was fitted with a 'VC Tune' variable preselector like the FTdx101D, I believe that it would rank very close to the top of the list.

The FT-710 is slightly smaller than the FTdx10, with a 4.3-inch touchscreen rather than a 5-inch screen. It is slightly lighter as well. There are fewer buttons, with more features relegated to the touchscreen menu. It has most of the same functionality, including an antenna tuner, a parametric microphone equalizer, and VOX. You get the same spectrum and waterfall or 3DSS (3-dimensional spectrum stream) display. There are voice and CW keyers, but it lacks the CW, PSK, and RTTY decoders.

This is no big loss in my opinion because they are poorly implemented in the FTdx10. Also, RTTY and PSK are less popular than FT8 these days.

The received radio spectrum is sampled by two ADCs located after the selectable attenuator, bandpass filters, and RF preamplifiers. An FPGA (field programmable gate array) carries out filtering and frequency translation before the signal is passed to a DSP (digital signal processing) stage. The transmitter is also an SDR design. The audio signal is converted into a digital data stream, modulated and upconverted to the final frequency in the DSP and FPGA stages, and converted with a high-speed DAC (digital to analogue converter) to an RF signal. It is then amplified in the RF power amplifier and sent through low pass, or bandpass, filters to the antenna connector.

## FLEXRADIO

FlexRadio has been producing software defined radios for the commercial and amateur radio markets longer than any other company. They were the first to introduce a true SDR transceiver to the amateur radio community with the introduction of the SDR-1000 transceiver and PowerSDR software in 2003.

The company ceased production of the 6000 series in May 2024 and has replaced it with the new FLEX-8400 and FLEX-8600 transceivers. The change introduces a faster processor and a bigger FPGA capable of supporting future projects. New features include a built-in GNSS (GPS) receiver for very accurate frequencies. The front-end preselector filters are now solid state switched, and a sampler has been installed to allow for adaptive transmitter pre-distortion. FlexRadios have great technical performance and SmartSDR has some excellent features that are not available on other amateur radios. Users are kept in touch with developments via 'The Flex Insider' email newsletter, there are many FlexRadio videos on YouTube and the FlexRadio website, and there is a thriving user forum on the company website. The 'Signature' 6000 and 8000 series radios work on the 'thin client' model where the DSP functions are performed by a DSP chip inside the radio.

Models with an M suffix include the Maestro front panel controls and display. All models support remote operation using an external Maestro terminal, or a computer running SmartSDR. You can also operate the radio using a tablet, iPad, or iPhone. The radio can be operated remotely over the Internet and it is even possible for two people to operate the radio from different locations at the same time. An interlock ensures that only one person can transmit at any given time.

### Maestro

Maestro is a portable user device, or console, for the FlexRadio 6000 and 8000 series of radios. It is a stand-alone computer terminal running the SmartSDR client software. Not just a box supplying the controls that are missing from the front panel of the radio.

The device can connect to the radio directly over Ethernet, via your home LAN using WiFi or Ethernet, or worldwide over the Internet. As well as control knobs and some buttons, it features an 8-inch touchscreen so you can control the radio with your fingertips. The display can support two panadapters, which is all you could expect to use on an 8-inch monitor. You can be truly portable with a WiFi connection and the optional internal battery pack. The radio has jacks for electret or dynamic microphones, a Morse key or bug, PTT, speakers, and headphones. There is a mono speaker inside as well.

With a Maestro you can operate your FlexRadio from anywhere in the house or anywhere in the world that has a reasonably fast Internet connection.

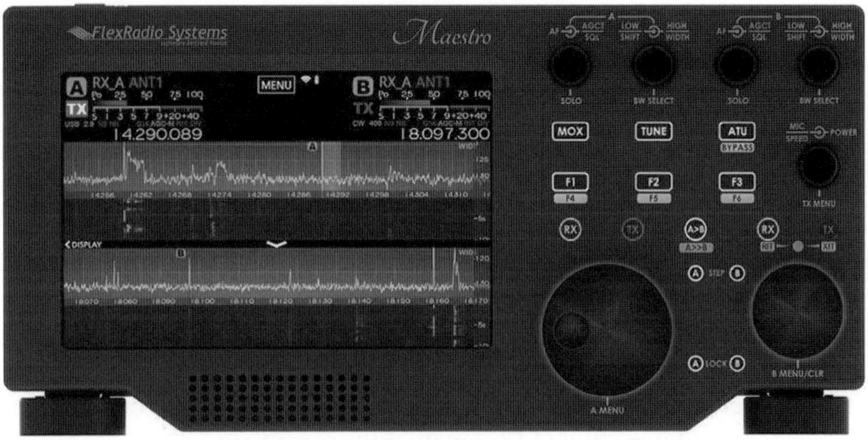


Figure 27: FlexRadio Maestro controller https://on5vl.org/maestro-flexradio/

### FLEX-6400, FLEX-6400M, FLEX-6600, and FLEX-6600M (2017)

FlexRadio released the 'Signature' range in 2017 with two black box radios and two with front panel controls and a screen. The front panel was kept deliberately simple with only the most used controls. The less-used controls and menu settings are accessed through the SmartSDR software. Any of the radios can also be used with SmartSDR running on a Windows PC, Mac, iPad, or iPhone. There is also a Mac client called 'dogparkSDR.'

The transmitters cover the 160m to 6m amateur bands, with receiver coverage between 30 kHz and 54 MHz. The FLEX-6400 and FLEX-6400M have a single ADC. FlexRadio calls it an SCU (spectral capture unit). The radio can display one or two receivers on one or two 7 MHz wide panadapters.

The FLEX-6600 and FLEX-6600M models have two ADCs (SCUs). When using SmartSDR running on a PC, they can display up to four receivers on up to four 14 MHz wide panadapters. The display on the radio, an external Maestro unit, or small screen devices like an iPad or iPhone shows one or two panadapters.

The ADCs are clocked at 245.76 Msps rather than the 122.88 Msps sampling rate in the FLEX-6400. Having two ADCs, the FLEX-6600 radios can simultaneously receive on two bands using different antennas, for example, 20m and 6m. The model also supports diversity reception using phase addition or cancellation between two antennas to enhance a wanted signal or reduce noise and interference. The FLEX-6600 radios have 7$^{th}$ order bandpass filters on the contest bands and the antenna tuner is supplied. It is an option on the FLEX-6400.

### FLEX-8400, FLEX-8400M, FLEX-8600, and FLEX-8600M (2024)

The FlexRadio 8000 series is expected to be available from late August 2024. Other than the model number, the radios look identical to the 6000 series. There are several key differences internally. The FPGA is much larger in the new radios which will allow FlexRadio the capacity to develop new features for the radios. The microprocessor is also much faster.

New features include a built-in GNSS (GPS) receiver for very accurate frequencies. There is still the option to install a GPSDO (GPS disciplined oscillator) if you want super accuracy, but you would only need that if you were using a microwave frequency transverter. The front-end preselector filters are now solid state switched. A sampler has been installed to allow for adaptive transmitter pre-distortion. I assume that this feature will be added to SmartSDR. There will also be improved spectrum display features and a new option to display a wideband view of several bands at once. Future developments include the ability to transmit and receive modes such as FT8 over a 40 kHz range without moving the VFO frequency. DX stations will be able to transmit multiple FT8 signals at the same time.

The differences between the 6400 and the 6600 remain the same between the 8400 and the 8600. FLEX-8600 radios will still have higher speed ADCs and better bandpass filters. An antenna tuner is supplied with the FLEX-8600. It remains an option on the FLEX-8400, and having two ADCs the FLEX-8600 supports diversity reception.

## APACHE LABS

Working in conjunction with members of the OpenHPSDR group, Apache Labs has been producing SDR transceivers since 2012. The early ANAN-100 and ANAN-10 models used the 'Hermes' SDR board and 'Alex' filters developed by the OpenHPSDR group. Apache Labs commercialized the designs and financed the development of newer more powerful designs with dual ADCs.

Apache Labs radios do not ship with any SDR software. They are compatible with the free, open source, Thetis software created and maintained by the OpenHPSDR group. They are also compatible with PowerSDR mRX PS, piHPSDR, SDR Console v3, and SDR++ (receive only).

The piHPSDR console units designed by John Melton G0ORX/N6YLT provided knobs and a panadapter display for the ANAN black box radios.

### Andromeda (2020)

The Andromeda 100 watt HF transceiver is no longer available. It included a Windows i5 or later an i7 PC which ran the Thetis SDR program and other non-SDR software. The radio featured two ADCs, diversity reception, and Pure Signal transmitter pre-distortion. The oscillator had an impressive 0.1ppm stability and -149 dB of spectral phase noise. The receiver was capable of an RMDR of 116 dB at a 2 kHz offset. I cannot find any technical data or even a user guide for the Andromeda transceiver on the Apache Labs website. But there is a sales brochure at, https://apache-labs.com/productPdf/1058_ANDROMEDA_v1.00.pdf.

### G2 (2022) and G2 Ultra (2024)

The ANAN G2 and G2 'Ultra' direct digital sampling HF transceivers use the 'Saturn' SDR board. I am a bit disappointed they didn't call the radio the Saturn.

The G2 'Ultra' has replaced the original G2 model with the screen and front panel controls. It has more buttons and an additional optical encoder, plus a larger screen. 8" on the Ultra and 7" on the G2. An Arduino board has been added to manage the knobs and buttons, passing the control signals on to the Raspberry Pi4 compute board that is in all the G2 models. The controls on the Ultra version can control Thetis running on your PC. There is also a 100 watt black box version and a 1000-watt black box version that has a display for the power amplifier, but not the radio.

The G2 Ultra (with knobs) is competitively priced and it has excellent specifications. It can be used standalone with the piHPSDR software running on an embedded Raspberry Pi4 compute module. Or, you can use an external monitor, keyboard, and mouse. There are several external software options. Thetis running on your PC is the best choice of external software. It includes features like diversity reception and Pure Signal. piHPSDR will run on a Linux computer or a piHPSDR console, but the consoles are no longer available. SDR Console v3 is another option. Finally, SDR++ works with the ANAN radios on a variety of platforms but it is 'receive only'.

The G2 has two HDMI outputs and GB Ethernet. The spec says that the hardware supports 10 digital down converters, but I don't know of any SDR software that can display ten panadapters. The oscillator has an impressive 0.1ppm stability and -149 dB of spectral phase noise. The receiver is capable of an RMDR of 116 dB at a 2 kHz offset. The radio has two ADCs, it is capable of diversity reception, and Pure Signal transmitter pre-distortion.

*TIP: I like the ANAN radios. They have excellent specifications and good build quality. However, Apache Labs is a small company and you may have to wait a few months for your radio to be shipped. There is a dealer in the U.K. or you can buy a radio directly from Apache Labs.*

## OTHERS

There are a number of Russian and Chinese designed QRP radios with knobs, being sold by the usual online stores. Some of them are very good, but you do need to read the fine print. If the advertisement does not explicitly say something is included, then it isn't included. Even if there is a photograph showing the item in question. Also, these websites often have different 'colours' with various buying options. Sometimes there is a cheap leader such as a cable, but the full radio costs a lot more. I even saw one advert that was selling an empty case at a much lower price. I only have first-hand experience with two models, and I was impressed with both of them.

### Malahit DSP2

The Malahit DSP2 is a compact receiver with great features and a huge frequency range, from 10 kHz to 380 MHz and 404 MHz to 2 GHz. Unlike most 'dongle,' and 'black box' SDR receivers, the DSP2 is completely self-contained. It has a colour LCD touchscreen, a built-in speaker, and an internal battery. The radio is portable making it ideal for plane spotters, and for use while walking or on holiday. My first impression when I opened the box, was the high quality of the radio and the nice travel case. The radio is surprisingly heavy, no doubt due to the metal case and the 5000 mAh 3.7-volt Lithium battery. I particularly like that you can charge the battery using the supplied Type-C USB cable from any USB power supply. This radio has both American FCC and European CE, type approval which is very unusual on a Chinese radio. It is a Russian design, built under license by a Chinese company. The fully licensed firmware is supplied on the Russian MalahiTeam website. It has the usual AM, SAM, DSB, FM, SSB, and CW modes, plus wideband FM stereo with RDS. There are decoders for CW, RTTY, and FT8.

### Xiegu X6100

The X6100 is a software defined transceiver with 5 watts of transmitter power when running on the internal battery and up to 10 watts when using an external DC power supply. The radio is well constructed and quite heavy at 880g. It has an excellent 4" (100mm) display which is very crisp and bright, although sadly it is not a touchscreen. The five 'screen control' buttons along the bottom of the display provide access to a variety of menu items that change when you press one of the six function buttons above the MFK (multi-function knob). I believe that the receiver is a QSD (quadrature sampling detector) design with a QSE (quadrature sampling exciter) in the transmitter.

It was a big surprise to discover that not only does the radio have a voice keyer and a CW message keyer. It also has built-in CW, PSK, and RTTY decoders just like the IC-7300. The CW messages work for PSK and RTTY while the decoders are active.

# ON-BOARD OR EXTERNAL DSP?

Digital signal processing functions like filtering, demodulation, modulation, and audio signal processing are software functions performed on the digital signals after the ADC sampling process. The DSP software can be implemented in several ways:

- Dedicated DSP chips inside the radio: These are the same specialized DSP chips as the ones used in many conventional amateur radio transceivers and receivers. The chips are specialized 'microprocessor like' devices, which can be updated periodically by uploading new firmware to the radio. Examples of radios that use dedicated DSP chips include the FlexRadio Systems Signature 8000 series and most of the 'SDR with knobs' radios. It is the approach used in the Yaesu, Icom, and Elecraft transceivers and the IC-R8600 receiver.

- Software running on a computer chip inside the radio: This could be a 'softcore' microprocessor emulation or Verilog code running on an FPGA. Or it can be software running on a microcontroller, an ARM (Advanced RISC Machine) chip, or a microprocessor. The difference is that this approach uses a computer or an FPGA rather than a 'commercial off the shelf' DSP chip. This type of DSP is also updated by uploading a firmware file to the radio. The Xiegu X6100 is an example that uses an STM32 microcontroller to perform the DSP functions.

- Software running on a computer or single board computer mounted inside the same box as the radio. The internal computer might run the Linux, Windows, or MAC operating system, but Linux is the most common. A Widows or macOS PC in the box works just like your desktop or notebook PC. The SDR software would be updated from the Internet. The Expert Electronics MB1 transceiver is an example of this approach. Embedded single-board computers such as the Raspberry Pi compute module in the ANAN G2 can run other software, but generally, they are dedicated to running the SDR software.

- Software running on your PC as a part of your favourite SDR program. The SDR software can be updated from the Internet. With some radios, you can choose from a selection of SDR programs. This is the traditional 'thick client' approach used by most SDR equipment manufacturers. The radio hardware acts as a front-end and the DSP is done by SDR software running on your PC. Examples include most of the black box receivers, such as the SDRplay models, Airspy, RTL dongles, RF Space, and loads of others.

If the radio is a stand-alone device without the need to connect it to a PC, the DSP must be done inside the radio. But even in the, 'no knobs' radios which do need a computer, there is a strong trend towards moving the DSP functions into the radio.

Splitting the DSP 'server' activities from the control and display 'client' activities has many advantages. The most important is the dramatic reduction in the amount of wasted data being sent from the radio to the PC over the Ethernet connection. With these 'thin client' onboard DSP radios, only the wanted data like the panadapters that you want to display and the receivers that you want to hear are sent to the PC. Not all of the data required for the entire HF spectrum or the whole bandwidth of a hybrid radio.

Some direct sampling SDRs with knobs, like the Expert Electronics MB1 use a built-in Windows computer. This internal PC is set up to run PC-based SDR software usually specifically designed for the radio. The architecture is convenient but expensive because you have to pay for a computer as well as a radio. It is a good idea if you want an "all in one box" high-performance SDR without the need for network connections or USB interconnection cables. The internal PC can run other programs at the same time as the radio, such as your station log, contesting, or digital mode software.

There are some good arguments for performing all of the DSP processes inside the PC software. The best one is that it makes the radio hardware cheaper. The introduction of faster interfaces like Gb Ethernet and USB3 makes it much easier to maintain a fast continuous data flow between a radio and a computer. So, it is now possible to have 'thick client' wideband radios connected to computers running the SDR and DSP software. But overall, I think that the benefits of the 'thin client' model with, DSP being performed inside the radio, are more important. The lower data rate required allows for remote operation over the Internet, using lower-power devices like phones and iPads, and multiple users. This can be two operators at different locations sharing a contest radio, or many people sharing a web receiver. For example, the four panadapters and receivers on a FLEX-8600 can be shared by two operators who can each use two receivers on two bands with separate antennas. The 'multi-operator single radio' contest station can activate four bands simultaneously using one radio with the single transmitter changing bands as appropriate. It can work very well since the radio operates in full duplex mode. One operator can listen or tune across a band while the other is transmitting on a different band.

Cheap radios that support a single panadapter may continue to send the data to an external computer performing the DSP because the low cost is one of the big advantages of those radios.

High-performance HF transceivers and the wideband VHF/UHF/SHF radios will probably all end up with DSP onboard. With the VHF, UHF, and SHF radios, you want very wide panadapters because the bands are bigger and signals are further apart. For HF transceivers, you want to display two or more panadapters on different bands. Either option requires huge amounts of data to be passed from the radio to the PC.

# SDR SOFTWARE

The SDR software running on your computer or another computing device is a very important component of any black box radio. It is the interface between the user and the radio, presenting the controls and the panadapter display. SDR software offers a user experience that is quite different from a conventional radio and you have to get used to operating in a different way. But essentially this is really no different to learning how to operate any new receiver or transceiver. Using a radio with external software has become so popular that many of the SDRs with knobs can also be operated from a computer, an iPad, a phone, or a Raspberry Pi. It gives the owner great options. You can operate your radio over WiFi while enjoying a summer evening outside, or away from home for work or on holiday.

There are dozens of different software defined radio receivers and transceivers available. All the models without knobs need a computer of some kind to provide a way to control the radio. The computer screen or 'window' that displays the radio controls and the panadapter display is known as a graphical user interface (GUI) or sometimes a 'console.' The software application provides a simulation of the knobs and controls that are missing from the front panel of the radio. It also displays the panadapter spectrum and waterfall displays. It often provides digital signal processing (DSP) functions such as noise and notch filters, modulation, and demodulation, but sometimes that is done inside the radio.

Most ham radio, transceivers produced over the last 20 years or so can be controlled using PC software. Computer control is handy for things like operating digital modes, logging, and CW keyers. If you use CAT (Icom CI-V) or digital mode software you should have no trouble adjusting to using an SDR. But the PC software for an SDR is much more than just an interface to the radio. Other computer programs may be able to remote control a conventional radio, and your digital mode program may include a spectrum or a waterfall display of the audio frequencies it is using. Only SDR software can display an RF panadapter or perform the DSP functions.

One of the big advantages of using an SDR is the ability to try different software with the same radio. New software can completely change the way the radio looks and operates, with new modes and features. Trying a different SDR application can be like getting a whole new radio and it helps to keep your radio up to date.

## SOFTWARE TYPES

There are many excellent SDR programs available. A lot of it is freeware, usually developed under open-source software rules. Other programs are commercial products intended for a specific SDR. Some software defined radios will only work with software developed by the manufacturer. Others may work with a variety of different applications.

Depending on what radio you purchase it can be a real challenge working out what software is compatible with it. Luckily, help is never far away, with a multitude of online forums on the Internet.

Open source software code is made available for anyone to modify, update, and reissue as long as the open-source rules are followed. Some open-source SDR software has variants, known as 'forks', which have been developed for different radios or generations of SDR hardware. PowerSDR was initially developed for the FlexRadio Systems 'Flex' series of radios, and later the Apache Labs ANAN series. It is still available as legacy software for the FLEX-1500, FLEX-3000, and FLEX-5000 radios. The Power SDR mRX variant can be used with Hermes-based radios and the Apache Labs radios using either 'protocol 1' or 'protocol 2.' The 'protocols' are different data formats between the radio and the computer. Protocol 1 supports multiple panadapters and receivers, but the panadapters must all have the same bandwidth. Protocol 2 supports panadapters with different bandwidths. Hermes was a very advanced direct sampling HF transceiver board, with a single 16-bit ADC sampling at 122.88 Msps. It was the main board in the ANAN-10 and ANAN-100.

PowerSDR and the Hermes board were developed as open-source projects by the OpenHPSDR group. Without this group of dedicated people who donated many thousands of hours of their time and expertise for free, it is unlikely that you would be using an SDR today. The software development team pioneered major advancements in the DSP software code, and adaptions for radio astronomy, receiver diversity, beam steering, and transmitter adaptive pre-distortion. Many of the things we now take for granted such as the spectrum and waterfall display were developed or refined for PowerSDR.

The Thetis program was developed by the same group. There are two forks of Thetis. The OpenHPSDR group's version and a fork by Reid Campbell MI0BOT which has enhancements for the Hermes Lite 2 QRP transceiver. Both versions can be used with the Apache Labs ANAN radios.

SDR software whether commercial or open source is usually developed for a particular model or type of software defined radio. Broadly, these align with the generations of SDR. There are programs aimed at Generation 1 'soundcard' software defined radios where the PC soundcard performs the analogue to digital and possibly digital to analogue conversion. Other applications are configured for the 'dongle' and other small box radios using a USB interface. Finally, there is software for radios, which use an Ethernet interface. Another consideration is the computer operating system. There are applications for Microsoft Windows, Apple Mac, and Linux. There are also applications for single-board computers and smartphones.

For some reason, there is a wealth of online information about how the SDR hardware works. But there is relatively little information about how the PC software works, even at a block level.

Table 3: SDR Software and Hardware Compatibility

| *Software* | *System* | *Hardware* | *TX/RX* | *Interface* | *SDR type* |
|---|---|---|---|---|---|
| CubicSDR | Mac Win Linux | SDRPlay, HackRF, BladeRF, AirSpy, NetSDR+, Red Pitaya RTL | Receivers | USB | DDC |
| CuteSDR (minimal) | Mac Win Linux | RFSpace, Afedri | Receivers | USB | DDC |
| GQRX | Mac Linux | FCD, USRP, RTL Airspy HackRF, RFSpace Red Pitaya | Receivers | USB Ethernet | DDC |
| GSDR (PowerSDR fork) | Win | Genesis, SR, SC | Receivers | USB SC | QSD |
| HDSDR Used with extio.dll for a variety of radios. Was Winrad. | Win | FCD, RTL, FDM-S1, SC, FDM-S2, SR, Hermes, ANAN, Perseus, Afedri, BladeRF, HackRF, Andrus Icom R6800 Xiegu SDRplay | Receivers | USB SC Ethernet | QSD and DDC |
| Kiss Konsole (code still available) | Win | Hermes, Apache Protocol 1 | Transceivers | Ethernet | DDC/DUC |
| Linrad | Win Mac Linux | RTL, RFSpace, Perseus | Receivers | USB | DDC |

| TABLE 3: SDR SOFTWARE AND HARDWARE COMPATIBILITY | | | | | |
|---|---|---|---|---|---|
| *Software* | *System* | *Hardware* | *TX/RX* | *Interface* | *SDR type* |
| PowerSDR | Win | FlexRadio 1000, 1500, 3000, 5000 | Transceivers | FireWire USB | QSD/QSE |
| PowerSDR IQ | Win | SR, SC, RTL | Transceivers | SC USB | QSD and DDC |
| PowerSDR mRX PS (replaced by Thetis) | Win | Hermes, Apache, Hermes Lite2 | Transceivers | Ethernet | DDC/DUC |
| PowerSDR UI | Win | FlexRadio 1000, 1500, 3000, 5000 | Transceivers | FireWire USB | QSD/QSE |
| QtRadio (SDR client used in other software) | Win Linux | RTL, Hermes, Apache, SR, SC, Perseus, RFSpace, HiQSDR, USRP | Receivers | USB SC Ethernet | QSD and DDC |
| Quisk (old) | Linux | Afedri, SR, SC | Receivers | USB SC Ethernet | QSD and DDC |
| Rocky 3.8 (old) | Win | SR, SC | Receivers | SC | QSD |
| SDR Touch | Android | RTL | Receivers | USB | DDC |
| SDR# "SDR sharp" | Win Mac | Airspy FCD, RTL, SR, SC, RFSpace, HackRF | Receivers | USB SC Ethernet | QSD and DDC |
| SDR Console V3 | Win | ANAN, ELAD, Hermes Lite 2, Lime, Pluto | Transceivers Receivers | USB, SC, Ethernet | DDC and QSD |
| SmartSDR (not open source). | Win | FlexRadio Signature 6000 and 8000 series | Transceivers | Ethernet | DDC/DUC |

TABLE 3: SDR SOFTWARE AND HARDWARE COMPATIBILITY

| *Software* | *System* | *Hardware* | *TX/RX* | *Interface* | *SDR type* |
|---|---|---|---|---|---|
| SodiraSDR | Win | RTL, SC,SR, Elektor, RFSpace, FCD, Pappradio, Afedri | Receivers | USB | QSD and DDC |
| Studio 1 (not open source). | Win | Perseus, RFSpace, RTL, SR, SC, FCD, FDM-S1, Apache, QS1R, USRP, Airspy Hack RF | Receivers | USB, SC, Ethernet | DDC |
| Thetis (replaces PowerSDR) | Win | ANAN Hermes Hermes Lite2 | Transceivers | Ethernet | DDC/DUC |
| uSDR (Delphi) | Win | RTL, Airspy, Blade RF, Hack RF, Lime | Receivers | USB | DDC |
| Wavesink + DAB and FM-RDS | Android | RTL | Receivers | USB | DDC |

This list is not exhaustive and some software may work with SDR hardware that is not listed. It is as accurate as possible at the time of writing. My apologies to any I inadvertently miss out. Inclusion or exclusion from this list does not imply any endorsement or criticism of any particular hardware or software.

FCD = FUNcube Dongle (all versions)

Hermes = OpenHPSDR

RFSpace may include SDR-14, SDR-IQ, SDR-IP

RTL = RTL Dongles USRP = Ettus Research radios

SC = Soundcard. It may include PC soundcards and USB soundcard devices

SR = Softrock and most other generation one soundcard SDRs

Win = Microsoft Windows operating system

Mac = Apple Macintosh operating system

Linux = Linux operating system

GNU = GNU Radio design (usually Linux-based)

Android = Android smartphone software

## THE GRAPHICAL USER INTERFACE

The graphical user interface (GUI) "Goo E" is the on-screen representation of the knobs, buttons, and displays that you use to control the functions of your software defined radio. Many conventional radios have menus or popup windows for setting lesser-used options. Most software defined radio programs do as well.

Many of the latest 'server type' SDR programs use an Internet web browser as the main display. The radio is connected to a single board computer which runs the server application to send the data out to the local network or the Internet. Any web browser can run the GUI to display and control the radio remotely.

Older SDR software tends to have a user interface designed to look like a hardware radio. Settings that are adjusted often appear on the main screen or the 'face of the radio.' Slider controls replace the volume control and other knobs because they are easier to use with a mouse. All the usual controls, the VFO frequency, and an S meter are displayed, as well as the panadapter spectrum and waterfall. Newer software like SmartSDR has a less cluttered look with a larger panadapter and waterfall displays. The user controls pop-up when you move the computer mouse to control areas on the computer screen.

*TIP: There are often two or three ways to adjust settings in PowerSDR and Thetis. Settings in the setup menu may change if you make adjustments on the main screen and vice versa. This is normal, but you might wonder why a menu item has changed status.*

The on-screen controls will be familiar to users of digital mode software, but for computer 'Newbies' here is a quick guide on how to set the main kinds of control. If you are left-handed the mouse buttons may be reversed.

- 'Click' means to move the mouse over the control that you want to change then press and release the left mouse button.
- 'Click and hold' means to move the mouse over the control then press and hold down the left mouse button. You will be able to adjust the control by moving the mouse sideways and/or vertically.
- 'Right click' means to move the mouse over the control then click or click and hold the right mouse button. As with your word processor and most other programs, the right click is used to display additional menu items. A pop-up window will list the options that you can then select with a left mouse click. If the list of available options is longer than the display size of the pop-up window, moving the mouse vertically will scroll through the list.
- 'Mouse wheel' is the wheel located between the two mouse buttons. It might not be an option on MacOS computers. Click on the control then use the mouse wheel to change the value. In most SDR software, clicking on the panadapter will allow the mouse wheel to fine-tune the frequency.

*Slider – click and hold to adjust. Or click to the right or left of the knob for a coarse adjustment.*

*Text box – click inside the box and type in a new value. Often the mouse wheel and the up and down keys can be used to change numeric values.*

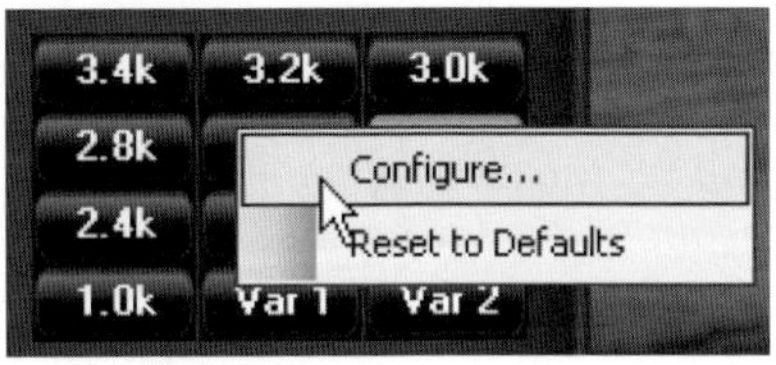


*Right click – may display a list of options. Then left click on an option to make a selection. Sometimes a new window will open to display choices that are more complex.*

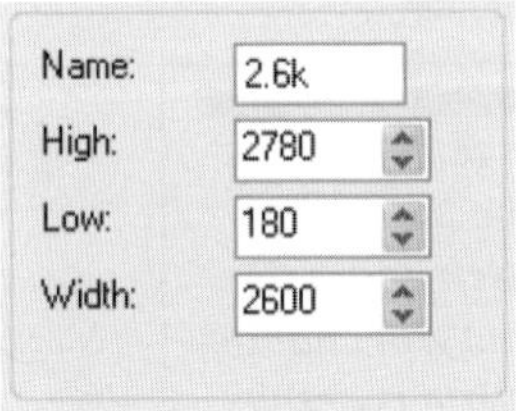


*List Box – click the up or down arrows to adjust the value, or click in the text box to type in a new value. New entries may be restricted to pre-set values.*

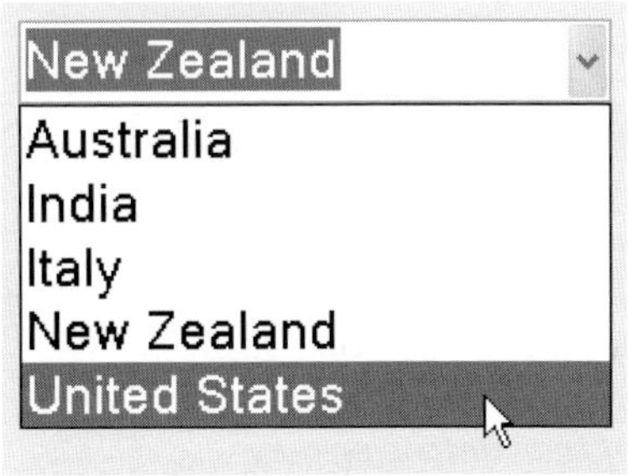


*Drop down box – click on the down arrow to open a list of options. Sometimes the list has scroll bars on the right to show more choices. Then click on an option to change the selection. Normally an unopened dropdown box shows the default, or the last chosen option.*

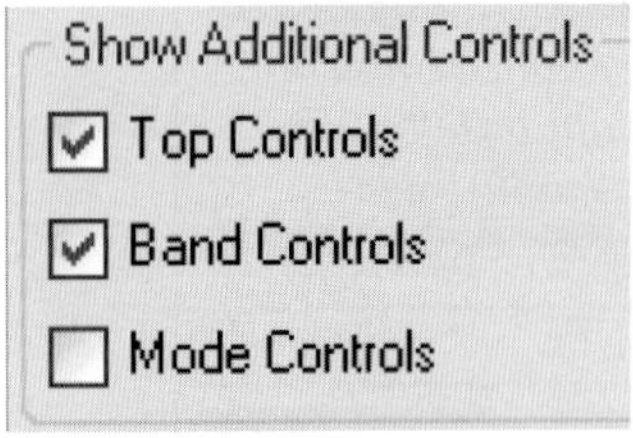


*Check Boxes – click in the square to place a check mark 'tick' beside options that you want to activate. Click again to unset. Multiple options can be selected.*

*Radio Buttons – click in the circle to activate an option. Like an old fashioned car radio, you can select any one of the available choices.*

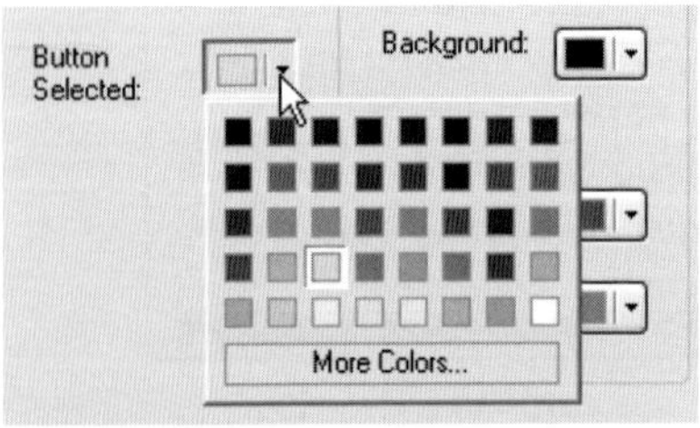


*Color chooser – a drop down box that lets you set the color of an item on the main display for example the text in the frequency read out.*

*Button – just click on it. They usually latch. Click the button to turn a function on, click again to turn it off. The graphic will change to show when the button is active or depressed.*

*Mouse hover – moving the mouse over controls on the screen, may cause pop-up windows to appear. These may display 'tool tips' such as information relating to the function of the control, or they may show a list of hidden setup options.*

Figure 28: Common SDR program controls

## SOFTWARE DISTRIBUTION METHODS

**Commercial** – you pay for the software when you order or download it. The software is usually downloaded from the company website. The vendor will provide an activation 'key' code by email. Linux software is often downloaded from a Linux repository. Cell phone 'apps' are downloaded from an 'App Store.' These 'paid apps' require a credit card or other electronic payment before the program can be downloaded. Commercial source code is usually proprietary and not available to the 'general public.'

**Shareware** – you can download (or order) the software and pay for it after you have tried it out. Usually, there is an incentive to register the software and send the author or distributor a payment. Either the full functionality is not available in the unregistered version, or the software will stop working after a few weeks. Normally the source code is proprietary and not available to the 'general public.' But software can be both open source and shareware.

**Freeware** - you can download (or order) the software and use it for free. However, the source code may not be available to the public.

**Open source** – applies to software developed and distributed under the open-source rules. Very often you can download (or order) the software and use it for free, but charges can apply. Not all open-source software is freeware. Even if the product is available free of charge there may be a charge for the CD and postage. The source code must be made available to other developers.

## OPEN SOURCE SDR SOFTWARE

A lot of the software that is available for use with your SDR hardware has been produced under the 'open source' rules. Even when the software is a commercial proprietary product, open-source .dll files or APIs may be used to allow you to use your SDR with their software. Firstly, I had better explain what 'open source' means.

*"Open source software is software that can be freely used, changed, and shared (in modified or unmodified form) by anyone. Open source software is made by many people and distributed under licenses that comply with the open source definition."* *http://opensource.org/*.

The open-source rules allow software developers to use software code developed by others in their own software programs. This is tremendously beneficial as it saves everyone from having to 'reinvent the wheel.' A programmer using someone else's open-source code can choose to give away the resulting software application as freeware, make it shareware, or sell it, but the rules state that they must make their revision of the source code available to others under the same terms as the license of the original software. There are other rules mostly relating to fairness and non-discrimination. See http://opensource.org/definition for the full definition.

Open-source software is terrific because, depending on the SDR that you are using, you may be able to try several different programs with your radio. It also means that the software evolves and extends as other developers contribute new ideas or upgrades to the original software code. You may be able to suggest improvements or cool new ideas, which might eventually be incorporated into new revisions. The radio astronomy and Pure Signal transmit pre-distortion upgrades to PowerSDR mRX PS and Thetis are very good examples of this type of collaboration.

This spirit of sharing makes SDR development much faster and cheaper, allowing software developers to concentrate on interesting new ideas and everybody wins.

*"A report by the Standish Group (from 2008) states that adoption of open-source software models has resulted in savings of about $60 billion per year to consumers." http://en.wikipedia.org/wiki/Open-source_software.*

Most open source software is developed by amateur developers for free and there is no guarantee of any future support or bug fixes. Surprisingly, given the challenges for developers who have only their leisure time for programming, open source software programs such as Thetis, SDR#, HDSDR, and many, many, others are extremely well written and perform very well indeed. In most cases, the developers are very approachable, interested in extending the software, and responsive to constructive criticism and requests. Don't demand new features or bug fixes. Open source developers are working hard to provide you with free software.

*"Open-source software (OSS) is computer software with its source code made available with a license in which the copyright holder provides the rights to study change and distribute the software to anyone and for any purpose. Open-source software is very often developed in a public, collaborative manner."*
*http://en.wikipedia.org/wiki/Open-source_software.*

The developers of many SDR programs have reused open-source DSP code, which is responsible for the filtering, modulation, and demodulation functions. The spectrum display and some filters rely on very complex mathematical fast Fourier transformations (FFT). Luckily, the software for performing FFT calculations is provided free of charge, as open-source .dll files.

Many SDR computer programs use the FFTW.dll file.

*"The Fastest Fourier Transform in the West (FFTW) is a software library for computing discrete Fourier transforms (DFTs) developed by Matteo Frigo and Steven G. Johnson at the Massachusetts Institute of Technology.*

*FFTW is known as the fastest free software implementation of the Fast Fourier transform (FFT) algorithm (upheld by regular benchmarks). It can compute transforms of real and complex-valued arrays of arbitrary size and dimension…"*
*http://en.wikipedia.org/wiki/FFTW*

One downside of using open-source software is that because you didn't pay for the software, you can't expect individual support or any guarantee that it will even work on your particular computer and radio setup. However, most of the widely used packages have online forums where you can ask for help.

Most open-source developers are enthusiastic about adding new features and extending the ways that SDR technology can be used. Sometimes they are less enthusiastic about mundane things like providing a user manual.

It can be frustrating when new revisions of the software introduce new features and setup options that I don't understand or know how to use. This is less of a problem with simple SDR programs as they don't have a lot of setup options. Pretty much all SDR programs do have a 'Help' menu covering at least the major functions.

## OPEN SOURCE .DLL FILES

A .dll is a Windows 'dynamic link library' file. It contains one or more software programs or subroutines, which can be accessed or 'called' from another piece of software. They can also contain data files or other resources like constants. The intention is to avoid re-work. Rather than 'cut-n-pasting' the same pieces of software code into every application that a developer writes, they can put a lot of useful code blocks into a .dll file. The routines inside the .dll file can be called from the main software program. Also, the .dll file can be made available to other software developers.

*TIP: The Linux equivalent of a .dll is a .so (shared object). The macOS equivalent is a .dylib (dynamic library).*

SharpDSP by Phil Covington, WDSP by Warren Pratt, and FFTW by Matteo Frigo and Steven G. Johnson are all examples of open-source software that is made available to developers in the form of .dll files.

The .dll file needs to be in a location where the calling program can find it. Traditionally they are placed in one of the Windows System directories where any program can find them, but they might be in the same directory as the SDR program files. In that case, only the calling program will be able to find the .dll file. The extio.dll file usually resides in the SDR program directory because there may be several different versions of it on your PC.

### extio.dll

One of the 'spin-offs' of the open software approach is the ability to add 'patches' or 'add-ons' to existing software. An example of this is the various flavours of extio.dll, which allow various software defined radios to work with existing SDR applications. The extio.dll provides an interface between the data format or 'frame structure' that the radio is sending out and the data structure that the software is expecting to see.

This patch approach only works if the original SDR program developer has incorporated the possibility of it being present, into their code. HDSDR is a good example. The program must do something like; "Go and look in the program directory for a file called extio.dll. If it exists, run the code inside it. If it does not exist, carry on without it." That is why there are several versions of extio.dll with the same file name but intended for use with different radios. You have to download the one that is intended for your radio.

## API (APPLICATION PROGRAMMING INTERFACE)

An API is a set of rules, published as a standard or document, that are intended to formalize a method of interfacing with a software application. It works like a kind of 'back door' allowing software developers to add features to a program, or interface a different kind of radio with it, without changing any code in the original program.

A modern SDR has several layers of computing, which means that there may be more than one level at which a software developer can interface with the radio hardware. The radio normally uses a USB or Ethernet interface to connect to the PC, so that is the first point that a PC software developer can access the data and control signals. They can write a completely new software application that reads the data from the Ethernet or USB port and writes back control information.

If the radio is a transceiver, the audio for transmission can also be sent to the radio hardware. The new application could interface with the 'native' SDR program or completely replace it. But first, the software guru needs to understand the way that the data is being carried over the USB or Ethernet cable, which is known as the 'protocol,' and they need to know how the data stream is formatted or packaged, which is known as the 'frame structure.' The protocol usually follows a known data standard, for example, UDP or TCP-IP, but the data format may be unique to the particular type of radio.

At this basic hardware level, an API would consist of tables describing the data frame structures and the control information. As well as information about how the radio will respond to commands.

Alternatively, there may be a higher level of interface, where the developer can query and receive data or control information from the SDR software application.

I guess you could say that the CAT (computer-aided transceiver) commands that are sent from digital mode and logging programs to the radio are a kind of high-level API. A CAT implementation uses a table of commands that can be sent to the SDR program and a list of responses and data that will be returned. If you know the CAT command structure you can write a program to control the SDR program and hence the radio. The difference is that a real API allows you to control the radio in the same way as the client program. You get full access to the radio rather than the restricted commands that CAT offers.

API access is much faster than CAT because CAT is done over a slow serial connection and is a low priority for the microcomputer in the radio or the SDR program. CAT only provides controls like changing a band or frequency. An API includes ways to send or receive audio or set the characteristics of a filter, not just enable or disable it.

You might use the API interface to replace the existing GUI so that your new program changes the way that the SDR looks and operates. Or you might just add a new filter or mode. In the future, the most common use of the API will probably be to add an interface between the SDR application and logging or digital mode applications. In other words, replacing the CAT interface with a more integrated and faster way of getting status and control information to and from the radio. One developer is using the FlexRadio API to add the FreeDV (Codec II) digital voice mode written by David Rowe, to the Signature series radios. Another possibility would be to add FT8 or a built-in CW, PSK, or RTTY decoder and memory keyer. There is more about the FlexRadio API in the next chapter.

## SOFTWARE TRENDS

It has never been easier to write software for SDRs. I'm not sure that is a good thing. Several factors are making it easier, but is it better?

There is a move towards the server/client architecture which places the difficult-to-write DSP software inside the radio. An external software developer only needs to write the client display and control software. Increasingly the DSP is being performed by embedded STM32 or ARM processors rather than an FPGA. The programming languages are more familiar to developers than the Verilog or Xilinx languages used to program an FPGA. The new processors are fast enough to manage the load, but the main advantage of using an FPGA is lost. FPGAs are very good at parallel processing, so they are much better when you want to create multiple panadapters. DSP performed by a computer processor is normally limited to one or perhaps two panadapters.

GNU Radio means that people can create SDR software by putting a series of graphical code blocks together, with very little understanding of how an SDR works and no understanding of how the code is written. Some developers are using Internet web browsers as the graphical user interface. It results in slower code because the web browser is acting as an interpreter.

It is great that people are experimenting with writing applications. But there doesn't seem to be as much innovation. People are creating software, but they are not creating anything new. Ten years ago the SDR scene was expanding. Exciting new technologies, like new DSP code, adaptive transmitter pre-distortion, CESSB, running DSP code on CUDA core video cards, and 'direct Fourier conversion' were being invented or developed. Now the emphasis is 'multi-platform' software running on single-board computers, web browsers, and cell phones.

# SMARTSDR

SmartSDR is the software package that is provided with the FlexRadio Systems Signature 6000 and 8000 series. The interface is very modern with most of the screen dedicated to the spectrum panadapter and waterfall displays. Many of the controls are hidden until needed. The most used controls and indicators are displayed in a little popup box, which attaches to the normal receiver indication. This information stays with the receiver when you move it across the panadapter. Each receiver is shown as a vertical line indicating the nominal carrier frequency and a shaded zone indicating the receiver bandwidth. FlexRadio calls this a 'slice receiver.'

## SPECTRUM PANADAPTER

Each SmartSDR panadapter can be extended out to a massive 14 MHz bandwidth on the FLEX-6600 and FLEX-8600. Or 7 MHz bandwidth on the FLEX-6400 and FLEX-8400. Normally you would operate with the panadapter set to cover a band of interest, for example, the 350 kHz wide 20m band, or perhaps just the CW segment of a band.

If you have a FLEX-6600 or FLEX-8600, SmartSDR can support up to four panadapters and four slice receivers using one, or two antennas. You can have two slice receivers on each panadapter. For example, you could have a receiver on CW, one on PSK, and also monitor both sides of a split SSB operation. If you just want to look at the band activity, you can display a panadapter with no slice receivers on it. These transceivers have two ADCs (FlexRadio calls them Spectral Capture Units) so they can perform diversity reception by receiving signals from two antennas at the same time. The FLEX-6400 and FLEX-8400 support up to two 7 MHz panadapters and two slice receivers. They only have one SCU so they can't do diversity reception.

A neat feature is that you can change the scale of the display in either direction by simply dragging the axis with the computer mouse. To make the panadapter wider just drag the display to the right. To change the level of the displayed signals just drag the vertical axis. This sounds very logical but most SDR software can't do it. You can change the receiver slice frequency by dragging it left or right and you can move the whole panadapter higher or lower in frequency by dragging that. Finally, you can move the divider between the spectrum and the waterfall to make a big waterfall and a narrow spectrum or the other way around.

## THE WATERFALL DISPLAY

Each SmartSDR panadapter can display a spectrum and/or a waterfall display. Most people prefer to run both at the same time because they each contribute useful information. This mode is called a 'Panafall' display. The SmartSDR waterfall display adds some new features that are not included in the majority of SDR software packages.

Other programs like Thetis store the waterfall information as a graphics image like a photo. After each line on the spectrum display is drawn, the levels are added to the waterfall display as a complete line. As each line is added to the top of the picture, the oldest line disappears from the bottom. If you change the panadapter frequency, the coloured vertical lines showing active signals, curve across to the new frequency. If you reduce the panadapter bandwidth, the newest waterfall lines will suddenly transition to a wider trace. In other words, the waterfall shows the history of the spectrum exactly as it was when it was captured. However, if you change the panadapter width on SmartSDR the waterfall display adjusts to compensate. It recalculates the entire displayed waterfall at any time that you change the frequency or bandwidth. This may not sound like a big deal, but it is a very nice feature. If the panadapter frequency is changed, the whole waterfall moves with it, so you don't get the curved trace lines. When the panadapter is dragged wider, the whole waterfall adjusts to a wider or narrower trace, not just the top line. This can create black voids where the spectrum information has not been captured, but overall I think it looks better.

Another innovation is that you can stop the waterfall and look back into the history. Depending on the update speed that you are using, you can scroll back and see waterfall images that occurred up to 10 minutes earlier. While you are doing this, the waterfall is still being recorded, so you don't miss anything when you return to normal mode.

*Technical TIP: When you drag the panadapter very wide, the radio switches to a wideband mode by switching out the front-end receiver pre-selector filters. The receiver will be more prone to interference signals if the panadapter is showing a bandwidth that is wider than the normal bandpass filter. A wide view also reduces the resolution of the image. Each pixel on the panadapter represents a wider selection of frequencies, i.e. the bins are wider. Normally wider panadapter bins would mean a higher receiver noise floor, but FlexRadio has stated that they are addressing that problem. I expect that the faster processor and FPGA in the 8000 series will allow for a wider FFT so that the same resolution can be maintained at any level of panadapter zoom.*

## TRACKING NOTCH FILTERS

Notch filters are implemented on the audio signal after the demodulator. They are used to eliminate interference carriers and 'birdies' within the audio passband.

The SmartSDR tracking notch filters remember the RF frequency of the interfering signal, rather than the audio frequency within the receiver passband like the automatic or manual notch filters on other radios. When you change the slice receiver frequency, the notch filter stays on top of the interfering signal instead of moving with the receiver. If you tune to a different part of the band or even another band and then return later, the notch will still be in place, filtering out the interference signal.

The notch filter eliminates signals that are fixed or changing much more slowly than the wanted speech signal. Some SDR software lets you adjust the depth of the notch, its width, and the centre frequency. Automatic notch filters can find an interfering carrier and move a notch onto it. Some can notch out multiple interfering signals. Sometimes the interference is a bunch of closely spaced carriers or a modulated signal. Setting a wider notch should eliminate any 'long-term' signal. A standard notch filter continues to filter the same audio frequency even when the radio is moved to a new frequency and the filter is no longer required. You have to remember to turn it off. The SmartSDR notch stays on the interfering signal so you can leave it activated.

## SMARTLINK® TECHNOLOGY

SmartLink is a registered trademark of FlexRadio. It is the SmartSDR tool that allows FlexRadios to be used over the Internet. A wired Ethernet connection is preferred, but it can work over a fast WiFi connection. You have to register for SmartLink and create a login for remote access. With SmartLink, you can have a local user and a remote user connected to the radio at the same time. It is also possible for SmartSDR users to connect to a choice of FlexRadios, ideal for contest or club stations.

## CWX (CW EXCHANGE)

CWX is the CW keyer included with SmartSDR. It is fully interactive, unlike the message keyers that are included with the Yaesu and Icom radios. You can use a keyboard to send as you type, or type ahead and then send. It has twelve text macros that are easy to change. The macros are mapped to the keyboard function keys. They can be sent either by clicking the mouse or pressing a function key. CWX supplies everything you need from an auto-keyer. But it does not include a CW decoder for reading Morse code. You could add external software such as CW Skimmer or MRP40 to decode the received Morse code.

## MULTIFLEX

MultiFLEX allows two operators to use the radio at the same time, each with a panadapter screen and controls. They may be operating within the same LAN, or remotely via SmartLink. The intention is to allow a single FLEX radio to be used as a multi-operator contest station. The resources are shared and only one operator can transmit at a time. This fits with the rules of a multi-operator single radio station. A FLEX-8600(M) or FLEX-6600(M) can have four panadapters and four receivers, so each MultiFLEX operator has access to two panadapters and two receivers. Since the radio has two SCUs two operators can work on different bands without having to sacrifice the front-end bandpass filters. The FLEX-8400(M) or FLEX-6400(M) can only support two panadapters and two receivers, so each MultiFLEX operator has access to one panadapter and one receiver. The operators can still work on different bands but the radio would have to switch to the wideband mode.

## DAX (DIGITAL AUDIO EXCHANGE)

DAX is used to connect audio signals to other applications, mostly digital mode software. You can also send IQ signals to an external program so that you can run external SDR software, for example, CW Skimmer or a DRM receiver. A virtual soundcard is created for each DAX audio output channel to provide receive audio to your digital mode software. All of the DAX outputs can work at the same time. So if you have a radio with four slice receivers, they can all send audio to the PC.

A single DAX soundcard channel is created for routing audio from digital mode programs into SmartSDR. Only one transmit audio channel is required since the radio can only transmit one signal at a time. The transmitter will use audio from the software connected to the receiver slice associated with the TX indicator, or as set in the DAX transmitter setup.

Two or four IQ output channels are available, depending on the radio. The DAX IQ channels can be used in any combination, which does not exceed a total of 384 ksps. The options for a four slice receiver radio are; 1x 384 ksps, 2 x 192 ksps, 1 x 192 ksps plus 2 x 96 ksps, or 4 x 96 ksps.

SmartSDR can send audio or IQ data to multiple digital mode programs at the same time. You can receive digital modes simultaneously on different bands with automatic transmit switching as required. This could be used to operate PSK on 20m and FT8 on 40m or to operate the same mode across several bands at the same time. You can even run 'full duplex,' receiving on one band while transmitting on another. For example, you can hold two FT8 conversations at the same time, transmitting FT8 on even periods on one band and odd periods on another band. How cool is that?

*Technical TIP: Windows defaults to an audio sample rate of 44.1 kHz, the standard data rate for CD audio. But the connection works best when set to a multiple of the 24 kHz sample rate used by SmartSDR. It is suggested that you change the sample rate on all of the DAX audio 'soundcards' to the 48 kHz sample rate.*

## API (APPLICATION PROGRAMMING INTERFACE)

An API is an Application Programming Interface. It is a set of instructions that are built into a computer program that lets a second program control aspects of the first. For example, you might write a digital voice program that takes and sends audio to SmartSDR. The SmartSDR API is a novel feature of SmartSDR. Even though the SmartSDR application is not 'open source,' the API allows external software developers to contribute functions and extensions.

The API enables the Open Source development community the opportunity to write applications and features for the radios. I am sure some great features will result from the collaboration. There is already a FreeDV 'digital voice' implementation available for use with SmartSDR, which uses the Waveform API.

## SmartSDR API – Technical description

The FlexRadio Systems radios use a client/server arrangement. The server is running as a Linux application on the ARM processor in the radio. The client is the SmartSDR Windows application running on your PC.

The client software has three distinct layers. At the bottom, there is a data handling section, which manages the interface with the server software in the radio. The communication is UDP and TCP-IP packets carried over the Ethernet interface. The UDP packets carry all the panadapters, waterfalls, IQ streams, audio from the PC to the radio, and the audio streams from each active Slice Receiver. The TCP-IP packets carry control and status information. TCP-IP uses a standard socket connection to port 4992. The data format of the UDP packets follows the ANSI VITA49 extended standard. Control over the TCP-IP is in the form of easy-to-read text strings, so it is easy to develop applications to remote control or read information from the radio. You can ask the radio to send your application an IQ stream or audio from a receiver via the UDP packets.

The next layer is a dynamic link library called FlexLib. It converts the VITA49 UDP data into complex and real streams, which can be used by high-level software.

The top layer is a Windows-based .NET graphical user interface.

SmartSDR offers three levels of API. One for each of the layers described above. This means that if you are writing a program that will talk to a FlexRadio there are three ways you can arrange for your program to interface with the radio. They are called the 'Waveform API' the 'Ethernet,' or 'network' API, and the 'FlexLib API.' The API that you choose to interface with, depends on what you are trying to achieve.

### The FlexLib API

The FlexLib software library provides access to SmartSDR commands, control signals, radio status, panadapter, waterfall, panel meter, and data samples. If you are writing an application that will run on a Windows-based PC and you like using a high-level .NET based programming language like C or C# then you would interface to SmartSDR using FlexLib. It is the best option for people who want to link data from the radio with a logging or digital modes program that they have written for Windows. It is effectively the same level of access as the SmartSDR GUI client has. The FlexLib library provides a way to create much faster and more closely integrated connections between the radio software and digital mode programs than using traditional CAT control. I expect that the developers of software such as Fldigi, MixW, and HRD will be interested in using the FlexLib API to add direct FlexRadio support to their digital mode applications.

The FlexLib .NET library will only work on a Windows platform. You can't use FlexLib to interface a FlexRadio with an application designed to run on a Linux, macOS, iOS, or Android computing device.

If you want to interface a FlexRadio with a program running on a 'non-Windows' device, you have to use the Ethernet API.

### The Ethernet, (network) API

You could use the Ethernet API if you were writing a completely new client program designed to connect with the server software running inside the radio. This level of API concentrates on receiving and sending UDP and TCP-IP data to and from the radio. It can operate without SmartSDR, but you would have to duplicate a lot of the functionality of SmartSDR in your new program. The added complexity makes writing code that uses the Ethernet API level of access to the FlexRadio a much more challenging programming task. The most logical use of the Ethernet API would be to create an SDR client application, which will run on Linux or macOS.

The interface between your software and the radio will be faster since your application is directly connected to the server inside the radio. Commands and data streams sent over the Ethernet do not have to be passed through SmartSDR.

### The Waveform API

The third type of API is different because the code you write runs on the server in the radio instead of the client computer. You download your firmware code and run it on the ARM processor inside the radio. The SmartSDR client can access Waveform API functions. If you add a new operating mode, SmartSDR will automatically create a button on the GUI for it. This is very cool indeed. Your open-source software can add new functions to the FlexRadio server software, and SmartSDR will accommodate the change. Wow!

The Waveform API is primarily intended for new modes. For example, you could add digital voice, FT8, WSPR, or DRM as a standard mode. Or perhaps a new noise filter, or a CW decoder. These new functions could be added by interfacing with the high-level APIs on the PC, but this way, the new function is built into the radio. The advantage becomes apparent when you operate the radio using different computers. You would not need special software patches or additional software. It also means that there is no additional Ethernet bandwidth required between the radio and the client computer, to run the new mode or function.

# THETIS

The Thetis SDR software is compatible with all of the Apache Labs models, the original OpenHPSDR Hermes board, and the Hermes Lite 2. It uses an Ethernet interface to communicate with the radio. It is a full SDR program that includes the DSP, modulation, demodulation, filters, etc., not a client/server application.

*THETIS was a Greek goddess of the sea and the leader of the fifty Nereides. Like many other sea gods, she possessed the gift of prophecy and the power to change her shape at will. [https://www.theoi.com/Pontios/NereisThetis.html]*

I think that the name of this SDR software is well chosen. Being able to change "shape at will," such as the way the radio looks on your computer screen or the functions that the radio can perform by adding new modes or filters, etc. is an important benefit of the radio being 'software defined.'

This section explains some of the innovative features of Thetis and it provides useful information about some of the neat things that people may not be aware of.

## A SHORT HISTORY

Thetis is a direct descendant of PowerSDR mRX PS. I believe that the multitude of additional features and variants known as 'forks' was starting to make PowerSDR cumbersome, so Thetis was designed to make use of new software code and the Protocol 2 data format. Protocol 2 allows multiple panadapters with different spans.

### PowerSDR

PowerSDR was originally called 'SDR Console.' It was first developed in 2002 by Gerald Youngblood, K5SDR, for the FlexRadio Systems FLEX-1000 transceiver. SDR Console and the first version of PowerSDR were written using the Visual BASIC computer language, but this proved to be too limited, and future software development was moved to the C computer language. According to the FlexRadio website, www.flexradio.com/amateur-products/flex-series/powersdr/ the current version for use with FlexRadios is written in a combination of ANSI C and C#. The 2016 OpenHPSDR 'PowerSDR mRX PS' version was written and compiled, using Visual Studio 2010 (Visual C++).

PowerSDR has always been open source software and many people have helped to extend the program. It has been extensively modified and improved over the years. As new technology arrived, PowerSDR was modified to support it. There are variants for all generations of SDR and different interfaces. Some of the many variants or 'forks' include versions for connecting to the I.F. output from a receiver, and there are Genesis SDR and SoftRock versions. There is a version for the FlexRadio Systems FLEX-1500, FLEX-3000, and FLEX-5000 models, and there is the PowerSDR mRX PS version used with the OpenHPSDR boards and the Apache Labs ANAN transceivers.

Most of the development work for the mRX PS version was done by Doug Wigley W5WC and Bill Tracey KD5TFD. The 'PS' stands for Pure Signal and the 'mRX' indicates that the software supports multiple receivers. Innovations included a special long-term recording functionality for radio astronomy and the excellent Pure Signal adaptive transmitter pre-distortion algorithm, developed by Warren Pratt NR0V. Warren also developed new noise reduction and noise blanking filters. The software developers work closely with the OpenHPSDR hardware developers including Phil Harman VK6PH and the team at Apache Labs. My sincere apologies if I left anyone out!

PowerSDR mRX evolved fast! There were 24 software releases between January 2014 and April 2016. Some software upgrades required firmware updates for the FPGA on the transceiver boards and correspondingly some firmware upgrades required modifications to PowerSDR.

Early 'forks' of PowerSDR, including the I.F. version, the Softrock version, and the FlexRadio versions, use the DttSP DSP library written by Francis Brickle AB2KT and Robert Gwier N4HY. This .dll file is called by the main program to perform the modulation, demodulation, and filtering tasks. The main program takes care of the user interface and display functions. Versions of PowerSDR mRX after V3.2.7, use a new DSP library called wdsp.dll, developed by Warren Pratt NR0V. WDSP has different code for FM modulation and demodulation, new noise blankers and noise reduction, new equalizers, FM squelch, AM squelch, and metering. The introduction of the new DSP library resulted in some changes to the DSP options on the setup screens.

### Thetis

Thetis represents a development of PowerSDR, focusing on HPSDR's new 'Protocol 2' radio interface. The development was led by Doug Wigley W5WC. It provides a familiar environment to existing users but includes a new interface and DSP technology to accommodate current and future radio products. Thetis was originally designed for the 'Protocol 2' data format between the radio and the computer, but a later version added 'Protocol 1' for backward compatibility. The software automatically changes to the data protocol being sent from the radio.

## THETIS RELEASES

These programs are open source and new versions are released regularly. There are three different versions available.

1. The last OpenHPSDR version was v2.10.0 (19th June 2023).
   https://github.com/TAPR/OpenHPSDR-Thetis

   This software was designed for the Apache Labs radios. It will work with Hermes Lite 2 but the MI0BOT version takes into account some hardware differences and it has some extra features for HL2.

2. The current stable Ritchie (Ramdor) MW0LGE version for the Apache Labs and Hermes is v2.10.3.5 December 2023. There is a more recent v2.10.3.6 'development' release. https://github.com/ramdor/Thetis/releases

   This software was designed for Apache Labs radios. It will work with Hermes Lite 2 but the MI0BOT version takes into account some hardware differences and it has some extra features for HL2.

3. The current stable Reid Campbell, MI0BOT version for the Hermes Lite 2 is v2.10.3.4 February 2024. This version takes into account some hardware differences between the HL2 and the Apache Labs radios, and it has some extra features for HL2. Unfortunately, it is not a standalone release. You must install the Ramdor version first and then install the MI0BOT version over the top. That sets up the skins and resources files although the database file is separate. I found that running the Ramdor version and changing settings like the skin caused the MI0BOT version to fail. Reinstalling the program didn't help. I had to delete the C:\Users*name*\AppData\Roaming\OpenHPSDR\Thetis-HL2-x64 folder and let the program rebuild it, including re-running the FFT calculations. https://github.com/mi0bot/OpenHPSDR-Thetis/releases

## USING A MOUSE ON THE THETIS PANADAPTER

You can change the panadapter frequencies by clicking on the horizontal display axis and dragging the scale. This requires a click and hold of the left or right mouse button. The 'zoom' slider control adjusts the span of the spectrum and waterfall display. You can also use the mouse to adjust the dBm vertical scale of the panadapter.

Mouse operation of Thetis can be a little confusing. I have assumed that you will be using the 'Panafall' display, which has a spectrum display at the top and a waterfall at the bottom. The Panafall display is the most popular mode because it provides you with frequency, signal level, bandwidth, and a history at the same time.

*TIP: Technically when* **CTUN** *is turned on you are seeing a panadapter display. When* **CTUN** *is turned off it is a bandscope display.*

- If **CTUN** (click tune) is turned off, clicking and holding the left mouse button anywhere on the spectrum display allows you to drag the panadapter centre frequency higher or lower. Move the mouse right or left on the screen. The frequency numbers at the top of the panadapter will change to indicate that you are moving it up or down the band. The VFO A frequency display changes as well because the slice receiver frequency is changing. The receiver slice stays in the same position on the screen as the panadapter is moved underneath it.

If the **CTUN** (click tune) button has been turned on, the receiver slice will jump to the place on the panadapter that you clicked. Click and hold the left mouse button to move the selected receiver slice without changing the panadapter. The panadapter will scroll a little if you drag the receiver to the right or left edge. You can still drag the whole panadapter by clicking and holding the frequency numbers at the top of the panadapter and dragging the mouse left or right. The **CTUN** mode is excellent when the band is busy as it allows you to jump directly to each station.

*Technical TIP: Position the line just to the left side of an upper sideband SSB signal or waterfall trace, to minimize the amount of fine-tuning required. For lower sideband signals click a fraction to the right side. Click directly on the signal for CW, AM, or FM. Fine-tune the frequency with the mouse wheel. I normally have the mouse wheel* **Tune Step** *set to 1 kHz or 100 Hz for SSB mode and 10 Hz for CW mode.*

- If you hover the mouse over the edge of the shaded receiver pass-band furthest from the red centre frequency line, the mouse pointer will change to a horizontal arrow. Click and hold and you can drag the receiver bandwidth wider or narrower. You can also set the receiver bandwidth **width** and **shift** using the slider controls, or select one of the bandwidth buttons on the lower right side of the main window. Two small numeric displays show the high and the low cut frequencies. They become adjustable if you move either the **width** or **shift** slider control.
- Changing the receiver bandwidth, either by dragging the shaded receiver bandwidth or by using the slider controls changes the selected bandwidth button to the **VAR 1** variable filter. You can click the bandwidth buttons to reset the receiver bandwidth to one of the 10 preset values. Right-click any filter button to change any of the filter button settings.
- A right-click on the spectrum display will create a vertical yellow line. Place the line where you want the receiver to be. A left click instantly moves your receiver to the frequency that the line was on.
- When **MultiRX** (sub-receiver) or **split** mode operation is in use, a second right click will create a red vertical line, allowing you to position the second receiver or the split transmitter on the panadapter in the same way as the yellow line on the main receiver. It changes the VFO B frequency. Successive right clicks cycle through, the yellow line, red line, and normal display mode.
- Two vertical yellow lines on the shaded area indicate that the **TX FIL** button is turned on. The lines indicate the transmitter passband. I find them distracting and leave **TX FIL** turned off. The TX passband is normally set with the Transmit Profile dropdown, but you can change it with the **Low** and **High** controls above the **TX FIL** button.

## ADC OVERLOAD

My Hermes Lite 2 indicates an ADC overload due to high out-of-band interference signals. To fix this problem, go to **Setup > General > Options > Options 1 > Hermes Lite Step Attenuator.** Check the **Enable RX1** checkbox and ignore the number. Check the **Auto Att** checkbox and leave the delay set to 100. After a while, you should see a red 'adc linked' indication. The attenuator in the bottom left of the main screen will change to A-ATT to indicate automatic attenuator control.

## THE THETIS PANADAPTER

The horizontal axis of the panadapter display indicates frequency. The span of the display can be zoomed using the **Zoom** slider control below the panadapter. The vertical axis is calibrated in dBm, so it is a logarithmic scale. It is very accurate. The scale can be dragged by holding the left or right mouse button. The left moves the panadapter up or down and the right changes the scale. You can change a setup option if you want to indicate a timescale in seconds on the waterfall display.

There are some useful numbers displayed on the line below the waterfall. Check that there is a 1/2 indication next to the **PEAK** button.

### Largest signal on the panadapter

Three numbers will be displayed on the right side of the bar. They relate to the largest signal currently within the displayed span. The first number indicates the frequency offset above or below the panadapter centre frequency. The second number is the level of the signal in dBm and the third is its actual frequency. Note that the largest signal might not be the one that you are listening to. You can use this feature to find and measure the strongest signal on the band.

### Current cursor position

When the mouse is over the panadapter and crosshairs are visible, three numbers which relate to the current mouse position are displayed on the message bar below the waterfall. The first number indicates the offset between the frequency that the mouse is on and the panadapter centre frequency. The second number is the level of the received signal at that point in dBm and the third is the actual frequency. You can use this feature to measure the frequency and level of any signal on the waterfall by moving the mouse pointer to its peak level. If the mouse is over the waterfall image the dBm figure changes to the time scale of the waterfall.

Click on the message bar to change to the 2/2 display. Five new measurements are displayed.

### RBW resolution bandwidth

The first is the resolution bandwidth (RBW) of the spectrum and waterfall display. This is determined by the sampling rate for the panadapter, **Setup > General > F/W Set > RX1 Sample Rate**, and the FFT size and Windowing function set in **Setup > Display > RX 1 > Fast Fourier Transform**.

The first RBW number is the width of each FFT bin, taking into account the windowing function being used. The number in brackets is the difference in the noise floor between the level in the bin and the level that would be in a bin that was 1 Hz wide, measured in decibels. If you set the 1 Hz/BW checkbox the panadapter will be mathematically offset to show the noise level in a 1 Hz bandwidth.

### PB (Passband)

The PB measurement shows the selected receiver bandwidth. It is usually set with one of the **bandwidth buttons**, but it could be set with the **width** or **shift** slider, or by dragging the passband on the receiver slice. The number in brackets represents the difference in noise level between the receiver slice bandwidth and a 1 Hz bandwidth.

### NPSD (spectrum display noise power)

The noise power on the spectrum display shows what the spectrum display noise floor would be in a 1 Hz bandwidth. If you add the number in brackets on the RBW() measurement to the NPSD, you should be very close to the noise floor showing on the spectrum display when the 1 Hz BW checkbox is not selected.

Spectrum display noise floor in dBm = RBW (dB/Hz) + NPSD

> On my ANAN-100 the noise floor = 9.4 + -134.4 = 125 dBm spectrum noise floor when the antenna is connected.
>
> On my ANAN-100 the noise floor = 9.4 + -155.4 = 146 dBm spectrum noise floor when the antenna is not connected.

### PBNP (passband noise power)

The passband noise power is the noise level you will hear in the receiver slice. It is not shown on the panadapter display, but it affects the signal to noise ratio of a signal from a signal generator. The noise level will be much higher with an antenna connected.

PBNP= NPSD + 10*log(BW in Hz/1Hz). For a 500 Hz CW bandwidth.

> On my ANAN-100 the PBNB = -134.4 + 10*log(500/1) = -134.4 + 27 = -107.4 dBm with the antenna connected.
>
> On my ANAN-100 the PBNB = -155.4 + 10*log(500/1) = -155.4 + 27 = -128.4 dBm when the antenna is not connected.

### PBSNR (passband signal to noise ratio)

The passband signal to noise ratio shows the signal to noise ratio in the receiver passband. A wide passband has more noise so the SNR will be lower. I don't know how this is calculated. When you have blobs turned on, the peaks show the signal level and signal to noise ratio. Note that this is the panadapter SNR, not the signal to noise in the receiver slice.

## SETTING THE AGC GAIN CONTROL

The easiest way to set the AGC gain threshold is to select the automatic control option. But you can adjust the 'G line' maximum AGC threshold manually. Set it just above the noise floor, or set it by ear. It is important to set AGC gain threshold control correctly. It should not be left at maximum.

### Automatic adjustment of the AGC gain threshold

I use the automatic setting for both panadapters. Right click the **AGC Gain** slider control, or **Setup > DSP > AGC/ALC > Auto AGC RX1** and **Auto AGC RX2.** With **shift** set to 20 and **attack** set to 2000.

### Manual adjustment of the AGC gain threshold

If you choose to set the AGC gain manually, you should adjust the **AGC Gain** slider control, or drag the G line on the panadapter every time you change bands, because the noise floor will be different on each band.

The AGC Gain control works like the RF Gain control on a conventional non-SDR receiver. If it is set too low, the receiver will seem to be insensitive. Do not yield to the temptation to leave the slider set to maximum all the time. Using more gain than you need will make the radio sound noisy.

If you have the G line turned on, you can drag the green square icon or adjust the AGC Gain slider so that the green line is a little above the noise floor. If the G line is not turned on, you can adjust the setting by ear. Tune to a frequency that only has band noise. As you adjust the control, listen for the 'knee' where the audio level starts to reduce rapidly around 5 dB above the noise level. The AGC Gain control should be set just before that point. If you are listening to a strong signal, you can reduce the AGC gain further to make the signal sound less noisy. You can turn on the AGC G Line by checking the '**Display RX1 Gain Line**' checkbox on the **Setup > DSP > AGC/ALC** setup tab.

## THETIS AGC MODES

Thetis offers six options for AGC, rather than just the fast, medium, or slow options on most conventional receivers. The AGC in a conventional receiver performs two roles. It limits the audio output on strong signals and it protects the receiver from being overloaded, by decreasing the I.F. stage gain when strong signals are received.

This extends the limited dynamic range of the receiver. The AGC setting in Thetis reduces the audio output on strong signals but there is no need to compensate for a lack of dynamic range. The dynamic range of an SDR receiver is large and the audio output level is directly proportional to the RF signal level. This makes big signals sound too loud, so the AGC limits the audio level when you are listening to a strong signal.

According to the excellent paper written by NR0V (23rd Nov 2012), the AGC in PowerSDR and now Thetis is a single attack, dual decay system. This means that the AGC responds quickly to large receive signals and it has two decay rates. If the input is a short burst of noise, the decay time is made short in order to restore the receiver output to a normal level quickly. If the input has a longer average time constant, for example, normal speech, the decay is longer, and the threshold level is controlled by the AGC gain setting that you have selected on the main SDR console dropdown list.

The **AGC** dropdown box on the console sets the AGC mode.

- Fixed – AGC is off, there is no limitation to the audio output on strong signals.
- Long – AGC attack time is 2ms, the decay time is 2000ms and the hang time is 2000ms.
- Slow - AGC attack time is 2ms, the decay time is 500ms and the hang time is 1000ms.
- Med - AGC attack time is 2ms, the decay time is 250ms and the hang time is not enabled. The H line will not display on the panadapter.
- Fast - AGC attack time is 2ms, the decay time is 50ms and the hang time is not enabled. The H line will not display on the panadapter.
- Custom - AGC times are as per the settings on the **DSP AGC/ALC** setup tab.

Another unique function in Thetis and PowerSDR mRX, is that the gain changes made by the AGC action lead, i.e. occur in advance of, the changes in the RF input signal, so there is never any AGC undershoot or overshoot. This can be done because the AGC is calculated numerically and applied to the digital signal. We tend to forget that all of the digital signal processing is performed by mathematically manipulating the data using computer software.

The AGC changes are not performed instantly because that would cause a choppy distorted audio output from the receiver. They ramp up and down following exponential curves, in much the same way as analogue circuits with capacitors would in a conventional receiver. A 'hang' function is included to delay the onset of AGC action when AGC is set to long or slow modes. This prevents the AGC from ramping up the audio level during pauses in SSB speech or CW words.

Without the hang function, pauses in the received speech could be accompanied by a 'swishing' sound as the noise level rises between sentences. 'Slope' allows strong signals to sound a bit louder than weak signals, without blowing you out of the room.

You can configure Thetis to display the max AGC and the Hang threshold right across the panadapter or just within the light grey receive pass-band. I usually select the right across the panadapter option.

The green G line indicates the maximum AGC gain. Dragging the G Line on the panadapter has the same effect as changing the AGC Gain slider control on the console. Note that dragging the G line or changing the slider turns off the automatic mode. For general listening, you would normally set it so that the line is just above the noise level on the panadapter. If you are listening to a particular station, you can reduce the noise level by placing the AGC Gain until just before the wanted audio signal drops. As you reduce the AGC gain, the background noise level should reduce but the wanted signal should stay much the same. At a point usually about 5 dB above the noise floor there is a noticeable 'knee' and after that, the wanted signal level will begin to reduce at the same rate as the noise. The best AGC Gain setting is just before the point where the wanted audio begins to reduce. The background noise has been reduced, so the signal to noise ratio is at its best. If the signal is strong, you can raise the G line even more and this will reduce both the signal's audio level and the background noise.

The yellow H line is the hang threshold. It is only available if the AGC is set to 'long,' 'slow,' or 'custom.' If the signal is strong enough to cross the hang threshold line, the 'hang mode' will activate. If the hang line is set above the peak signal level on the panadapter the hang mode will not be active.

The AGC is calculated using the AGC settings on the Setup tab.

- Slope dB - allows strong signals to sound a bit louder than weak signals. It is best not to be too aggressive with the settings I use the default slope of 0. You could try anything from 0 to 20 dB, but I would suggest a setting of not more than 10 dB
- Max gain dB – this is the same as the G line on the panadapter (if selected) and the AGC Gain slider control on the console.
- Attack time ms – Is not adjustable, it is fixed at 2 ms.
- Decay time ms – sets the normal (no noise spike) decay. 50 ms is fast 2000 ms is very slow. It can only be adjusted when the 'Custom' AGC setting has been selected.
- Hang time ms – delay before the hang allows the AGC to activate, usually 1 to 2 seconds. It can only be adjusted when the 'Custom' AGC setting has been selected.

- Fixed gain dB – stage gain when AGC is set to off (Fixed). This level is the same as the Fixed Gain slider that appears in place of the AGC Gain control if you select 'Fixed AGC.'
- Hang threshold dB – same as the H line. It sets the threshold that a signal has to exceed before the hang mode activates. The H line is only active if you have selected long, slow, or custom AGC.

*Technical note: the AGC attack time constant is always 2ms. The decay time is a constant as well, but the time is different for each AGC mode. The time constants are used mathematically by the very complex AGC algorithm in the Thetis software. The AGC routine is one of the largest code blocks in the software, mostly because there are so many options available. Hang time is not a 'time constant' calculated from the data rate it is an absolute time interval.*

## NOISE BLANKING AND NOISE REDUCTION

Thetis has very good noise reduction and noise blanking filters. This is one of the big strengths of software defined radio and DSP based systems in general. Each of the noise filters works in a different way. They are designed to eliminate different types of noise. Noise blankers are primarily for impulse 'ignition' noise and lightning crashes. The noise reduction filters are aimed more at wideband atmospheric noise especially on the low bands. To get the best results from the noise reduction systems it is important to have the AGC gain control set correctly (see above).

By introducing a very small delay, the filters are able to look ahead and modify the digital data streams to remove noise and interference before you hear it. This is not easily achievable with analogue circuitry. It would require analogue delay lines. Thetis has two different noise reduction filters. I found the NR2 filter to be amazing when listening to weak signals on 80m. On my radio, I do hear a slight Dalek / Darth Vader distortion to the audio, but it cuts the noise completely. When you turn it off or onto NR1 which is an older design, you realize just how well it is working.

Noise blanking occurs very early in the receiver DSP process. It is performed on the wideband spectrum before any demodulation or other filtering takes place. Noise reduction is performed on the filtered signal i.e. within the receiver passband indicated by the grey shaded area on the panadapter. You can select whether it is applied before or after the AGC section. Thetis also has a spectrum noise blanker (SNB) which performs noise blanking across the whole panadapter bandwidth.

## ADJUSTING THE NB AND NR SETUP CONTROLS

All of the filters are adjustable. My advice is "don't adjust them." Or at least export your database before you make any changes so you can return to the previous settings.

The developers, most recently Warren Pratt, have done a lot of work on these filters and they have decided that the default settings work well. If you have an on-going noise problem that's been driving you nuts, then maybe you can tweak the settings a bit.

It is important not to make the threshold adjustment too low. If you do, the filter will act on the speech and not just the noise and you will make the received audio sound horrible. On the other hand, if you think that the current settings are already distorting the audio, you can raise the threshold a bit and make the filter less aggressive.

Sometimes it is a trade-off. I find that the new NR2 filter does add a slightly electronic Dalek sound to the speech, but it also completely eliminates the background noise. Which do you prefer and which is less tiring to listen to? Some people leave it permanently enabled.

The following is an extract of the setup instructions recommended by the man who wrote the filter algorithms, Warren Pratt NR0V.

*"THIS IS CRITICAL; I'LL SAY IT FIRST: Begin with a very high Threshold value, high enough that the blanker is having NO impact on the noise. Gradually reduce the threshold just until the blanker is catching the impulses. If you do not do this, you are very likely to introduce low-frequency distortion into the signal stream.*

*Recommended adjustment steps:*

- *If using NB2, select (one of the 5) modes of operation.*
- *Set all delay values to their minimum values.*
- *Adjust the threshold per the guidance above.*
- *Try increasing the Lead to determine if any improvement can be obtained.*
- *Try increasing the Lag to determine if any improvement can be obtained.*
- *Try increasing the Slew to determine if any improvement can be obtained."*

## BLOCKING NOISE BLANKER (NB)

Depending on the threshold setting, the blocking noise blanker can completely remove noise spikes, which are bigger than the average signal level, but it has no effect on noise spikes that are below the average signal level. The NB noise blanker does not act as a clipper. When it detects a large noise spike, it mutes the signal completely. Using NB introduces a very small, two sample, delay (latency) to the received signal. When it acts, it blanks the signal to zero for around six data samples. Hopefully, by then, the noise impulse has passed. At higher sample rates, the data stream is faster so the latency and the mute period are shorter. You will not be able to hear the blank space, it is much too short. What you will notice is the elimination of the impulse noise.

*The basic concept of a noise blanker is to replace the signal during the impulses with either zero or some value much more representative of what the signal should be. The noise blanker algorithm depends upon the fact that the impulses are much larger than the signal to be able to detect them. It compares the signal at each sample with the long-term average value across a wide bandwidth. If the samples exceed a "Threshold" referenced to the average, it concludes this is an impulse. (Warren Pratt NR0V)*

- The software looks ahead to see if a noise burst is coming.
- When an impulse a preset amount larger than the average signal is detected, the signal level is decreased over a few data samples to zero. The time from full signal to zero is the 'Slew' rate.
- The signal is held at zero during the pulse and for a 'lag' time after the pulse. After the lag period, the signal ramps back to a normal level over the 'transition' time.
- The slew, lag, and transition times are all adjustable on the **Setup > DSP** > **NB/SNB** setup tab. The default is 0.1 ms for each. You can increase the times if you are experiencing longer noise bursts.
- The threshold setting adjusts the amount above the average signal level that the impulse noise has to be before the filter will operate. Too high and the filter won't work, too low and the blanker will try to work on voice peaks.

## AVERAGING NOISE BLANKER (NB2)

There are dropdown options on the **Setup > DSP > NB/SNB** setup tab, for the NB2 noise blanker. At present NB2 supports five modes of operation. The differences relate to the way that the software manages an impulse when it is detected. Rather than muting the audio to zero amplitude when there is a noise pulse, the different routines attempt to mimic what the signal might have been without the noise.

The different settings fill the gap where the noise impulse was removed with an average of the incoming signal, the level just before the impulse was detected, the level just after the impulse was detected, or an average of what the level was before and after the impulse.

You can try the methods and decide for yourself what works the best for the particular noise that you are experiencing. The five methods yield very similar results. The noise impulses are very short and you might not be able to tell if the gap is filled with an average level or left at zero.

- Zero: when an impulse is detected, the output signal is set to zero amplitude for the duration of the pulse. This is the method used in the blocking noise blanker.
- Sample-Hold: the output is set to an average of the incoming sample for the duration of the impulse.

- Hold-Sample: for the duration of the impulse, the output signal is held at the value occurring just after the impulse.
- Mean-Hold: for the duration of the impulse, the output signal is held at the average of the values occurring just before and just after the impulse.
- Linear: the output signal is a linear interpolation of the signal value occurring just before the impulse and the signal value just after the impulse.

## SPECTRUM NOISE BLANKER (SNB)

The spectrum noise blanker is aimed at wideband interference, often right across the band you are listening to. The type of noise that is often caused by switch mode power supplies, TVs, or LED lighting systems. There are two threshold controls on the **Setup > DSP > NB/SNB** tab. The Thetis manual does not describe what they do and states that it is best to leave them at the default settings. It then goes on to say that if you can achieve better noise blanking by adjusting them, then you can. I have found the SNB excellent using the default settings. It eliminates a lot of low frequency rumble. Note that although the SNB removes the effects of wideband noise from the signal that you are hearing, it does not remove the noise spikes from the panadapter display.

## NOISE REDUCTION (NR)

The noise reduction filters work to reduce annoying background noise. If you are experiencing noise impulses or bursts it is probably better to use a noise blanker. The noise reduction filters work best when the received signals have a good signal to noise ratio. Using the NR filter will probably not help you to pull an otherwise unreadable signal out of the noise. However, the NR2 filter, which uses a different type of noise reduction calculation, is very effective with weak signals.

Noise reduction is performed on the narrowband signal after the receive filter, rather than on the full panadapter. In other words, the NR filter only affects the approximately 3 kHz bandwidth you are listening to. You have a choice of applying the noise reduction before or after the AGC. I find that the NR2 filter works better if applied before the AGC, but you may prefer it the other way around. Post AGC noise reduction might be more effective in some situations such as when there is fast fading. It is best to set the AGC Gain line about 10 dB above the noise floor so that the AGC action doesn't undo what the noise reduction is trying to achieve. Or use the auto gain feature.

If you have persistent noise problems or just like the way it sounds, you can adjust the filter settings in the **Setup > DSP > NR/ANF** tab. I suggest that you don't make large changes. It is better to make small changes and evaluate the effects that you are hearing before proceeding further. This is because the NR filters are adaptive filters. They take some time to settle. The noise level gradually reduces as the filters average the noise levels over time. I leave everything at the default settings.

The recommended default settings for a 192 kHz sample rate are:

- Taps = 64. 'Taps' sets the number of LMS (least mean square) stages used in the calculation. You can increase the number of taps to expand the frequency range and reduce the noise. A higher number of taps increases the CPU load on the host computer while the NR filter is in use because more computations are being performed.
- Delay = 16. Delay changes how many samples backwards in the data stream the system works. Increasing the delay makes the filter act earlier, which should reduce the noise.
- Gain = 100. Decrease the gain to reduce noise the noise further. Lower values may work better on CW signals.
- Leak = 100. The leak is the amount of signal that is applied to the next iteration of the software, i.e. how much the past noise level affects the current calculation. More 'leak' should reduce noise. But it also means the filter takes longer to respond to changes in the received noise level.

### Noise reduction systems - technical description

The noise filter is an adaptive interference canceller. It finds a periodic (repeating) signal like speech and cancels out incoherent signals such as noise.

The data stream is passed through a delay line and an adjustable FIR (finite impulse response) filter, which is a bandpass filter that automatically moves to the signal we want to keep, (or eliminate in the case of a notch filter). The delayed signal becomes a kind of reference signal at the periodic frequency.

It is subtracted from the current samples creating an 'error' level. After some time the filter coefficient automatically adjusts in order to progressively, reduce the 'error' level, meaning that the delayed signal is as close as possible, in phase with the current samples. When the 'error' signal is at its smallest, the wanted periodic speech is enhanced and the unwanted non-periodic noise is suppressed. The technique uses a least mean square algorithm developed by Widrow and Hoff in the late 1950s. These adaptive filters work in the time domain, no FFT is required.

## ADVANCED NOISE REDUCTION (NR2)

The NR2 noise reduction uses a 'noise predictive equalizer' (NPE) algorithm. The filtered audio is reported to be less tiring to listen to than the typical LMS noise reduction used in the NR filter. It is also much better at bringing a weak signal out of the noise.

It is important to adjust the AGC gain correctly before turning on the NR or NR2 filters. AGC is the enemy of noise reduction algorithms because it tries to increase the gain, thus increasing the noise level while the noise reduction is trying to reduce the noise level. I have found it better to use NR2 in the Pre-AGC mode so that it is not competing with the AGC.

I recommend using the default 'Gamma' gain mode, either OSMS or MMSE as the NPE method and leaving the AE-Filter checkbox checked. I don't know what it does, but the audio sounds weird if the AE filter is not selected.

I found that the NR2 filter is excellent on 80m. It cuts the background noise completely, but it does introduce a strange electronic sound to the speech. The NR2 noise filter is enabled by pressing the NR button, then pressing it again for NR2. It is certainly worthwhile experimenting with the filter settings. It's always a good idea to write down the default settings before making adjustments, so you can remember what they were.

*The NPE adapts to achieve a minimum 'error signal' calculated using either the MMSE or OSMS calculation methods.*

*NPE = Noise Power Estimation*
*OSMS = Optimal Smoothing and Minimum Statistics*
*MMSE = Minimum Mean Square Error*
*AE = Artifact Elimination*

## AUTO NOTCH FILTERS (ANF)

Thetis includes an automatic notch filter (ANF) and multiple tracking notch filters (MNF). The ANF is extremely effective at removing constant interference such as a carrier signal within the receiver's passband. This is a very common occurrence on the HF bands. ANF is not effective when you are listening to digital signals like RTTY or PSK because it will attempt to notch out the fairly constant wanted signal. It is not useful when receiving CW for the same reason. In both of those cases, you can usually use a narrow receiver filter and avoid the interfering signal that way.

The LMS algorithm used for ANF is very similar to the code used for the NR noise reduction although the aims are different, and different time constants are used. In the noise reduction filter, the aim is to keep the steady speech signal and eliminate relatively random noise signal. In the ANF filter, the aim is to eliminate the steady interfering carrier and keep the relatively random speech signal. The same type of filter is used, but the output of the filter is taken from a different part of the code. Using the automatic notch filter is CPU intensive. The CPU load increases with the number of taps and with higher sample rates.

Increasing the gain allows the ANF filter to lock onto offending carriers more effectively. But if you push the gain too high, you will begin to distort the wanted speech. I leave the ANF adjustments at the default settings.

For a 192 kHz sample rate the defaults are:

- Taps = 64, increase taps to expand the effective frequency range
- Delay = 16, lower to improve the maximum effective frequency
- Gain = 100, increase to more effectively lock onto offending signals
- Leak = 100: there is probably no point changing this. Less leak means it looks at less of the audio when working out what is audio and what is an offending carrier, so more leak should be better.
- Pre-AGC checked. Post-AGC may be better for fast fading or on AM when you are using the fade leveller.

## MULTI NOTCH FILTER (MNF)

Version v3.3.6 of PowerSDR mRX introduced a new Multi Notch Filter (MNF). It replaced the FlexRadio 'Tracking Notch Filter,' which was included in FlexRadio versions of PowerSDR but not the versions intended for other radios, due to licensing issues. The PowerSDR software was open source, but the tracking notch filter code was owned by FlexRadio.

*"The, Multi Notch Filter (MNF), allows you to specify up to 1024 notches. The notches are specified by RF frequency and width and will be invoked, as needed when they overlap the passband. This feature is useful for those who have interference that consistently appears on specific frequencies. To avoid phase distortion, the notches are implemented with linear phase. Also, they introduce no additional processing delay nor do they consume any additional CPU cycles once you're on frequency, and the notches are set up. This is all accomplished by simply "cutting" the notches into the existing bandpass filters rather than adding additional filters."* [Doug Wigley, W5WC and Warren Pratt, NR0V].

Turn the feature on with the MNF button and set each notch with the +MNF button. A vertical yellow line will appear on the panadapter. You can drag it over the offending signal. While MNF is turned on, the notch will stay on the offending signal even if you move the receiver slice or the panadapter frequency.

Right clicking the filter line opens a popup window where you can change the width of the filter. There are four preset widths and a slider control. The maximum width is 1000 Hz. The filter might not achieve the full attenuation of >100 dB if the notch width is too narrow. Adjust the slider or use the buttons to eliminate the interference. As with all filters, the interference only disappears from the signal you are hearing. It remains visible of the spectrum and waterfall display.

Leaving the 'auto increase width' checkbox checked on the **Setup > DSP > NR/ANF** tab, will automatically increase the notch width to achieve 100 dB of attenuation when the notch is cut in.

## DOLLY FILTERS

'Hello, Dolly!' The Dolly Filters are 'double lobe' twin peak filters designed specifically for receiving RTTY. RTTY transmissions send a signal that alternates between two 'tone' frequencies that are usually 170 Hz apart.

These filters are very narrow and it is important that they are set to cover the same audio offset frequencies as the DIGU transmit profile that you use for RTTY. I have found them to be extremely effective. They allow me to decode RTTY signals that are completely illegible without the Dolly Filter. The **Setup > DSP > Audio** tab includes three groups of Dolly filter settings, one for each receiver.

One annoying downside is that there is no front panel control to turn the Dolly filters on and off. Because they are very narrow, I find it nicer to turn them off while tuning to another signal. It would be great if the software developers could allocate a button on the GUI for the Dolly filters. It would also be handy if these filters were only enabled for the DIGU and DIGL modes. At the moment, they work on any mode setting. If you are using CW mode and you select both the CW APF and the Dolly filters at the same time, both filters will be in the circuit and you will hear nothing.

## CW AUDIO PEAKING FILTER (APF)

The CW audio peaking filter is a very sharp adjustable filter great for pulling weak Morse code out of the noise. There is a handy 'RX1' button to turn it off when you are tuning around the band. You can change this to the sub receiver or the receiver on the RX2 panadapter if you need to. The CW APF bandwidth is very narrow. It is adjustable from 10 Hz to 150 Hz. The frequency is centred on the standard CW tone offset and is adjustable plus or minus 250 Hz.

Right click the RX button to access the setup controls for this feature. They are located on the **Setup > DSP > Audio** setup tab. There are three groups of APF settings, one for each receiver.

Selecting the CW APF filter causes a duplicate set of controls to appear in the 'transmit audio settings' block of the main screen. You can control the filter gain, filter bandwidth, and fine-tuning of the frequency relative to the standard CW pitch.

## ZERO BEAT (0 BEAT)

The '0 Beat' button automatically pulls the frequency of the radio so that you are receiving CW signals with exactly the same tone offset as the transmitter will use when you are sending CW. This feature makes netting the received CW signal with the transmit frequency much easier.

I like to receive CW at a pitch of 750 Hz. If you operate CW regularly, I am sure you will have a favourite CW tone. The CW pitch offset is set on the **Setup > DSP > CW** tab, or on the front panel in the transmitter section.

## THE DOWNWARD EXPANDER (DEXP)

The downward expander makes your transmitted signal sound better and it is something that most radios don't have. It has been a feature of PowerSDR and Thetis since the early days. However, many users don't know what it does or how it works.

The function reduces the amount of noise that you transmit between words and syllables. When you key the microphone in SSB mode without talking, you will notice some RF signal on the spectrum display. This is background noise from things like; fans, thermal noise, and other noise in your shack modulating your transmitter. The downward expander works like a noise gate, but it is much smoother and less aggressive. It effectively increases the dynamic range of the microphone by decreasing the noise level when you are not talking.

Traditional noise gates do not decrease the gain. Below the threshold, they just switch off the audio and they tend to 'chatter' on and off which causes a choppy sound on your transmitted signal.

The downward expander has adjustable controls on the **Setup > DSP > VOX/DE** setup tab. **Threshold** sets the level below which the microphone audio will be attenuated and **Exp ratio** sets how aggressively the signal will be attenuated. The threshold control is the same as the DEXP slider control on the front panel.

*Technical TIP: The threshold control sets where the downward expander stops operating. When the microphone audio level is below the threshold, the algorithm will reduce the microphone gain. When it is above the threshold, the downward expander has no effect.*

## THE COMPRESSOR (COMP)

If you plan to work DX, it is worthwhile using some compression on your transmit audio. The compressor increases the quiet levels in your speech to increase the average power of your signal. This reduces the dynamic range of your audio, but it makes your signal easier to copy. It stops the quiet parts of what you are saying from disappearing into the noise, especially when the received signal strength is weak. Your signal will be more 'punchy' and more effective at cutting through the noise and QRM.

You don't want your signal to sound overly compressed and unnatural. I usually operate with 6 dB of compression. Forum reports say that the signal will start to sound odd and unnatural at about 10 dB. I don't recommend using more than 10 dB of compression.

## CONTROLLED ENVELOPE SINGLE SIDEBAND (CESSB)

Controlled envelope SSB is a relatively new method of getting a higher average transmitter power from your SDR transceiver. It is another excellent example of innovative software development making your SDR transceiver even more effective.

Dave Hershberger, W9GR, published interesting articles on CESSB in the November/December 2014 issue of QEX and in the February 2016 edition of QST. The intent of his algorithm is to increase "talk-power" on SSB by bringing the average power/speech level closer to the peak level. Part of this method is used in the compression (COMP) routine in Thetis and the other part; **CESSB overshoot control** is selectable on the **Setup > Transmit** tab. Apparently, the same algorithm is used in the FlexRadio SmartSDR software. FlexRadio has reported an average increase in talk power of 2.56 dB (x1.8). This increase is equivalent to stacking another identical antenna top of your main antenna!

David Hershberger, states that overshoot is inevitable in SSB modulators. You have to either reduce the audio modulation to a level where the audio peaks do not drive the RF power amplifier to a level where clipping distortion occurs, or use ALC (automatic level control) to dynamically reduce the level of audio peaks. Both of these methods reduce the average talk power of your signal and ALC is only partly effective in reducing the effects of overshoot in the modulator. Traditional ALC systems reduce the amplitude of an SSB signal in response to overshooting envelope peaks. Slow ALC may result in clipping and splatter because it has failed to act fast enough. Fast ALC will significantly reduce the average transmitted power. The advantage of DSP systems is that the peaks can be measured and reduced before they get to the modulator. David's CESSB algorithm increases the average transmitted power without letting the signal go into distortion.

*"Controlled Envelope SSB envelope control is accomplished by 1) Pre-filtering and peak limiting the audio input signal. 2) Baseband "RF clipping" of the SSB signal to reduce Hilbert transform overshoots. 3) Overshoot compensating the remaining envelope peaks resulting from baseband "RF clipping." By accurately controlling SSB envelope peaks at the point where the SSB is generated, ALC is unnecessary. ALC, even with look-ahead, reduces transmitted power when it does not have to — before and after an envelope peak. Average transmitted power can be significantly higher, without introducing "speech processor" type artifacts. An SSB signal with well-controlled envelope peaks makes more efficient use of the RF power amplifier and produces higher average power for a given peak envelope power."* [Dave Hershberger W9GR]

To start using CESSB, enable the **CESSB overshoot control** checkbox on the **Setup > Transmit** tab. CESSB will not operate unless the compressor **COMP** is also enabled. However, the compressor will still work, without CESSB control, if the CESSB offshoot control has been not selected.

Note that, *"CESSB is intended for use with speech signals. Although it is a nonlinear process, the nonlinearity has a negligible effect upon speech. For non-speech SSB applications, however, such as digital modes (PSK, JT65, SSTV, and others), CESSB should be tested to be sure that any average power increase is not offset by nonlinear distortion."* [Dave Hershberger W9GR]

## DIVERSITY MODE

Diversity reception is very useful if you operate on bands where wideband noise is a problem. Thetis includes a diversity mode, which will work with radios that have two synchronized ADCs and receivers. As far as I am aware the only dual ADC radios compatible with Thetis, are the Apache Labs ANAN transceivers, and possibly the Afedri AFE822x SDR-Net (Dual Channel) receiver. The Hermes Lite 2 only has a single ADC. The SDRplay RSPduo can do diversity reception with its SDRconnect or SDRuno software.

The diversity mode allows you to combine signals received from your main antenna with signals from a reference or 'noise' antenna in such a way that much of the wideband noise is cancelled while the wanted signal remains almost unaffected. You can use it to emphasize a wanted signal or reduce noise and other interference. A 'radar' control lets you adjust the phase and amplitude of one receiver in relation to the other receiver. It is very quick and easy to adjust while watching the noise floor decrease on the panadapter display, or listening for a signal improvement. The system works best if the noise being received at the same or greater level as the levels received on the main antenna. If you don't have enough noise being received off the reference antenna, the subtraction effect will be diminished and so the noise cancellation will be ineffective. There is no point in putting the reference antenna near power lines or machinery in order to, deliberately introduce noise that is not affecting your main antenna.

You will need an outside reference antenna. It can be close to your main antenna, but not so close that a lot of power from your transmitter will be coupled into it. The 'noise' or reference antenna does not need to be directional. It can be a long wire, dipole, or a multiband vertical, used in conjunction with your directional antenna.

If the noise problem is within your house such as your own plasma TV, low voltage lighting, ADSL modem, PC switch mode power supplies, etc. Then the diversity system may work better with a small antenna inside the shack or up in the ceiling space. You may have to experiment with different options.

The diversity mode works best in situations where the interference signal is fairly constant, but it doesn't necessarily have to be wideband. It will not be effective against lightning crashes, static, or vehicle ignition interference. Setting the diversity level and phase angle is a manual adjustment, so it is less effective against changing or non-repetitive noise sources.

There is an excellent online video by Nige Coleman G7CNF illustrating how effective it is, https://www.youtube.com/watch?v=qdQm17udCSo. He demonstrates up to 20 dB of reduction in the noise floor. There is another great video at https://www.youtube.com/watch?v=ZTtTa6EQxgg by "The AM Window" about diversity reception using Thetis.

## PURE SIGNAL – TRANSMITTER PRE-DISTORTION

Amplifier nonlinearity results in poor IMD (intermodulation distortion) performance, which is bad. Adaptive transmitter pre-distortion modifies the audio modulation signal to correct for the nonlinearity of the amplifier so that the net result is a flat linearity response. This improves the transmit IMD performance a lot. Maybe it sounds easy, but it is not! Both phase and amplitude linearity errors need to be corrected and the nonlinearity characteristics change dynamically with changes in frequency, antenna load, temperature, power supply voltage and all sorts of other factors.

Pure Signal is an innovation for PowerSDR mRX PS (now Thetis) that makes use of the full duplex capability of the software defined radios. It also requires the wide bandwidth capability of the transmitter and the calculating power of the PC application.

*Technical TIP: Full duplex is the ability to receive and transmit simultaneously. Most conventional architecture HF radios are not able to operate in a duplex mode. They tend to use the same local oscillator and I.F. amplifiers for transmitting and receive and their receivers may not be able to cope with the high RF levels radiating from the transmitter.*

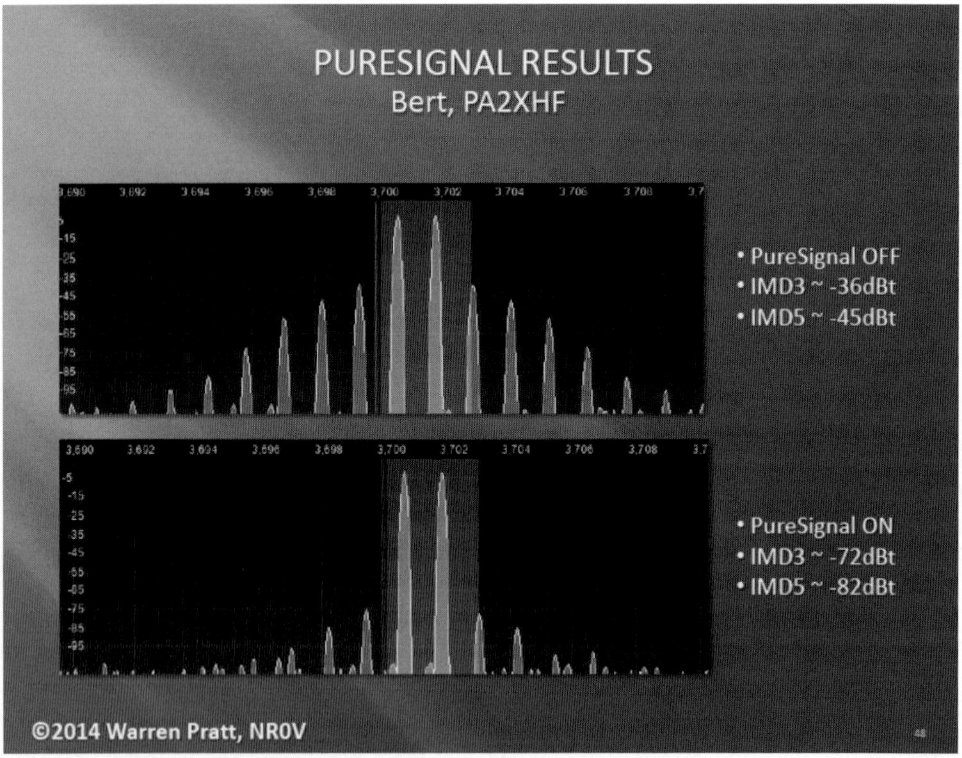


Figure 29: PureSignal IMD improvement

The image above is the result of a two tone transmitter IMD performance test performed by Bert Meijer, now PA0XHF. The two tone IMD test is the standard measurement of intermodulation distortion caused by the nonlinearity of the RF power amplifier in a transmitter. It uses two test tones to modulate the SSB transmitter, but speech causes a similar spreading of the transmitted signal. If the intermodulation products are large, they can be heard as noise or distorted speech "splatter" when you are listening to a station working close to the same frequency.

*TIP: So far, there are very few ham radio transceivers with the ability to use adaptive transmitter pre-distortion. The Apache Labe ANAN transceivers and Hermes Lite 2 transceivers use Pure Signal with Thetis. FlexRadio is planning to introduce transmitter pre-distortion in SmartSDR for the new FLEX-8000 radios.*

Pure Signal was developed by Warren Pratt NR0V. It dramatically improves the transmitter's intermodulation performance by adjusting the modulated RF signal going into the RF power amplifier so that the signal at the output of the amplifier is a more accurate but amplified representation of the input signal. It is so effective that when Pure Signal is employed, it currently gives the ANAN radios better transmit intermodulation performance figures than any other HF ham radio transceiver tested by ARRL labs, (based on published QST data). The ANAN-8000DLE claims a 'third order IMD' typically -72dB below PEP @ 200W output on the 20m band.

*Technical TIP: Warren's "adaptive pre-distortion Pure Signal" algorithms operate in real time rather than using the LUT (lookup table) approach. Illustrations in the introduction to Pure Signal document available from the Apache Labs website indicate a typical improvement in IMD performance of 20 dBc, which is very impressive indeed.*

**How does it work?** A low-level sample of the transmitter's output power is taken, usually from a coupler at the antenna port. It is received and demodulated by the receiver, which in an ANAN or Hermes radio is always running because the radio is full duplex. It is extremely important that this feedback signal is attenuated to between 0 dBm and +10 dBm, to avoid damage to the sensitive receiver front end.

The demodulated sample signal is compared with the audio input signal being sent to the modulator. Phase and amplitude changes are applied to the modulated signal in order to dynamically reduce the IMD products. This is not straightforward because the amount of adjustment required changes continuously with the changing input signal. The level and phase adjustment need to adapt. Also, the amount of nonlinearity caused by the amplifier will be different at different operating frequencies and when the transceiver is connected to different antenna impedances.

If the radio was just sending an un-modulated carrier signal the phase and level adjustment would be easy. But then life would be boring because you can't convey any information with an un-modulated carrier signal. No, not even CW, which is an amplitude modulated digital mode, "carrier on, carrier off".

Inside the SDR software both the audio signal being used to modulate the transmitter and the audio from the received feedback signal are digital signals. The Pure Signal adjustments can be calculated mathematically and applied by changing bits in the 'transmit' audio data stream. Over a short period, the software adjusts the modulation signal to compensate for the IMD created by the amplifier. It continuously compares the transmitted signal with the modulation and adapts to changing conditions.

The idea of using pre-distortion to improve the transmitter's linearity has been around for quite a while, but it is tricky to implement with conventional electronics. In a software defined radio, the problem can be resolved using a mathematical algorithm and that allows fast, dynamic, adjustment of the modulation signal. The receiver needs to be working at the same time as the transmitter because you need to continuously receive and compare samples of the transmitted signal. The process needs to be able to apply adjustments over a wider frequency range than the normal SSB transmission bandwidth so that harmonics of the modulation frequencies are included. That is another problem if you are trying to implement adaptive transmitter pre-distortion on conventional architecture radios.

Adaptive transmitter pre-distortion could be implemented in the DSP chips of conventional transceivers. But, they do not usually have enough transmitter bandwidth and most of them are not able to transmit and receive on the same band at the same time. They reuse circuit blocks such as the local oscillator which prevents operation in a duplex mode. Future designs may use an embedded SDR receiver or a sub receiver to acquire the required feedback signal.

**What is transmitter nonlinearity?** All RF power amplifiers introduce some nonlinearity into the transmitted signal. 'Nonlinearity' just means that the relationship between the input power and the output power is not entirely flat. In other words, the output signal is never a perfect but louder copy of the input signal. The power amplifier in your transceiver may have a little less gain when the input signal is small and it might not put out as much power as it should when the input signal is large. This is usually due to the power supply voltage drooping under a high current load. Components heat up under high current conditions. This can cause them to have more resistance or to change their capacitance or inductance, which might de-tune the transmitter slightly. All of these factors combine to introduce two kinds of nonlinearity. Amplitude nonlinearity where the gain of the amplifier does not remain the same for all input levels and phase nonlinearity when the phase of the output signal is being affected by the input level or frequency.

Using a 'linear' amplifier after your transceiver adds an additional source of nonlinearity. If you place the feedback coupler at the output of the linear amplifier instead of the output of the transceiver, Pure Signal can compensate for the overall nonlinearity of the transceiver and linear amplifier combination.

**Why do we want good transmitter linearity?** When we transmit a digital mode or SSB HF transmission, the last thing we want is complaints that our transmission sounds bad, or that it is causing splatter interference across the band. We are legally required to ensure that we do not cause interference to other band users and this includes making sure that our signal is mostly contained within the wanted bandwidth, which is usually less than 3 kHz for an SSB signal.

SDR transceivers are in general no better and no worse than any other kind of ham radio transceiver. Hopefully the 3rd order IMD products, measured using the standard 'two tone' test, are at least 30 dB below the level of the two test tones. Even if this requirement is met, this level of interference might be heard at the receiver's location. For example, if you were transmitting USB on 14.200MHz and a station was receiving your signal at 30 dB over 9, then a station in the same locality listening on 14.203 MHz could hear an interference signal at around S9. Improving the transmitter IMD performance of your transmitter is a way of being "nice to the neighbors."

Transceivers with higher voltage 50V output stages tend to have slightly better 3rd order transmit IMD performance than 12V transceivers. The improvement is typically 6 dB to 8 dB.

Transmitter pre-distortion can improve the 3rd order transmit IMD performance by a massive 20 dB.

*TIP: The ITU says that the 3rd order product should be at least -35 dBc for an SSB transmitter, (-41 dB PEP for the ARRL method). Many transceivers don't meet this benchmark.*

*The ARRL 'Clean signal initiative – published in May 2024' recommends that transmitter manufacturers should ensure that the 3rd order IMD product in a two tone IMD test should be better than -42 dB PEP (-36 dBc). At present, vary few amateur radios meet that specification. The ones that I know of are ANAN radios running Pure Signal, the FLEX-6000 series, the IC-7300, TS-890S, TS-590SG, and the FTdx101D.*

**The technical background:** Amateur radio transceivers use linear power amplifiers because the voice signals that we want to transmit vary in amplitude, causing the transmitted RF signal to range from zero to the full output power of the radio. You see this as the peaks on your transmit power meter during SSB operation. It is important that the output power of the amplifier accurately follows the changing level of the input signal. This is called the amplifier's amplitude linearity.

During AM or SSB transmission, the phase of the output signal should not be affected by changes in the level or frequency of the audio input signal, this is called the amplifier's phase linearity. Phase nonlinearity can be caused by components changing their inductance or capacitance when they are stressed due to high voltage or current causing heating.

The changed value of capacitance or inductance causes de-tuning of the RF power amplifier resulting in phase shift. Phase nonlinearity effectively causes a small amount of phase modulation when we only want amplitude modulation.

Severe nonlinearity causes the different frequency components of the RF signal to mix together creating unwanted sum and difference frequencies. The result is known as 'intermodulation distortion.' The amount of degradation is dependent on the design and tuning of the output stage of the amplifier, the output filters, load impedance, drive level, the frequency of operation, modulation signal, the stiffness of the power supply, components like toroids, and the output transistors.

'Linear' amplifiers tend to be less linear at low input levels and the output power tapers off near full drive level. If you plot the output power vs input power on a chart, you would see a slight S shape rather than the desired straight line.

Intermodulation products mixed into the wanted transmitter signal sound bad because they interfere with the wanted speech. Intermodulation products that are created outside of the wanted transmit bandwidth are heard as 'splatter' and make the transmitted signal wider than intended. If your transmitter is only used for CW and digital modes like PSK and RTTY the audio modulation level does not vary. The modulation is either on, or off. With these modes, the RF amplifier does not have to be linear. They generate much less IMD, but Pure Signal can still be used.

**How well does it work?** The normal way to measure a transmitter's IMD performance is to apply two audio test tones to the SSB modulator at a level that causes full transmit power. The transmit signal is measured using a spectrum analyser and the level of the 3$^{rd}$, 5$^{th}$, 7$^{th,}$ and 9$^{th}$ order intermodulation products compared with the two tone level is recorded. You might have seen the IMD products on pictures in transceiver reviews. A 3$^{rd}$ order Transmit IMD of -30 dBc compared to the two tone level is good. A 3$^{rd}$ order Transmit IMD of -36 dBc compared to the two tone level meets the new ARRL standard. Pure Signal can produce a 3$^{rd}$ order Transmit IMD of better than -50 dBc, (-56 dB PEP ARRL method).

**What it can't do:** Pure Signal can improve amplitude and phase linearity problems but is less successful with 'memory effects' because they are related to past events, not the input signal that is currently being compared. 'Memory effects' are phase or amplitude distortions caused by earlier signal events affecting components in the radio. For example, a large input signal could cause the RF transistors and heatsink to get hot resulting in an amplitude and/or phase change in the amplifier output signal. When the input signal level drops, the devices take some time to cool so the distortion persists even though the input signal has reduced. Another example is a high input level causing the amplifier to run at near full power, drawing maximum DC current. This may cause the power supply voltage to droop and the lower supply voltage will affect the amplifier linearity. When the input signal level reduces, the DC supply voltage takes some time to return to full voltage, so again the distortion persists even though the input signal has reduced.

Adaptive pre-distortion can only improve IMD caused by nonlinearity that is attributable to the input modulation signal. It may not be able to adjust for fixed sources of IMD such as saturated Toroids in the output filter, badly tuned linear amp stages, or IMD occurring in the antenna system after the feedback coupler.

**The feedback coupler:** The feedback level from the transmitted signal is quite critical. It has to be quite large but not so big that it causes the receiver ADC to overload. If you place the coupler on the output of your linear amplifier, you will need additional attenuation to reduce the feedback to a safe level.

## SHED YOUR SKIN

One of the very cool things about Thetis is the ability to change what the operating interface 'GUI' looks like. This is called changing the skin. The buttons and controls all remain in the same place and have the same functions, but you can change what they look like. You can also change the background image of the radio and the panadapter. There are several skins bundled with PowerSDR. They can be selected using the **Skins** drop-down box on the **Setup > Appearance > General** tab. You can download other skins using the controls on the **Setup > Appearance > Skin Servers** tab. You can replace the image files with other pictures of a similar size. The skins are a collection of small graphics files. There is a separate graphics file for every button and control on the console and larger image files for the background pictures.

Figure 30: Change the skin to customise the look of Thetis

# OTHER POPULAR SDR SOFTWARE

## HDSDR

HDSDR has become popular because it can be used with a wide variety of SDRs, using an extio.dll file to convert between the radio's data format and the format expected by HDSDR. You can also use the program with IQ audio data from a soundcard type SDR. The Icom IC-R8600 and the Malahit DSP2 receivers are both self-contained radios which can interface with HDSDR.

I have extio.dll files for the RTL dongles, IC-7610, IC-R8600, OSA spectrum analyser, and the SDRplay RSP2 receiver. I didn't have to use an extio.dll file for the Malahit DSP2 because it was sending audio data from the radio's built in soundcard.

*TIP: The USB driver for the IC-R8600 receiver includes the extio.dll file for HDSDR. You will have to find it on your computer and move it into the HDSDR directory.*

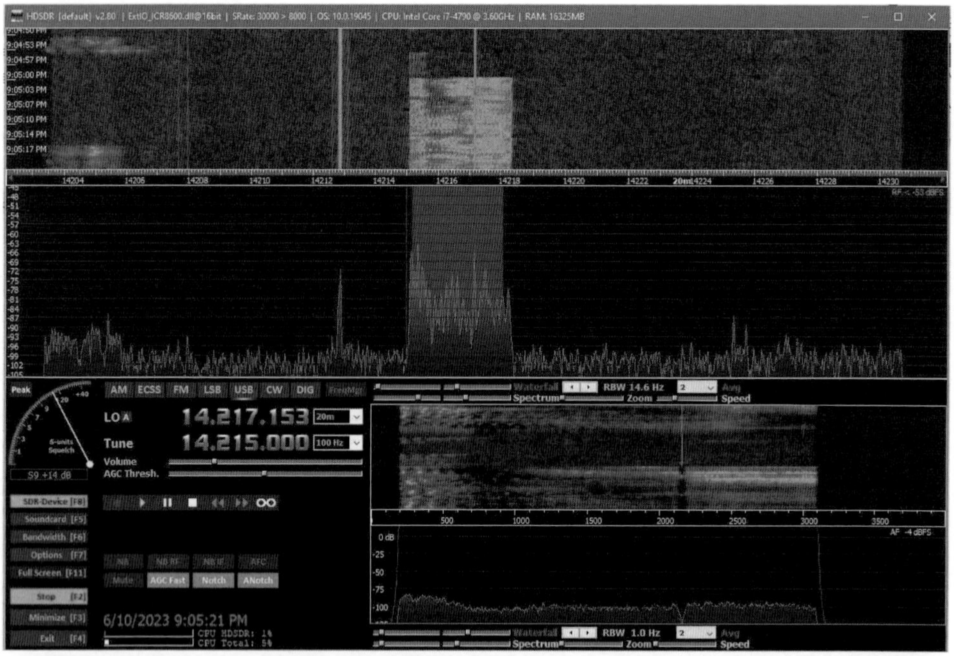


Figure 31: HDSDR. An SSB signal on a narrow panadapter

HDSDR is for receivers only. It works very well. One very useful feature is the lower spectrum and waterfall display which shows the signals within the receiver passband. You can see the effect of noise and notch filters and you can change the audio passband width very easily. With soundcard radios, you will see a noise spike in the centre of the panadapter. There is a filter which helps. The noise spike can be avoided by listening to a 'Tune' frequency away from the L.O. panadapter centre frequency. Some extio.dll files add extra control features.

## KISS KONSOLE

Note that Kiss Konsole only supports OpenHPSDR type radios using the Protocol 1 data format. It will work with Hermes Lite 2 and ANAN radios that have Protocol 1 firmware ("not many, if any").

Kiss Konsole is one of the few SDR software programs for the OpenHPSDR and Apache Labs ANAN radios that support the ability to transmit. It was never designed to be a full-featured SDR application. The intention was for it to act as a learning tool for budding software developers. It is an open source project and the software code is extensively commented so that it is easy to understand. Kiss Konsole was written using the C# computer language. All of the code is available from the OpenHPSDR downloads site. When it is running, the program looks and works like a very simple version of PowerSDR. It only supports a single panadapter and waterfall, but it works well. I even had a go at modifying the code by making a small change to slow down the display of the received signal strength meter.

After reading a lot of information from many websites, magazine articles, and papers I gained a fairly good understanding of how the hardware side of software defined radios work. But the hardware is only half the story. I also wanted to understand the SDR software. Strangely, my online research hit a wall when I tried to find out how the PC software works. I thought that this was rather odd because there are many SDR software applications available and most of them are open source. I expected there to be a lot of reference information, but explanations even at a block diagram level are pretty rare. I seemed to fall into a small group (of one?) who wanted to know how it works, but who did not want to create my own software. I spent a lot of time studying the C# computer code that makes up Kiss Konsole.

*Kiss Konsole uses the SharpDSP .dll written by Phil Covington. The rest of the Kiss Konsole code was developed over several years by a team of developers, notably Phil Harman VK6PH and Warren Pratt NR0V who are still actively developing Thetis.*

Kiss Konsole is an excellent learning resource if you want to see how filters, modulators, demodulators, and other SDR features like the panadapter and waterfall are coded. However, many SDR software developers now use 'GNU Radio' rather than using open source .dll files and creating software from the ground up. GNU radio uses software blocks linked together using a graphical interface to create software. Not just SDR software. It is good because you can reuse blocks coded by others, but I feel that this has stifled SDR software development.

## SDR# (SDR SHARP)

SDR# "SDR Sharp" was developed for the Airspy range of SDR receivers using the C# computer language. It rapidly became the preferred software for a range of USB and soundcard-based, software defined radios because it worked with extio.dll. Unfortunately, that connectivity option was removed in 2015.

The current list (on my installation) is Airspy, AOR IQ, RF Space, FunCube Dongles, RTL dongles (USB or TCP), HackRF, and soundcard IQ radios.

The software is for receivers only. It can demodulate; AM, FM narrow, FM wide (stereo) with embedded RDS text, USB, LSB, DSB, and CW. Not many SDR programs can demodulate FM stereo.

SDR# has extensive control over the waterfall and spectrum displays, digital noise reduction, and a noise blanker. It supports unlimited memories grouped into user-defined banks and recording of either the IQ signal or the demodulated audio. You can show all of the controls or hide them leaving only the panadapter and waterfall displays. The screen controls, (zoom, contrast, range, offset), frequency and volume control also remain visible. If the I.F. spectrum, FM MPX spectrum, or audio spectrum displays have been selected they show as well.

The program is extendable. There is an SDR# SDK for 'Plugin Developers.' The basic installation comes with some standard plug-ins and several others can be added.

The standard plug-ins are;

- Zoom FFT – zoomed IF spectrum display (I.F and FM MPX)
- Audio FFT – audio spectrum display
- Noise Blanker
- DSP noise reduction
- Baseband IQ or audio recording
- Frequency manager – memory channels in groups

SDR# can include several plug-ins contributed by other developers;

- Frequency scanner (by Jeff Knapp)
- Frequency scanner (by Vasili rtl-sdr.ru)
- CTCSS tone decoder (by Vasili rtl-sdr.ru)
- Baseband IQ or audio recording (by Vasili rtl-sdr.ru)
- Digital audio processor (by Vasili rtl-sdr.ru)
- DDE client for Doppler correction when receiving signals from satellites

I like SDR#. I use it with my FUNcube Pro+ dongle. Admittedly, it spends most of its time on the FM Broadcast band. For FM broadcast reception, you must turn off the low noise amplifier in the dongle, which you can do from within SDR#.

I have also used the SDR# and FUNcube Pro+ combination for monitoring local 2m and 70m repeaters, and for receiving aircraft ACARS, automatic air traffic control messages and maritime VHF AIS beacons. There is free software for decoding AIS and ACARS signals available online.

The SDR# developer also published ADSB# and ADSB Spy which are used to decode the signals from 1090 MHz ADS-B aircraft position beacons.

*AIS = automatic identification system (VHF ship ID, status, and position beacons)*

*ACARS = aircraft communications addressing and reporting system (basically a text messaging system for pilots, often used to report faults to the ground crew)*

*ADS-B - Automatic dependent surveillance-broadcast. It is a rebroadcast from the aircraft of satellite-sourced position information. The received and decoded transmissions can be sent to sites on the Internet, which plot aircraft movements on maps.*

## SDR++

SDR++ is a multi-platform (Windows, Linux, macOS, and Android) SDR receiver program that works with several SDR models including

- Adalm Pluto SDR
- Airspy - all models
- BladeRF
- CaribouLite - Raspberry Pi hat (planned)
- Ettus USRPs (in beta)
- HackRF (see troubleshooting section about removing soapy source)
- Hermes Lite 2 (experimental)
- Lime SDRs
- Per Vices Noctar (planned) 6
- PicoSDR (coming soon)
- RFSpace devices (in beta)
- RTL-SDR dongles
- SDRPlay receivers - providing the API is installed and running
- SPECTRAN V6 spectrum analyser (Coming soon)

The program has a built-in server which is started with a command line on Linux or Windows. I'm not sure about the other platforms. This lets you access the radio from any connected device running SDR++. You would probably have to open a port on your router if you want to access the radio from outside of your local LAN. You can plug a small box or RTL dongle into a USB port on your Raspberry Pi4 or 5 and listen to it on a desktop PC, tablet, or Android phone.

## SDRCONNECT

SDRconnect is a multi-platform, (Windows, macOS, Linux including RaspberryPi), replacement for the SDRuno program used with the SDRplay series of receivers. SDRuno is a modular program that opens several windows. SDRconnect operates in a single window. There is a spectrum and waterfall display of the receiver slice and a spectrum view of the demodulated audio. The software supports diversity reception with an RSPduo receiver which has two ADCs. A major complaint with SDRuno was managing all the separate windows. With SDRconnect you can separate windows or dock them differently, but normally the program operates on a single screen. The program supports receivers on multiple panadapters. I don't think that it can place two receivers on the same panadapter. You can also run SDRconnect as a server and connect to it from a remote computer.

You can now set your IARU region. This puts band markers on the panadapter and changes the labels on some of the band buttons. You can record the IQ data stream for the whole panadapter display, which means that you can listen later, even to a station that you were not listening to when the recording was made. It means that you can finish listening to a conversation and then go back and listen to a different signal on the recording. You can add multiple manual notch filters and see their effect on the receiver spectrum display. SDRconnect includes a memory bank system with tabs that you can name. It allows you to create banks of channels for the ham bands, FM broadcast stations, air band repeaters, etc.

I have not tried SDRconnect as I don't have an SDRplay receiver at the moment. It looks like a useful improvement over the old software which did work very well other than having too many windows. SDRuno and SDRconnect will only work with SDRplay receivers.

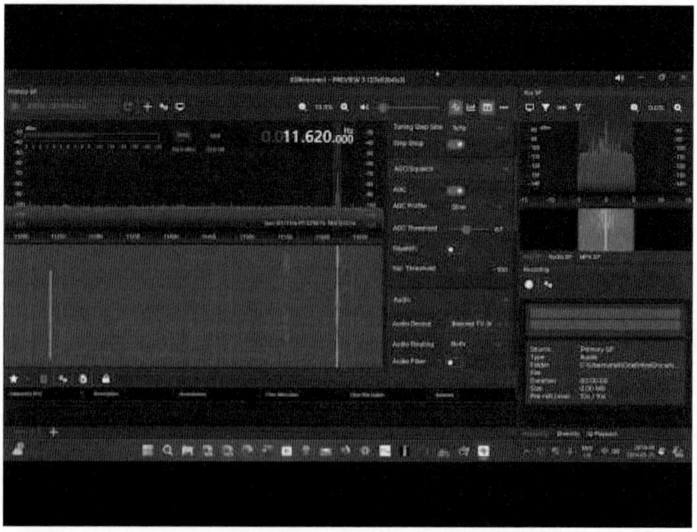

Figure 32: SDRconnect

## EXPERTSDR2

ExpertSDR2 is supplied free of charge by Expert Electronics, for the MB1 and SunSDR2 pro transceivers, and the Colibri receivers. It is closed-source software and as far as I am aware, it cannot be used with any other SDR hardware.

ExpertSDR2 can support transmitting and it has two panadapters. Each panadapter can have two receivers active, so you can operate split and hear both sides of the QSO. You can hear all four receivers at the same time if you want to. On each panadapter, you can shift the audio from each of the receivers to separate the signals left and right in the stereo audio field. I like to place the audio from the higher RF frequency on the right side and the lower RF frequency on the left side. There is some ability to customize the look of the software including changing the image behind the panadapter, the grid lines, spectrum display colours, and the waterfall. If you do change the background image, choose a darkish picture so that the spectrum display is clearly visible in front of the image.

One of the unusual features of ExpertSDR2 is that the wideband scope includes a waterfall as well as the spectrum display and you can zoom the wideband image so that it only shows the range of frequencies that you want. I believe that this is unique to this software. Another novel feature is that each of the two panadapters has its own VAC (virtual audio cable) connectivity and its own CAT command port. So you can route the panadapters to different digital mode programs and run them both at the same time. The program uses the Kenwood TS-480 CAT command set.

The 'quick memory' automatically remembers the last 16 frequencies. There is a memory bank as well which you can tag with channel names or descriptions. The main console shows the temperature of the radio, DC voltage applied, current draw, CPU loading, day and date, local time, and UTC. On transmit, the 'S meter' shows both transmitter power and reflected power along with a numeric indication of the SWR. While receiving, it shows the received signal strength in S points and dBm.

The program can automatically launch other applications when it starts. This could be used to launch your favourite digital mode program, a rotator controller, or a logging program. Or all of them! You can auto-start up to 10 applications.

*Technical TIP 1: Initially I had a problem with the CAT control. This was not a fault in ExpertSDR2 or the radio and it was quickly addressed by the manufacturer.*

*Here are the details of what happened, in case a reader has a similar problem. The symptom was that the frequency and band control could not be changed from the digital mode software. It worked fine from the radio to the digital mode software and other functions seemed OK. Every time I moved the frequency via the software, it would bounce back. I think that this was caused by the radio sending the frequency back to the digital mode program before it had settled on the revised frequency. I managed to fix the problem by changing the settings on the com0com virtual Com ports. Unselect the two 'emulate baud rate' checkboxes and the two 'enable buffer overrun' checkboxes leaving all options unselected.*

The wideband display, equalizer, S meter, and mixer controls all show in boxes that are initially docked to the main GUI display. They can be dragged off the GUI so that they are displayed in separate windows, which you can place anywhere on your desktop, or a second monitor. Once undocked, I was not able to get them stuck back onto the GUI again. An email to the helpdesk resolved the issue. If you move the window to the left edge of the spectrum display, it will dock on the left side. If you move the window to the bottom centre of the GUI, it will dock at the bottom of the screen.

The SunSDR2 Pro transceiver uses a static IP address, rather than accepting an address from your Internet router via DHCP like most Ethernet devices. This means that when you first plug in your transceiver and start ExpertSDR2, the program may be able to see that the radio is on the network but be unable to send and receive data. This happens because the radio is on the network but has an incorrect subnet address. The procedure to fix this is included in the equipment user guide. I found an alternate way that was easier for me.

I fixed this one myself. Setting the address of the radio only has to be done once and after that, it will work with no problem. The radio ships with an address of 192.168.16.200. If your computer uses a different subnet, the ExpertSDR2 software will be able to locate the radio but it will not be able to receive or send any data. You can find the subnet by looking at the IP address of your PC or a connected printer in the computer management settings or your Internet router LAN (not WAN) settings. The subnet number is the third set of digits in this case 16. It is more likely that your network's subnet address will be 1. There are a couple of ways to resolve the problem. You can connect the radio directly to the Ethernet port on your computer and use the setup page of ExpertSDR2 to set the address of the radio so that it is in the correct subnet, or you can take the approach that I did. I left the radio connected to my Ethernet switch, the same as the ANAN. Then I set up a static route on my Internet router to 192.168.16.200. It is easy to do. That lets the radio communicate with the PC over the network. I could have left it that way, but instead, I used the ExpertSDR2 software to change the address of the radio to 192.168.1.20, which is in my subnet. After that, I could disable the static route in the Internet router, and the radio continues to work fine.

*TIP: ExpertSDR3 was supposed to be released in 2021. There was a video of a Beta version in 2023, but the project seems to be dead since then.*

## GNU RADIO

What's GNU Baby? GNU Radio is a clever new way to write your own SDR software. It is an open-source project with hundreds of collaborators.

You can take advantage of software code blocks, which have been written in computer languages like C++ or Python and made available for other developers to use in their applications. Apparently, GNU is pronounced 'Guh New.'

Using GRC (GNU Radio Companion), the GNU graphical user interface, each software code block is represented on the worksheet as a rectangle. Each rectangle has one or more inputs called, 'sources' and outputs called 'sinks.' You can combine the blocks in the same way as building a Lego kit. This means that you can create your own SDR software application without writing any software code. You just link the graphics blocks together. If you do want to write your own Python or C++ code blocks, you can expand on the open-source examples to create whatever you want. You might design an improved filter, or a demodulator for a particular digital mode, and share it with the group.

There is a version of GNU Radio for Microsoft Windows, but most users use the Linux operating system. You can use compiled GNU Radio applications with a variety of SDR radios. It is very popular with people using experimenter's boards like Ettus USRP, Blade RF, and HackRF One.

*"GNU Radio is a free software development toolkit that provides the signal processing runtime and processing blocks to implement software radios using readily available, low-cost external RF hardware and commodity processors. It is widely used in hobbyist, academic, and commercial environments to support wireless communications research as well as to implement real-world radio systems. GNU Radio applications are primarily written using the Python programming language, while the supplied, performance-critical signal processing path is implemented in C++ using processor floating point extensions where available. Thus, the developer is able to implement real-time, high-throughput radio systems in a simple-to-use, rapid-application-development environment."*
http://gnuradio.org/redmine/projects/gnuradio

The GNU Radio software suite is not only used by SDR enthusiasts it is also used as a development platform by scientists, military, and commercial developers. Not all GNU applications are related to software defined radio.

I believe that GNU radio has led the move away from creating SDR and DSP in FPGAs, which requires difficult Verilog or similar coding skills, towards SDR and DSP running on Linux platforms such as single-board computers. Unfortunately, this approach is not as good, yielding slower and inferior SDR systems that are more prone to failure. Computers crash and they are not solely focused on the DSP process which can lead to variable latency issues and processing drop-outs. FPGAs are great at parallel processing such as creating multiple panadapters and receivers.

# I AND Q SIGNALS

In any technical discussion about software defined radio, it is not long before I and Q streams are mentioned. It is at this point that people's eyes glaze over and they decide that SDR is too complicated for them. Don't panic, it is not really that difficult.

The ADC in a direct sampling receiver converts the incoming RF signal from an analogue signal into a digital signal. After sampling, the data is converted into two data streams; one representing the original analogue signal and the other representing the same analogue signal with a 90-degree phase delay. The non-delayed digital audio signal is known as the 'I' or 'incident' stream, meaning that it is in phase with the input signal. The delayed digital audio signal is known as the 'Q' or 'quadrature' stream, meaning that there is a 90-degree phase difference from the input signal. If you have both the I data and the Q data, you can demodulate any kind of modulation and you can cancel out any image signals. For transmitting, the modulating signal is applied to I and Q data streams, which are then upconverted to the wanted frequency and changed from a digital to an analogue signal. Then the resulting analogue RF signal can be amplified by the RF power amplifier and passed through standard low pass filters to the antenna.

*Technical TIP: The I and Q signals both contain the same signal amplitude information. You can demodulate AM modes like AM, SSB, and CW from either individual stream because the mathematical algorithm only needs the signal's amplitude over time. You need the second stream to eliminate image signals using the phasing method and to demodulate phase or frequency modulated modes like FM and PSK. To demodulate modes based on phase or frequency changes you need to detect the phase difference between the streams, as well as the signal amplitude.*

The older QSD radios mix the RF signal down to I and Q audio streams, which are then converted, to digital bit streams using two analogue to digital converters. The two analogue to digital converters are often the left and right stereo channels of a computer soundcard. One of the resulting digital bit streams is the, I stream and the other is the Q stream. The phase and amplitude relationship between the I and the Q signals are used to demodulate the signals and the IQ phase difference is used to eliminate the image signals that arise from the single conversion mixing process.

With hybrid HF software defined radios and VHF or UHF SDRs, the RF signal is mixed down to an I.F. frequency which is sampled using an ADC. The resulting data stream is sent to the DSP stage of the receiver. Sometimes the ADC is inside the DSP chip. After that, the DSP process is the same as direct sampling software defined radios. In some hybrid receivers, the hardware mixer and oscillator, and the ADC are located inside a tuner chip that outputs IQ signals for processing in the DSP stages of the radio. The RTL dongle receivers and other radios using Mirics tuner chips are based on this type of arrangement.

So, in summary;

- Direct sampling receivers use a single ADC to sample the HF spectrum directly with no hardware mixing or down conversion. The resulting high-speed data is split into I and Q streams inside the DSP stage of the radio.

- QSD receivers mix the RF signal down to I and Q audio streams, which are then converted to digital signals using two ADCs, often a stereo ADC chip.

- Hybrid SDR receivers mix the RF signal down to an I.F. (intermediate frequency) that is within the range of the ADC. It can be at a radio frequency in the HF range in the case of a UHF/SHF receiver, at a low I.F. around 24-36 kHz for a hybrid HF receiver, or at audio frequencies.

  The signal from the hybrid mixer output is sampled with a single ADC and converted to I and Q digital signals in the DSP stage. Alternatively, some hybrid receivers send the mixer output to a fixed tuned QSD system which outputs I and Q audio streams to a stereo (2 channel) audio frequency ADC, that outputs I and Q data streams.

IQ signals are also used in the transmitter. The wanted modulation is coded into the two data streams and then they are mixed together in a quadrature mixer in the same way as the received signals are in the receiver. The output of the mixer is an RF signal at the final operating frequency. The signal is carrying the wanted modulation. All that remains is to amplify it and to filter out any harmonics using standard power amplifier and low pass filter circuits. In the older QSE quadrature sampling exciter, the mixer is a Tayloe detector. In a digital upconverting (DUC) transmitter, the mixer is software code running on the FPGA or DSP stage. A single output data stream is passed to a digital to analogue converter, which outputs the modulated RF signal.

If you look at the audio I and Q streams from a QSD detector on an Oscilloscope they look almost identical. The only differences are the result of the slightly delayed sampling of the Q signal. The signals can be represented using a vector diagram, which depicts the phase and amplitude of signals rather than the amplitude over time representation of an Oscilloscope.

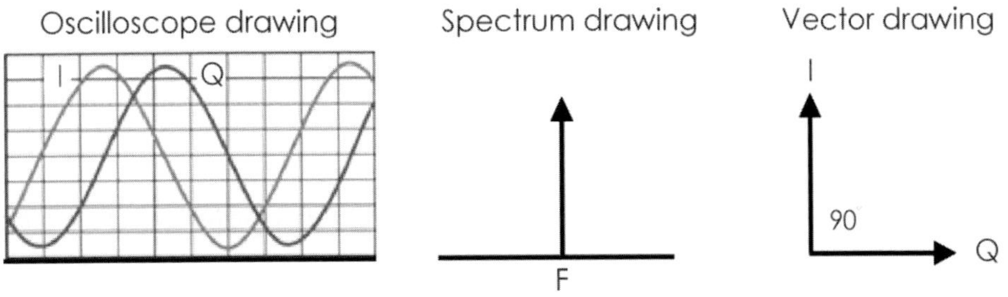


Figure 33: A vector diagram shows the phase relationship of the signals

# QUADRATURE SAMPLING

QSD (quadrature sampling detector) receivers usually use a Tayloe detector to detect the RF signal and turn it into two audio signals. Since it converts directly from RF to audio, it is acting as a single conversion receiver. The advantage of this is that it does not include the oscillators and mixers that can introduce noise and distortion to the wanted audio output. Also, it is very cheap to make, only requiring a few very cheap integrated circuits. The disadvantage is that all single or 'direct' conversion receivers are very prone to image signals.

The Tayloe Detector was patented by Dan Tayloe N7VE in 2001. The same design can be used in reverse as a quadrature sampling exciter (QSE) to make a signal which can be amplified to create a transmit signal. Gerald Youngblood AC5OG, the CEO of FlexRadio stated in a March 2003 QEX article that the Tayloe detector is the same in concept as modulator designs published by D. H. van Graas, PAØDEN, in 1990 and by Phil Rice, VK3BKR in 1998. However, he also says, "*Traditional commutating mixers do not have capacitors (or integrators) on their output. The capacitor converts the commutating switch from a mixer into a sampling detector (more accurately a track-and-hold).*" In other words, the earlier circuits did not have the capacitors on the mixer output, which act as the sample and hold mechanism. Anyway, Dan holds the patent and most if not all QSD receivers use the Tayloe design.

Tayloe detectors receive a band of frequencies above and below the nominal local clock frequency. In this case the clock frequency sets the centre frequency of the receiver. The bandwidth of the receiver is governed by the time constant of the circuit resistance and capacitors used in the detector.

*Technical TIP: Oscillators are used when you dealing with analogue signals, clocks are used when you are dealing with timing signals such as ADC sampling or passing data signals through registers. In the digital world, everything happens at distinct times, which are governed by clock signals. The time that the data signal changes state is usually when the clock signal transitions from low to high. This is the rising edge of the clock pulse. The time that the data is sampled is when the clock signal transitions from high to low. This is the falling edge of the clock pulse.* *See clock signals on page 25.*

*In the DSP part of the radio, you can represent an analogue signal such as a Sine or Cosine wave as a stream of digital numbers. Such a stream would still be an oscillator, not a clock since it will be acting in the same way that an analogue signal would in a hardware mixer. Combining the numbers representing the Sine or Cosine wave with the numbers from the ADC has exactly the same effect as using a hardware oscillator and mixer. You get an output stream of numbers representing the difference between the two signals. Unlike a hardware oscillator and mixer, you do not get sum products or either of the original signals. The digital process adds no noise or intermodulation products.*

## THE TAYLOE DETECTOR

The Tayloe QSD detector is very simple and cheap to make. It uses three very basic integrated circuits, a dual Flip Flop D type latch configured to divide the local oscillator signal by four, a multiplex switch, and a dual low noise Op-Amp.

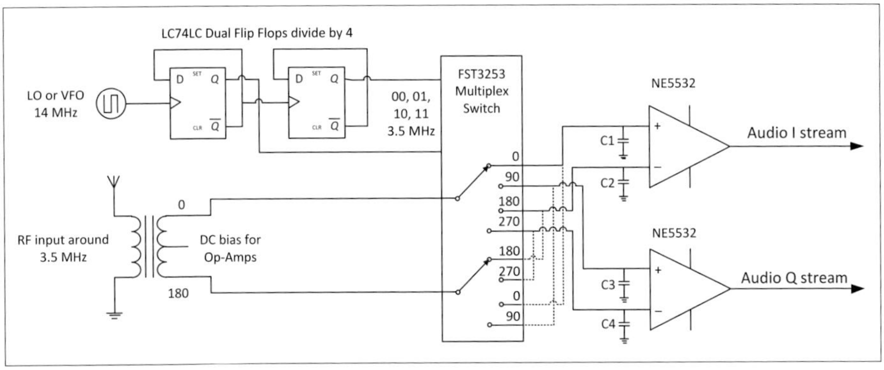


Figure 34: Balanced Tayloe quadrature sampling detector schematic

Take a look at the schematic diagram of the 'balanced' Tayloe quadrature sampling detector. The two flip flops in a 74LC74 chip are configured to divide the 14 MHz clock signal by four. The two outputs generate a 00, 01, 10, 11 binary pattern at 3.5 MHz which is used to switch the two multiplex switches in the FST3253 chip, to each of the four outputs in turn. If the VFO signal is from a variable device like a Si570 chip, the receiver can be made to cover a wide range of frequencies. The Tayloe design can be used up to around 1 GHz. In this example, the clock signals to the switch are at 3.5 MHz, so the receiver bandwidth is centred at 3.5 MHz. The signal from the input transformer is switched to all four outputs during every cycle of the input frequency. A pulse of the input signal voltage is applied to each of the four capacitors in turn. The capacitors store the voltage on the inputs to the Op-amps and end up averaging the input levels over time. This is called a 'sample and hold' circuit. It works as an envelope detector in the same way that the capacitor following the diode in a crystal set does. You get an audio frequency output. In a balanced Tayloe detector, the signal from the other side of the input transformer is not wasted. Because it is anti-phase it is used to top up the capacitor that is 180 degrees out of phase. So for each of the four switch positions two of the capacitors are charged up with the input signal.

As Dan Tayloe states in an article titled 'Ultra Low Noise, High Performance, Zero IF Quadrature Product Detector and Preamplifier,' "*... two separate detectors are driven with the inputs 180 degrees apart using an input transformer. The two detectors use a common set of four detector caps. Since the outputs of the two detectors are 180 degrees out of phase, the detector capacitors are now driven two at a time.*"

*This information was also published in RF Signal Processing magazine in March 2003 as 'A Low-noise, High-performance Zero IF Quadrature Detector/Preamplifier' by Dan Tayloe.*

A DC bias voltage for the Op-Amps is applied to the transformer centre tap. It passes through the switches to the Op-Amp inputs.

Because the 0° and 180° capacitors are charged up at the same time, (during switch positions 1 and 3), the signal on the 180° capacitor is the same as the 0° capacitor but with reverse polarity. This means that the 0° and 180° audio signals can be combined using the differential inputs of the Op-amp, providing a handy 6 dB increase in the I stream signal level without adding any noise to the signal. Likewise, the 90° and 270° audio signals are combined in the other Op-Amp to create the Q signal. Just like the 1950s phasing receivers, both audio outputs contain essentially the same signals, but the Q signal is delayed by 90 degrees. The audio output extends from 0 Hz up to the bandwidth of the receiver.

The bandwidth of a Tayloe detector is limited by the RC time constant (Ohms x Farads) of the capacitors and the overall resistance of the network. In the balanced version of the Tayloe detector the capacitors get topped up twice as often. They can have a smaller value as they only have to hold their charge for half as long. Less capacitance means that the bandwidth of the balanced Tayloe detector is wider than the original single balanced detector. The other factor limiting the bandwidth is the sample rate of the ADC (analogue to digital converters) following the detector.

In a 1st generation SDR the ADC is the PC soundcard. The I and Q signals are connected to the Left and Right line input on the PC. Don't use the microphone input because it may not be stereo.

Some QSD designs use the multiplex switch the other way around with four inputs feeding two LT6231 op amps. The design is still a balanced Tayloe detector, but it only requires two detector capacitors, and it uses a single DC power rail. https://fivedash.com/resource/RXTXEnsembleSchematic.pdf.

## THE QUADRATURE SAMPLING EXCITER

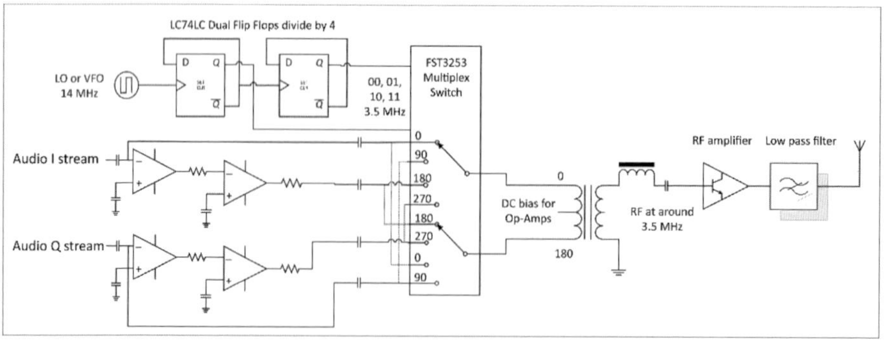


Figure 35: Balanced Tayloe quadrature sampling exciter schematic

The QSE or quadrature sampling exciter is the transmit version of a Tayloe detector. It is virtually the same circuit in reverse, and it works in much the same way.

Audio signals created from the data are sent to the QSE via the soundcard DAC (digital to analogue converter). The digital signals feeding the DAC carry the modulation added by the DSP software. In the QSE, the I stream is split into $0^{\circ}$ and $180^{\circ}$ signals. The Q stream, which already has a $90^{\circ}$ degree phase lag, is split into $90^{\circ}$ and $270^{\circ}$ signals.

The FST3253 is switched at the 3.5 MHz rate causing a 3.5 MHz RF signal to be created. This is amplified and passed through a low pass filter, then connected to the antenna. The QSE circuit is simple but effective. Similar designs are used in all generation 1, 2, and 3 SDR transmitters. The Elecraft KX3 and the SoftRock Ensemble use the circuit above. The Genesis design uses more Op-amps and eight 74HC4066N switches rather than a single FST3253 multiplex switch, but the design is the same at a functional block level.

## THE IMAGE PROBLEM

Direct conversion receivers are similar to superheterodyne receivers except the I.F. output is directly at audio frequencies extending from 0 Hz up to the bandwidth of the receiver. The local oscillator (LO) is at the same frequency as the receiver's centre frequency (Fo). In the mixer, an upper sideband signal that is 10 kHz above the local oscillator frequency becomes an upper sideband audio signal at 10 kHz. See the lighter coloured signal on the drawings below.

This is great except that the Tayloe QSD is a mixer. It works both above and below the local oscillator frequency. An RF signal that is 5 kHz below the local oscillator frequency cannot become -5 kHz. It becomes reflected into the audio range with a phase change of 180 degrees. See the darker coloured signal on the diagrams. So now, we have a wanted signal at 10 kHz and an unwanted 'image' signal at 5 kHz. The image signal becomes a lower sideband audio signal due to the 180-degree reflection. Problems arise when an image signal from below the local oscillator frequency appears right on top of a wanted signal.

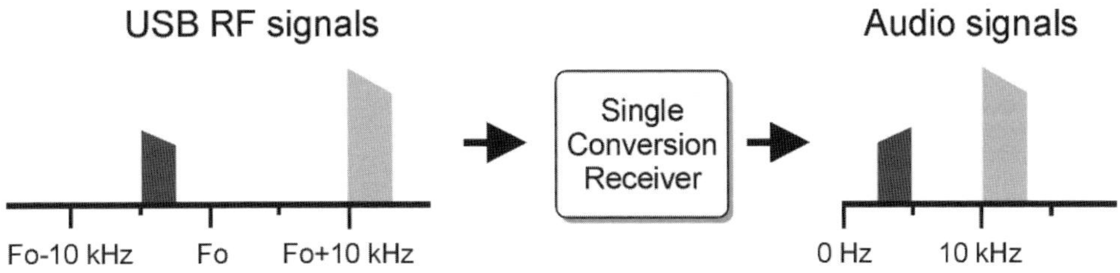


Figure 36: Single conversion creates image signals in the audio output

Traditionally this image frequency problem is managed in one of two ways; either the signals below (or above) the LO frequency are filtered out before the mixer, in the same way that image signals are filtered out before a mixer in a superheterodyne receiver, or the 'phasing method' is used to eliminate the image frequencies. SDR receivers use the phasing method for image frequency cancellation.

This is not a new idea. Direct conversion 'phasing' receivers such as the Central Electronics CE-100v were popular in the mid-1950s until improvements in filter design made superheterodyne receivers the preferred design. In the phasing method, the signal is split into two streams before the mixer and one stream is delayed by 90 degrees. This should sound familiar! After the mixer, the two streams are added back together which cancels the unwanted image signals.

The Tayloe QSD detector creates I and Q audio frequency signals in quadrature because the RF signal charging the capacitors on the Q stream occurs 90 degrees later. The two audio signals are converted to digital data using the PC soundcard or a dedicated analogue to digital converter (ADC) chip. The software in the PC applies a $+90^{o}$ degree phase shift to the I stream, or a $-90^{o}$ to the Q stream, to realign the I and Q streams, so they become 'in-phase.' The two data streams are then added together to make a new audio stream with twice the original amplitude. Subtracting the two data streams creates a second data stream containing the signals from below the local oscillator frequency and canceling the signals from above the local oscillator frequency. The image cancellation is achieved using mathematics applied to the digital data streams rather than circuitry. By combining these two streams into a single stream via a data buffer, the radio can display signals below and above the LO frequency while suppressing the image signals.

The process could be done in hardware, but in a software defined radio, everything after the analogue to digital converters is done using computer programming.

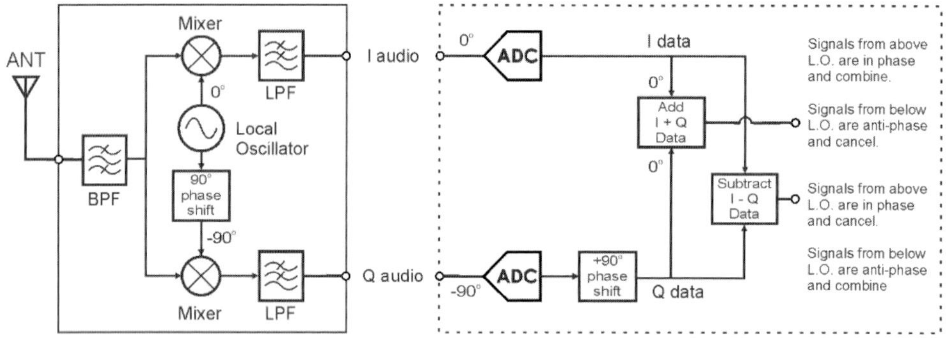


Block diagram of an SDR front end. This could be a QSD Tayloe detector.

PC soundcard (ADC) and SDR DSP software.

Figure 37: Typical QSD software defined receiver including image cancellation

## Creating a panadapter

The drawing on the next page shows how the panadapter display is created from the incoming RF signal. First, there is a single conversion receiver front end, which creates I and Q audio signals. Both streams are virtually the same and they both contain unwanted image signals. However, the phase relationships are different.

We want to create a panadapter that displays the signals above and below the nominal centre frequency Fo, with no pesky image signals.

The I+Q signal supplies the high end of the panadapter and cancels the image signals from below Fo. The I-Q signal provides the lower half of the panadapter and cancels the images from above Fo. But the panadapter is reversed and the sidebands have swapped.

By selecting the data from a different part of the FFT output buffer, we get a reversal of the I-Q part of the signal. This corrects the sideband reversal and places the signal at the proper frequency offset from the centre of the panadapter. Then the data can be combined to display a panadapter.

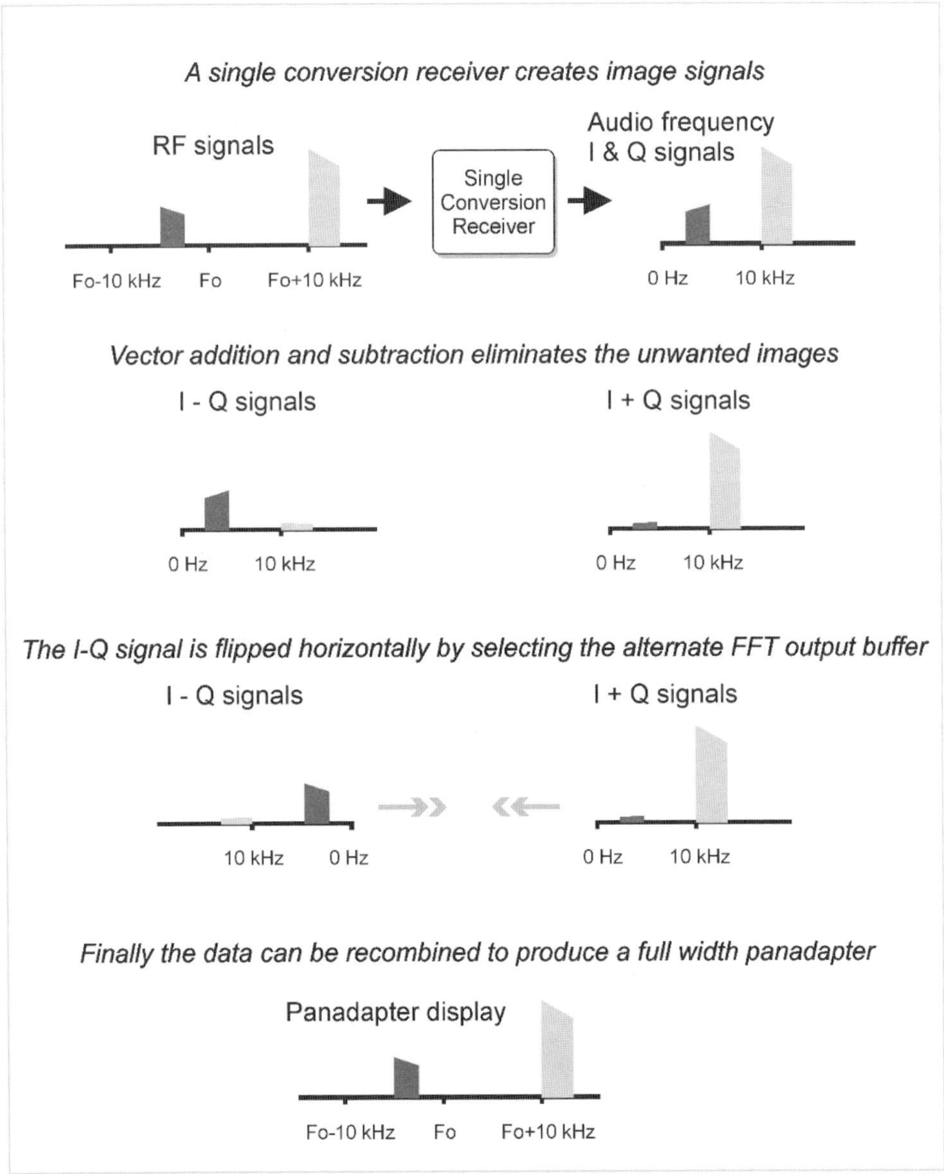


Figure 38: Using the I and Q streams for image cancellation

The image cancellation process can also be depicted using vector diagrams to show the vector addition and subtraction in action.

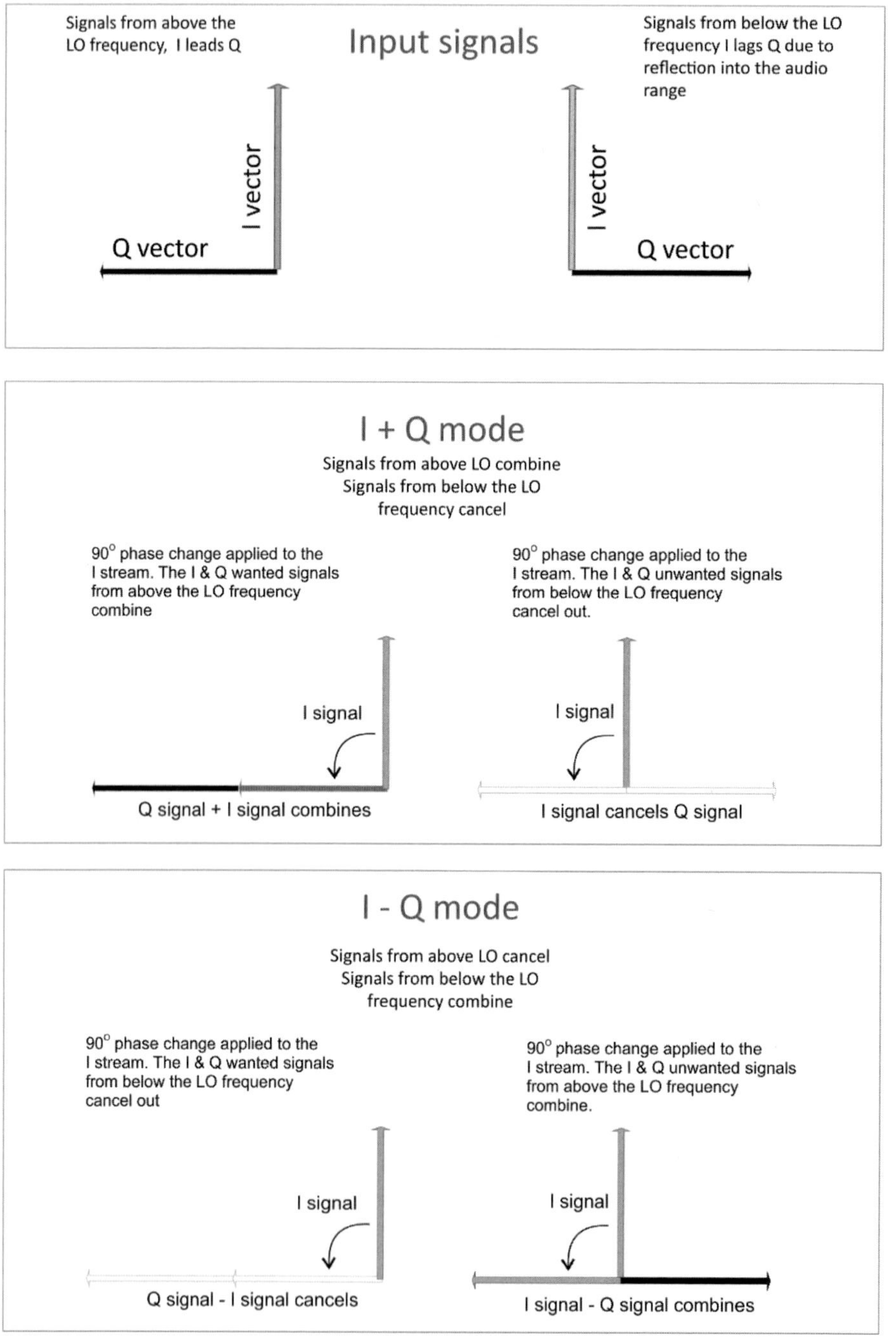


Figure 39: Vector diagrams illustrate the image cancellation process

To get good image cancellation the audio level of the I and Q streams must be the same and the phase difference must be exactly 90 degrees.

For 40 dB of image cancellation the difference between the I and Q levels must be within 0.1 dB and the phase error must be less than 1 degree. To achieve 60 dB of image cancellation, the levels must be within 0.01 dB and the phase error less than 0.1 degrees. The PC software is able to compensate for phase and amplitude errors to minimize the display and reception of image signals, but the correction is not perfect across the whole spectrum.

If you have a QSD type receiver, you can see the effect of the image cancellation by disconnecting or turning down the level of either the I or the Q stream, thus disrupting the IQ balance. With the streams unbalanced, each wanted signal will have a mirror image on the other side of the spectrum display, equidistant from the centre frequency. The mirror images will have the sideband(s) reversed, which is not a problem and not visible for CW, AM or PSK signals but obvious with SSB signals. Using a mono rather than a stereo soundcard input will have the same effect, i.e. no image cancellation.

## THE RECEIVER

As well as creating the data for the panadapter display, the IQ digital signals are used to determine the amplitude and phase of the received signals at the times they were sampled. The phase differences are used to decode FM or other phase modulated signals and the amplitude information is used to decode amplitude modulated signals like CW, AM, and SSB. If you have I and Q signals you can demodulate any type of modulation.

A 48 kHz or 96 kHz bandwidth is enough to display quite a few SSB signals or the entire CW or digital mode section of the band. You can click your computer mouse on any displayed signal and hear the QSO. Another function unique to SDR receivers is that you can record the full panadapter bandwidth and play it back later. You can listen to QSOs anywhere on the recorded band segment even signals you didn't listen to earlier.

Some other statistics for the Tayloe detector include a conversion loss of 0.9 dB. A typical conventional receiver mixer has a conversion loss of 6 to 8 dB. A low Noise Figure of around 3.9 dB. The NF of a typical HF receiver can be anything up to 20 dB. It also has 6 dB of gain without adding noise, and it is very cheap to make.

The Tayloe QSD SDR provides very good receiver performance at a much lower cost than conventional receivers and it achieves the goal of making the modulator and demodulator a part of the digital signal processing, bridging the gap between IF DSP and AF DSP. But it still has one mixing process, which is a potential source of intermodulation distortion. The next logical step is to eliminate the QSD mixer and sample the RF spectrum directly at the antenna. This is the basis of 4$^{th}$ generation SDRs. The process is known as direct digital sampling (DDS). It uses digital down conversion (DDC) receivers and digital up-conversion (DUC) transmitters.

# DIRECT DIGITAL SAMPLING

In a direct digital sampling (DDS) receiver, the incoming RF signals are directly sampled by a very fast analogue to digital converter. The oscillator and mixers in the FPGA perform digital down conversion (DDC) to select the panadapter centre frequency and create the I and Q data streams. The code in the filter stages reduce the data rate and maximum panadapter width down to the size that we want to display on the spectrum and waterfall display.

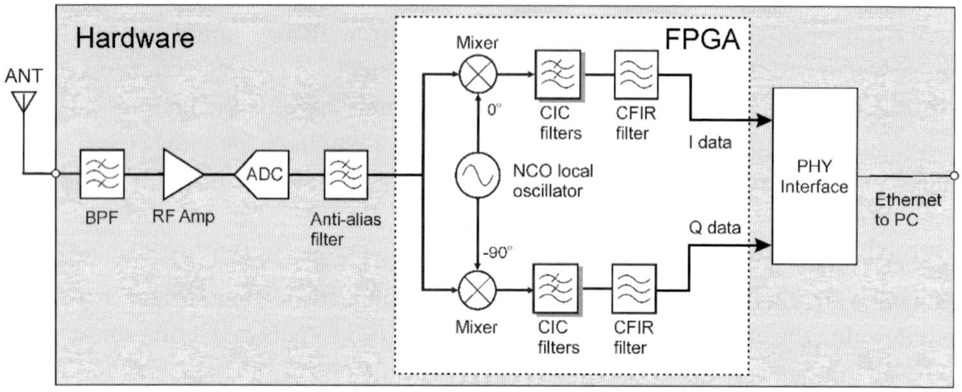


Figure 40: Typical digital down conversion (DDC) HF receiver (Hermes)

The block diagram of the oscillator and mixer in the FPGA looks similar to the mixer in a QSD Tayloe detector, but it is constructed purely in software code. The difference is that in a QSD receiver the mixer is before the analogue to digital conversion. In a direct sampling receiver, the mixer is after the analogue to digital conversion. It works by applying mathematical operations on the digital signals.

The NCO is a numerically controlled oscillator. It creates data streams of numbers that are equivalent to a Sine wave and a Cosine wave.

A CIC filter is a type of FIR (finite impulse response) filter that either interpolates or decimates. What? I hear you gasp! OK, CIC stands for cascaded integrating comb filter. They are multi-stage filters with the advantage that the mathematics governing the filter's operation is easy and efficient to code in software. The drawbacks are a certain amount of latency (delay) and they are not completely flat. They have a slight roll off at the top end. The more CIC stages that are used, the more roll off occurs and that is why the last filter in the I and Q data paths are CFIR (compensating finite impulse response) filters. These are trickier to implement in the FPGA because they use more FPGA elements and resources. The CFIR acts as a filter that has a rise at the high frequencies to compensate for the droop in the earlier filters. The result is a flat panadapter across the band. Without the CFIR filter, you would see a drop off at the far left and right ends of the panadapter. You sometimes see this drop-off in level at the edges of 'soundcard' SDR panadapters.

Decimation in the DDC receiver involves throwing away a whole lot of the ADC output bits. The combination of the digital mixer and decimation selects a subset of the input spectrum and sends it to the PC at a much-reduced data rate. The CIC filters in the receiver block diagram slow the data rate down from the ADC rate to the sampling rate that is sent to the PC. For example, in an ANAN-100 radio, the CIC filters slow the data from the ADC sample rate of 122.88 Msps to 1536 ksps or lower. Each time the sampling rate is reduced, you must apply a low pass filter at the Fs/2 frequency to stop alias signals appearing in the wanted pass-band. The CIC filters do both the filtering and the decimation. Not all direct sampling receivers use decimating filters. Onboard DSP radios may be able to process the entire output of the ADC without digital down conversion.

The digital up conversion (DUC) transmitter is quite similar to the receiver design. A modulated IQ signal is sent from the PC to the radio via Ethernet. The IQ data from the PC is at a much lower speed than the rate required for the DAC. It only has to be at least twice the highest modulating frequency. The CIC Interpolation filters speed the data up to the DAC sampling rate. Then the NCO and mixer converts the resulting data up to the RF frequency. The DAC converts the signal from a digital data stream into an analogue signal at the RF frequency. The analogue RF signal is amplified and filtered before being sent to the antenna. There will usually be several stages of amplification and filtering because the output from the DAC is only about 1 mW of RF power.

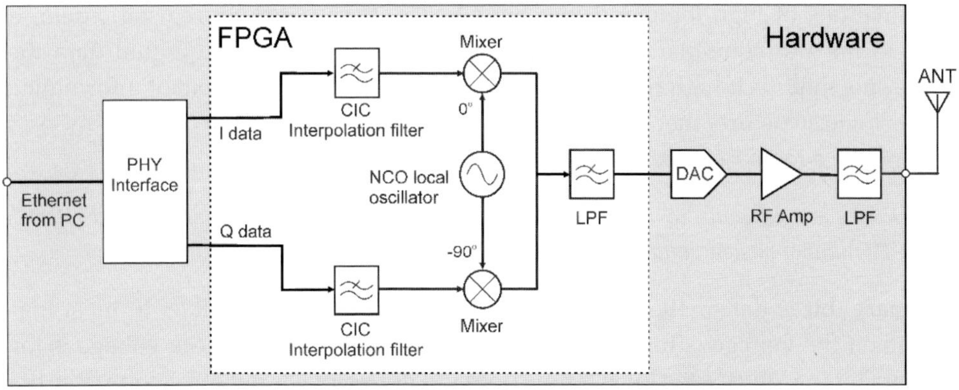


Figure 41: Typical digital up conversion (DUC) HF transmitter (Hermes)

Interpolation in the DUC transmitter is a mathematical way of filling in the gaps. The data rate of the IQ signals carrying the modulation is much slower than the data rate required by the DAC. This means that every 16-bit sample applied to the CIC interpolation filter results in a whole lot of 16-bit output samples. The output could just repeat each input sample two thousand five hundred and sixty times to interpolate between the 48 ksps audio IQ stream and the 122.88 Msps DAC input stream of an ANAN transmitter, but that would cause sudden signal level changes. The interpolation filter uses mathematics to smooth the rate of change.

# THE ADC

The heart of any software defined radio, or indeed the DSP stage in a conventional receiver or transceiver, is the conversion of the analogue signals that we receive on our antenna into digital signals that we can process with computer technology. We do it because manipulating the digital signal adds none of the noise and distortion that happens in the amplifiers and mixers of a conventional radio.

The analogue to digital converter (ADC) reads the voltage at its input, decides what the closest voltage step is, and then outputs a number representing that voltage step. This happens very fast. For example, the ADC used in the FlexRadio FLEX-8600 samples the input voltage 245.76 million times per second.

When we need to convert a signal from digital back to analogue, we use a digital to analogue converter (DAC). This integrates voltages between samples to create a smoothed analogue output signal.

## WHAT ARE BITS?

I keep talking about bits. But what are they exactly? Computers are stupid, but they work very fast. They do not understand decimal numbers or the text on this page.

At the particular time that the clock signal changes from a high state to the low state, the computer looks at the data lines or internal storage 'registers' and reads the combinations of high logic and low logic levels. It can also set registers, data, or address lines to combinations of high and low logic levels. The digital data that represents this, "is it high or is it low?" measurement, is called a 'bit' of information. In this logic structure, there are only two possible levels, so it is a 'binary' bit. A bit can only be a logic 1 or a 0.

*TIP: Binary is a reference to a base 2 number system. In the same way that decimal refers to a base 10 number system, and hexadecimal is a base 16 number system.*

A 'binary' bit is the smallest amount of digital information possible. In most logic systems if the voltage is high, the data bit is called a 'one' and if the voltage is low the data bit is a 'zero.' One bit at a time does not get you very far and a computer that could only handle a bit at a time would not be very useful at all. But, if we look at two bits side by side, the combination of the two bits can represent 4 different numbers. Four bits can represent 16 numbers, eight bits 256, and sixteen bits can represent 65,536 numbers.

| 2-bit Binary | 4-bit Binary | 8-bit Binary | Decimal |
|---|---|---|---|
| 0 0 | 0 0 0 0 | 0 0 0 0  0 0 0 0 | 0 |
| 0 1 | 0 0 0 1 | 0 0 0 0  0 0 0 1 | 1 |
| 1 0 | 0 0 1 0 | 0 0 0 0  0 0 1 0 | 2 |
| 1 1 | 0 0 1 1 | 0 0 0 0  0 0 1 1 | 3 |

Note that I grouped the 8-bit binary numbers into groups of four binary digits to make them easier to read.

Modern computers work with huge numbers that are 64 bits wide, but most analogue to digital converters are 8-bit, 14-bit, or 16-bit devices. With eight bits, you can use combinations of bits to instruct the computer to do something like add two other numbers together, or that a particular number is equivalent to a text character. With sixteen bits, you can do floating point mathematics.

### Quantisation error

An 8-bit ADC has eight output pins. The smallest number that can be output is zero, represented by 0000 0000 and the largest output number is 255 represented by 1111 1111. The eight data output lines of an 8-bit ADC can represent 256 different voltage steps because that is the maximum number of different numbers that can be fitted into 8 bits.

If the input range is an analogue signal with a minimum level of 0 and a maximum level of 1 Volt, then each sampling voltage step is 3.9 mV. An input level very slightly under a 3.9 mV step might create an output code of 1000 0001. An input level very slightly higher would create an output code of 1000 0010. That means that every digital sample could be incorrect by anywhere up to 3.899 mV. The input level could be any voltage between zero and one volt, but the output codes can only be one of the 256 steps. This sampling error is known as quantisation error because it is a mistake in quantifying the input level.

Quantisation errors cannot be recovered. When you convert the signal back to an analogue level, the output for that particular sample will be wrong by the amount of the quantisation error. This manifests itself as noise on the recovered signal. The best way to reduce quantisation errors is to make the voltage steps smaller. To do that you need to use more bits. But, that adds to the complexity and cost of the device. Using 14 or 16 output pins instead of 8 means a larger chip size with more internal buffers and your processing system has to cope with more data bits.

*TIP: Quantisation errors are also minimized by pipeline sampling in the ADC. This is a sort of successive approximation methodology.*

The output of a 16-bit ADC can represent 65,536 individual voltage levels because that is the maximum number of different numbers that can be fitted into 16 bits. If the input is in the same 0 to 1 Volt range, the voltage steps are now only 0.015 mV apart. This means much higher sampling accuracy and therefore much lower quantisation noise.

Adding more bits increases the data rate that we need to transfer over the serial USB or Ethernet interface between the radio and the SDR software running on the PC. Doubling the number of bits from an 8-bit data packet to a 16-bit data packet doubles the data rate required.

The actual serial data rate is not quite double because the number of overhead bits in each data frame remains constant. Radios with onboard DSP usually connect the parallel data lines from the ADC to parallel input lines on the DSP chip, or a computer device performing the DSP functions. That reduces the data processing rate because each of the data lines is only carrying data at the sampling rate, rather than at sixteen times the sample rate.

**What is serial and parallel data?** A 16-bit ADC outputs each sample as a 16-bit number, with each of the 16 bits appearing on a separate output pin. The pins are connected to other devices (digital chips) via tiny parallel wires or 'traces' on the circuit board. This is called a parallel data bus. Like a real bus, it carries many data bits (passengers) together. You can think of a parallel data bus as a highway with sixteen lanes in each direction. This works well on a circuit board, but it is less convenient when you want to send the data to your PC across a cable. You could use a cable with 16 wires, one for each bit, but that would need 16 data transmitters, 16 data receivers, and a fat cable. It is more usual to convert the data into a serial stream where each of the 16 bits is sent in turn. Then you only need one data wire and a ground return. But of course, you need to send the data 16 times as fast, so that you don't lose any of the information. Ethernet and USB data ports are both serial data interfaces.

Let's look at an example.

If we consider a 16-bit ADC clocked at a sample rate of 122.88 Msps, each of the 16 data output lines carries one bit of information for every ADC sample. The data rate on each data line is 1 x 122.88 Msps = 122.88 Mbits.

If we put the parallel data into a parallel to serial buffer and read it out as a serial data stream. The resulting data rate is 16 x 122.88 Msps = 1966.08 Mbits, (1.966 Gbits). This is not an intermittent or peak rate. The Ethernet or USB data connection and the computer processing software must be able to handle a steady continuous data rate of 1.966 Gbits.

*TIP: Mbits = millions of bits per second. Gbits = billions of bits per second.*

## ARE ADC BITS IMPORTANT?

Is the number of bits used by the ADC important? The answer is yes, but not as important as some people think. There is a large amount of misinformation on the Internet about how much the number of bits used by the digital sampling process affects the performance of software defined radio receivers. Some critics have claimed that SDR receivers cannot technically achieve the signal to noise and dynamic range statistics that are claimed in lab reports. They are dead wrong. I have taken it as my mission to try to clear this up.

The analogue to digital converter in the SDR receiver creates a stream of data bits which represent the signal at the ADC input at very specific instants in time.

In other words, the ADC looks at the input signal and outputs an 8, 14, or 16-bit number, which represents the voltage step that is nearest to the input voltage at that time. This is called sampling. An 8 bit ADC can describe 255 voltage levels and a 16-bit digital number can describe 65,536 voltage levels, although in practice internal noise performance reduces the number of usable bits. The noise performance of the ADC confuses the reading of very small input signals and this reduces the effective number of bits. For example, a 16 bit ADC may only be able to output 12 bits of usable data. This is known as the 'effective number of bits' or ENOB.

*It is the ENOB of the ADC, and the process gain from decimation, that determines the maximum dynamic range of an SDR receiver,* ***not*** *the number of bits in the output stream. That is why radios with 14-bit ADCs often have very nearly as good dynamic range as radios with 16-bit ADCs.*

## QSD RECEIVERS AND ADC BITS

"I have a QSD-based SDR and my PC soundcard uses 24-bit sampling. Does that make it better than a direct sampling receiver that only uses 16 bits?" The answer is no because the ADC for a QSD receiver is working on audio signals, not at RF.

It is true that the 24-bit soundcard has great accuracy and therefore low quantisation noise, but its sample rate is only 96 ksps. Or 192 kHz for a really good soundcard. The low sample rate limits the QSD receiver to a maximum panadapter bandwidth of 96 kHz, or 192 kHz in a few cases.

Noise near the centre of the panadapter and imperfect image cancellation are other limitations of the QSD design. The QSD receiver is a direct conversion receiver so the audio streams contain image signals, which must be cancelled by the DSP software. Direct conversion receivers tend to be noisy near the DC point, which is in the centre of the panadapter display and there is a risk of some intermodulation noise from the mixing process. In addition, soundcards have sampling limitations at very low audio frequencies. This often results in a pronounced noise spike in the centre of the panadapter display. Some SDR software has 'DC cancellation' to reduce the noise spike. It only masks the problem by removing the spike from the display.

If you are comparing the audio frequency ADCs used for QSD receivers, a 24-bit soundcard is better than a 16-bit soundcard, particularly if it can sample at 192 ksps. A 192 ksps sample rate will result in a panadapter display that is 192 kHz wide. This is a pretty useful size of panadapter display and of course, you can zoom in for more detail. Even though my radios can display wider panadapters, I find that 192 kHz suits my needs pretty well. A 96 ksps sample rate will result in a panadapter display 96 kHz wide and a 48 ksps sample rate means a maximum panadapter display only 48 kHz wide. I find a 48 kHz panadapter a bit too limited.

On VHF and UHF frequencies, the bands are much wider, and repeaters may be several MHz apart, so the ability to display wide panadapters is very useful.

Some direct sampling SDR boards can provide bandwidths of up to 20 MHz. The Icom IC-905 has a maximum panadapter width of 50 MHz.

## ADC SAMPLING RATE

The ADC sampling rate is very important because it sets how much of the radio spectrum the radio can cover in one chunk. In most cases the faster the ADC can sample, the better.

If the receiver uses a mixer to down convert HF, VHF, or UHF frequencies to an intermediate frequency, or from HF to audio in the case of a QSD receiver, the ADC sampling rate defines the maximum panadapter display bandwidth. For example, a receiver with an ADC sampling rate of 192 ksps can display a 192 kHz section of the RF spectrum. That might be from 14.100 to 14.292 MHz or from 144.000 to 144.192 MHz. You get a 192 kHz wide band centred on the receiver's nominated frequency.

A direct sampling HF receiver can cover frequencies from a few kHz up to a frequency set by the Nyquist limitation, (one half of the ADC sampling rate). For example, the ANAN-G2 uses a sampling frequency of 122.88 Msps, so the maximum frequency that can be sampled reliably is 61.44 MHz. The receiver is specified to cover up to 55 MHz and has a low pass filter to block any signals from above that frequency. The ADC used in the Perseus receiver is clocked at 80 Msps giving a maximum frequency of 40 MHz. The Perseus is specified to cover 10 kHz to 30 Mhz. You will note that the 6m band is not included due to the lower ADC sample rate.

With direct digital sampling, the maximum radio bandwidth is half of the sampling rate. If the DSP is in the radio the whole bandwidth may be sent to the DSP stage. If the DSP is happening in the SDR software on your computer, the data rate required to send the whole bandwidth may be too fast for the connection between the radio and the computer. The use of 1GB Ethernet and USB 3 or USB 4 has largely eliminated this problem.

Radios using Thetis do not send the full data bandwidth to the SDR software on the computer. CIC filters decimate the data down to a slower data rate such as 192 ksps, which results in a maximum panadapter span of 192 kHz. Thetis supports a maximum data rate of 1.536 Msps (24.576 Mbits) resulting in a 1.53 MHz maximum panadapter span.

- In SDRs with a mixer before the ADC, the maximum bandwidth in MHz can equal the ADC sampling rate in Msps, but it might be less. The ADC sampling rate is usually much lower than in direct sampling radios, typically 10 Msps to 20 Msps for hybrid receivers, or 48 ksps to 192 ksps for soundcard SDRs.
- In direct sampling SDRs with no mixer before the ADC, the maximum bandwidth in MHz is half the ADC sampling rate in Msps. The ADC samples at a very high speed to include as much radio spectrum as possible.

The ANAN and FLEX-8400 sample at 122.88 Msps. The FLEX-8600 samples at 245.76 Msps. The Perseus receiver samples at 80 Msps.

*Technical TIP: ADCs typically have a much wider frequency response than the Nyquist bandwidth. For example, the LTC2208 ADC used in the ANAN / Hermes receiver has a frequency response up to 700 MHz. You might wonder why anyone would design a chip with a 700MHz frequency response if the Nyquist frequency is only 61.44 MHz. The reason is to facilitate the deliberate use of the 'alias frequencies.' Using appropriate bandpass filters 61.44 MHz wide blocks of frequencies can be sampled all the way up to the 700MHz cut off. Each chunk of spectrum is called a Nyquist zone. You might elect to use the receiver on the 3rd Nyquist zone 122.88 – 184.32 MHz to receive signals on the 2m amateur band.*

*The LTC2208 ADC as used in the ANAN / Hermes receiver has a frequency response up to 700 MHz, so without filters ahead of the ADC, large signals anywhere in the 2nd to 11th Nyquist zones could be sampled by the ADC. Once sampled, they would end up being heard in the receiver and shown on the spectrum display. You will note that the 2nd Nyquist zone includes the FM broadcast band. In the Hermes design, a 55 MHz low pass filter is incorporated to attenuate alias signals and ensure they are not sampled by the ADC.*

*The idea of deliberately receiving frequencies above the first Nyquist zone is called 'undersampling.' It is used in some receiver designs. For example, the SunSDR2 pro radio has an ADC sample rate of 160 Msps. It uses its 2nd Nyquist zone to provide coverage of 80-160 MHz. The radio switches automatically between HF and VHF modes when you select any frequency above 80 MHz. You can only use one Nyquist zone at a time, i.e. HF or VHF.*

*If you use undersampling you must provide filters in front of the ADC to prevent signals from other Nyquist zones being sampled. If you try to cover more than one zone, the signals from both zones will be sampled and they will overlap on the Panadapter. Once signals are sampled, they cannot be removed. Without filters on the SunSDR2, a signal sampled at 10 MHz would appear right on top of a signal sampled at 150 MHz.*

## ADC NOISE PERFORMANCE

The number of ADC bits affects the achievable dynamic range of the ADC and ultimately the SDR. 8-bit ADCs have fewer voltage steps and therefore more quantisation noise. That makes the ADC noise floor higher and the achievable signal to noise ratio lower. 14 and 16 bit ADCs both have an ENOB of around 12 bits, so they have a similar ADC noise floor, although 16-bit ADCs are usually slightly better. The SDR receiver's noise performance is also affected by the ADC design. The noise floor of an ADC is measured over its entire frequency range. In the case of the 16-bit LTC-2208 used in the ANAN radios, that is up to 700 MHz. When this bandwidth is reduced in the radio down to the 3 kHz receiver slice bandwidth or the panadapter bin bandwidth or only a few Hertz, 'process gain' (page 203) improves the dynamic range and dramatically lowers the noise floor within the wanted bandwidth.

It is very important that the ADC used for a direct sampling radio has the ability to distinguish wanted signals from the background noise. This is such an important factor that the manufacturer's data sheets include tables of the results of several different measurements of the noise performance. The tests are done at several different input frequencies across the first Nyquist zone. If the ADC is being used in an under-sampling mode, the noise performance will translate up into higher Nyquist zones.

Unless you are building your own software defined radio you won't have any control over the choice of ADC, but it is worthwhile discussing these measurements because the ADC noise performance is the ultimate limitation of how well your SDR will be able to receive weak signals. And strong signals too. These measurements also shed some light on why 14-bit ADCs work very nearly as well as 16-bit ADCs.

Prepare to enter acronym heaven!

### Decibels (dB), dBFS, and dBc

All of the noise measurements on the ADC data sheet are expressed in decibels (dB). Measurements in decibels are always a logarithmic ratio. They have to be a value referenced to something else. You can say that a voltage is a certain number of decibels above or below another voltage reading or a standard value. The same goes for power measurements. If you are referring to an attenuator or an amplifier, the amount of loss or gain is expressed in decibels referenced to the level at the input of the device. An attenuator might have 10 dB of loss or an amplifier may have 20 dB of gain. Ohm's law gives us the relationship of Power to Voltage. $W = V^2/R$. So given the same Resistance:

> For power: $dB = 10 \times \log_{10}$ (power measurement 1 / power measurement 2)
> Twice the power is a 3 dB increase. Ten times the power is a 10 dB increase.
>
> For voltage: $dB = 20 \times \log_{10}$ (voltage measurement 1 / voltage measurement 2)
> Twice the voltage is a 6 dB increase. Ten times the voltage is a 20 dB increase.

Commonly used dB scales include; dBm, which is power compared to 1 mW, dBc which is power, or voltage compared to a 'carrier' level, and dBFS which is a level compared to the 'full scale' input level for the device.

dBFS is used to denote the maximum input level for the ADC because the actual maximum input level is adjustable. Typical input levels are 1 V peak to peak, 1.5 V peak to peak, or 2.25 V peak to peak.

The ADC data sheet usually has performance data compared to dBFS or a set value below the maximum level such as -1 dBFS. But your input signal may be much smaller than the full-scale value and you want to know how the ADC will work at the actual input level. So, the data tables also include measurements of noise and unwanted signals, using a dBc scale referenced to the input level.

## CLIPPING LEVEL

The clipping level is the maximum input signal that an ADC device can handle without catastrophic distortion. The full-scale figure of 0 dBFS is set a little under the clipping level so that it can be reached without distortion. The actual RF input level that your SDR receiver can manage without clipping is also affected by the loss in any front-end filters or attenuators and the gain of any pre-amplifiers.

Clipping level in a direct sampling receiver = ADC clipping level + losses before the ADC – preamplifier gain before the ADC. Obviously this doesn't apply to hybrid receivers that have tuner stages or hardware mixers before the ADC.

The clipping level is usually very high. Adam Farson (SK) A7OJ/AB4OJ, measured it at -10 dBm (about S9 + 60dB) on the IC-7300.

## ADC SPURIOUS FREE DYNAMIC RANGE (SFDR)

The dynamic range of an ADC is the difference between the noise floor and the 'full scale' maximum input signal of 0 dBFS. Nonlinearity in the front-end circuits of the ADC can cause harmonics of the input signals or intermodulation products, which will be sampled by the ADC. These harmonics and any other spurious signals such as interference resulting from clock signals show as spikes on the spectrum display. They are commonly referred to as 'spurs.'

The actual usable dynamic range of the ADC is affected by any harmonics or other spurious signals. This is recorded on the ADC data sheet as the 'spurious-free dynamic range.'

The SFDR quoted on the ADC data sheets is the dynamic range from the highest spurious or harmonic signal compared to either the level of the input test tone level (dBc) or to the 'full scale' clipping level (dBFS). Some data sheets quote SFDR at a variety of input frequencies measured against both dBc and dBFS. Typically, there are also separate tables showing SFDR relative to the larger 2nd and 3rd harmonics and SFDR relative to the smaller 4th and higher order harmonics.

The SFDR tends to degrade fairly linearly as the input signal level increases and also as the frequency of the input test signal increases.

For digital down conversion HF receivers, we are only using a small fraction of the overall frequency response of the device, so the slow roll off in dynamic range is not noticeable. The SFDR also gets worse at very fast sample rates, but with most of our receivers, the ADC sample rate is fixed and less than the maximum allowable rate, so this effect is not important either.

### ADC signal to noise ratio (SNR)

The maximum signal to noise ratio of a software defined radio receiver is directly related to the dynamic range. It is the range between the maximum 'clipping' level the ADC can sample and the noise floor of the device. The ADC data sheets show measurements of the SNR compared to the noise floor. The SNR on the data sheets excludes the first five harmonics of the input signal from the measurement. This is probably because the noise performance including the harmonics is covered by the SFDR test. Note that the ADC signal to noise ratio is not the same as the receiver's signal to noise ratio. It is measured over the entire frequency range of the ADC, not the tiny bandwidth of a receiver slice, or a panadapter bin. See the section on Process Gain below.

## ADC SINAD

SINAD is the Signal to Noise Plus Distortion Ratio [S/(N+D)]. It is the ratio between the input level of a test tone, and all of the noise, harmonics and distortion products in the band of frequencies below ½ of the sampling frequency (i.e. the first Nyquist zone). For HF receivers the 1st Nyquist zone is the band of frequencies we are interested in receiving, so SINAD is the best indicator of the achievable signal to noise ratio.

*Spurious-free dynamic range (SFDR) measures the dynamic range between full scale or the input signal and the level of the highest harmonic.*

*Signal to noise ratio (SNR) measures the dynamic range between full scale or the input signal and the noise floor, ignoring any large harmonics.*

*Signal to noise plus distortion ratio (SINAD) measures the dynamic range between full scale or the input signal and the average noise level, including any harmonics or spurious signals within the first Nyquist zone.*

## EFFECTIVE NUMBER OF BITS (ENOB)

The ADC cannot describe very small input signals due to the limits of its internal noise performance. Very low-level input signals might be recorded as 0000, 0001, 0010, or 0011. This makes the lowest two or three bits of the output data unreliable. ENOB is the number of bits, which contain usable information. It can be calculated from the quoted SINAD of the ADC. *ENOB = (SINAD-1.76)/6.02.*

- The ANAN / Hermes board uses a 16-bit LTC2208 ADC, which has a SINAD of 77.4 dB giving an ENOB of around 12.6 bits.
- The Perseus receiver uses a 14-bit LT2206 ADC with a SINAD of 77.1 dB giving an ENOB of around 12.5 bits.
- The AD9467 ADC used in the FlexRadio SDRs has a published SINAD of 76.2 dB and an ENOB of 12.4 bits.

You can see that the SINAD ratio of all three devices is pretty similar. Even though the Perseus SDR receiver uses a 14-bit ADC, the SINAD is almost the same as the 16-bit ADCs used in the other two radios.

The main difference between the three radios is the ADC sampling speed, which allows the FlexRadio to work to around 77 MHz, and the ANAN to work to 55 MHz, while the Perseus only works to about 40 MHz.

*Technical TIP: The number of bits that the ADC uses to describe the signal is not a reliable indicator of the receiver's dynamic range and noise performance. It is the ENOB that is important. A 14-bit radio may work just as well as 16-bit radios.*

## ADC GAIN COMPRESSION

Very large incoming signals appearing at the input of the analogue to digital converter can cause ADC gain compression. This temporary reduction in the dynamic range results in a higher noise level being displayed right across the panadapter. Typical indications of ADC gain compression are a high or jumping noise floor. If the noise floor is regularly jumping a few dB, the ADC is probably being affected by AM or FM broadcast signals. Adding a bandpass filter to the receiver antenna port or selecting internal high pass and/or low pass filters should fix the problem. Noise from lightning crashes will also cause the whole panadapter noise floor to jump, for the same reason.

An ADC with 16 bits has a wider dynamic range than an ADC with 8-bits, so it will be less susceptible to ADC gain compression. This is another reason that direct sampling radios with 16-bit ADCs are definitely better than radios with 8-bit ADCs.

You can sometimes see the effect of ADC gain compression when you switch out the receiver bandpass filters and operate the receiver in wideband mode. If the noise floor of the panadapter rises when you switch to the wideband filter, the dynamic range is being reduced by large signals that the ADC is processing. You may not be able to see the large signals on the panadapter. They may be outside of the displayed panadapter span.

*Technical TIP: ADC gain compression should not be confused with 'Digital Gain Compression', which is an optional feature built into some analogue to digital converters, (but not the LTC2208 used in ANAN radios). Digital gain Compression is an AGC action that is applied to the analogue signal inside the analogue to digital converter before the signal is sampled. It limits the input signal so that it always swings between 10% and 90% of the maximum input level. This ensures that the input signal can never get to the clipping level or into the fuzzy noise floor where the level might be incorrectly sampled, (see ENOB).*

## DITHER

Some ADC chips have a dither function built in. Others do not. If it is available, the SDR software will probably have a checkbox to select or de-select the function. The ADC used in the Apache Labs ANAN series has a dither option and the ADC used in the FlexRadio Signature 6000 series does not. I don't believe that dither is particularly relevant in a ham radio or SWL receiver, but it does no harm, so I usually leave it turned on. https://www.youtube.com/shorts/nwSpTr-8b1U

Dither improves the intermodulation performance of the ADC by ensuring the input signals are spread out across the analogue to digital transfer characteristic of the ADC. Pseudorandom noise is deliberately added to the analogue signal and then removed from the digital signal after the analogue to digital conversion.

The ADC samples everything that is presented to its input, so unless you have a particularly poor antenna there will usually be enough medium-size signals to ensure that the signals at the ADC input will transverse most of the ADCs input range. In that case, dither does not provide any advantage.

Dither could be helpful if you were using external narrow band filters for contest operation and it does improve the IMD performance when a two tone test signal is used, as part of lab testing because there are no other signals present. It could be a useful feature for weak signal work on VHF or UHF frequencies because the ADC will only be presented with very low strength signals right across the receiver's bandwidth. When you are using the SDR as an HF receiver there are many signals at varying levels and frequencies, so selecting the dither function is unnecessary. It will probably not provide any noticeable improvement and it may degrade the receiver noise floor by a couple of dB.

*Technical TIP: In the LTC2208 ADC the noise is generated as a pseudorandom digital code inside the ADC. The code is converted into an analogue signal and mixed with the incoming signal from the antenna. After the combined signal has been sampled and converted to digital data the original digital code is mathematically subtracted back out again. The result is better IMD performance with only a small degradation in the receiver's noise floor.*

## RANDOM

Some ADC chips have a 'random' function built in. If it is available, the SDR software will probably have a checkbox to select or de-select the function. The ADC used in the FlexRadio Signature 6000 series does not have a random function and the ADC used in the Apache Labs ANAN series does. I don't believe that the random is particularly useful in a ham radio or SWL receiver. At my house, I can't see any difference on the panadapter with random turned on or off.

"The ADC is a small device with a big problem." On the antenna side, there are very small sub microvolt signals which we want to hear in our receiver and on the other side, there are very high speed 3.3 V square waveform data and clock streams. Inside and around the chip both signals are in very close proximity. In the Hermes receiver, the ADC samples at 122.88 Msps and the same clock drives the FPGA. Capacitive and radiated RF coupling of the large 3.3 V signal into the sensitive analogue input stage is a problem even with very good circuit board design. The effect, particularly when there are a limited number of input signals such as during testing, is that the coupling could cause fixed signal spikes called spurs, which would be visible on the spectrum display.

The generation of spurs is a particular problem when the radio is being used in a situation where the receiver is being used to receive signals at fixed levels or frequencies such as on a fixed radio link or a repeater base station.

Spurs build up over time due to recirculating coupling from the digital signals to the ADC input.

*Technical TIP: The problem with fixed levels and frequencies on the ADC input is the repetitive nature of the data coming out of the ADC. Harmonics and related intermodulation products can be coupled into the ADC input. These unwanted signals and harmonics right up to the 700 MHz bandwidth of the ADC can be aliased back into the wanted bandwidth as they are being generated after the front-end anti-alias filters. The answer is to randomize the data coming out of the ADC so any coupling back into the input becomes noise rather than visible and audible spurs.*

*A 16-bit ADC does this by applying an exclusive or (XOR) function to bits 1 to 15 of the output signal based on the value of bit 0. This produces a randomized output bitstream resulting in a clean spectrum display with no spurs. As soon as the scrambled signal gets to the FPGA it can be converted back into usable data by simply applying another exclusive or function to bits 1 to 15 based on the value of bit 0.*

*The output of bit 0 is random due to the noise performance of the ADC so the scrambling function really does produce a randomized output. The number of output bits is 16 but the ENOB is 12 so the bottom few bits just toggle randomly.*

The creation of fixed spurs is not a problem on the ham bands or for shortwave AM reception. The random nature of the received signals means that the input signal at the ADC is fairly random and so the digital output signal is quite variable. Any coupling from the digital output of the ADC into the sensitive analogue input just degrades the overall noise performance. No fixed spurs build up because the digital signal is not repetitive.

As far as I am concerned the use of dither and random is a personal preference. You might notice some small improvement in the displayed spurs and IMD performance at the expense of a minimal degradation of the noise floor. On my ANAN radio, I see no improvement using the dither and random settings, but your experience may be different. Many software defined radios don't even have the ability to use dither and random.

## PROCESS GAIN

Process gain is a trade-off between the dynamic range and the bandwidth of the signal. The ADC might have an SFDR of around 95 dB at 30 MHz and a SINAD of 77 dB, but that is while sampling the entire 700 MHz bandwidth of the ADC. The noise level is spread out.

When you filter out the frequencies you don't want and just look at the signal in a narrow bandwidth, such as the 2.4 kHz receiver bandwidth required for an SSB signal or a 500 Hz wide bandwidth for CW, there is much less noise and so the dynamic range and SNR is much improved. The improvement in signal to noise ratio is directly proportional to the reduction in bandwidth.

Narrower filters result in a higher dynamic range because of process gain. It is why a CW signal is easier to copy with a 500 Hz filter setting than with a 2.4 kHz filter.

The process gain from changing from a 2.4 kHz filter to a 500 Hz filter is 6.8 dB.

Process gain is expressed in decibels, as a gain in signal level rather than a drop in the noise level, because an improvement in the signal to noise ratio is equivalent to an increase in the dynamic range.

$$\text{Process gain} = 10 * \log_{10}(BW1/BW2) \text{ dB.}$$

It is the reason that an SDR can deliver more than 120 dB of dynamic range from an ADC that might only have a SINAD of around 77 dB.

### Process gain on a panadapter or bandscope

The panadapter is much wider than the receiver bandwidth, and yet it displays a lower noise floor. This is due to the FFT (fast Fourier transformation). The panadapter is effectively made by plotting the levels of thousands of very narrow 'bins' each only a few Hz wide. Because each bin is so narrow, they each have more process gain than the much wider receiver bandwidth. By plotting the level of each bin side by side, the displayed noise level is lower than the noise level in the receiver bandwidth.

### Process gain mathematics

The theoretical maximum dynamic range of an ADC is (N bits x 6.02) +1.76.
For a 16-bit ADC, that is (16 x 6.02) +1.76 = 98 dB.

The typical 'real world' dynamic range of an ADC is (ENOB bits x 6.02) +1.76.
For a 16-bit LTC2208 ADC, the dynamic range is **(12.6 x 6.02) +1.76 = 77.6 dB**.

Process gain for an SDR is 10 x log(Fs/2 / B), where Fs/2 is the Nyquist bandwidth and B is the receiver bandwidth. [Fs is the sample rate].

For a 16-bit ADC sampling at 122.88 Msps using an SSB receiver filter of 2.4 kHz, the process gain is **10 x log (122880000/2 / 2400) = 44 dB**.

So, the expected dynamic range of the receiver in a 2.4 kHz bandwidth is **77.6 + 44 = 121.6 dB**. The real performance of an ANAN-100 is better than 120 dB in a 2.4 kHz bandwidth and 128 dB in a 500 Hz bandwidth. So its performance compares well with the theory.

## FREQUENCY FOLDING AND ALIASING

As discussed above, the ADC can sample signals well above the Nyquist frequency of Fs/2, where Fs is the sampling rate. This can be useful if it is managed properly because it can extend the frequency range of your receiver. However, you must ensure that you have bandpass filters before the ADC so that the ADC can only sample frequencies from one Nyquist zone at a time. Otherwise, the received signals will interfere with each other.

The way that the signals are aliased back into the first Nyquist zone depends on whether the sampled signal is in an even, or an odd-numbered Nyquist zone. Signals from odd-numbered zones are 'folded' back. This means that they will appear on the panadapter at an offset related to multiples of the Fs/2 frequency. Signals from even-numbered zones are 'reflected' back. The reflected signals will be inverted. An upper-sideband signal becomes a lower-sideband signal. In fact, the whole panadapter will be backward, with high frequencies at the left and lower frequencies to the right. This is easy to fix in the SDR software because the buffer at the output of the FFT stage contains both a 'normal' and a mirror image 'inverted' copy of the panadapter data. To show the 2nd (or other even-numbered) Nyquist zone panadapter you just use the data from the other half of the buffer. Software intended for receivers that deliberately use the 2nd or other 'even' Nyquist zones automatically selects the data that displays a correct panadapter. Signals received on frequencies in odd-numbered Nyquist zones are not inverted.

- For odd-numbered zones the alias frequency = MOD($F_{RX}$,Fs/2)
- For even-numbered zones the alias frequency = Fs/2-MOD($F_{RX}$,Fs/2)

In a spreadsheet, you can create the following simple calculator:

- Cell A1 "ADC Sample Rate," enter the sample rate into cell B1 e.g. 100
- Cell A2 "Nyquist Frequency Fs/2", type formula =b1/2 into cell B2
- Cell A3 "Received Frequency." Enter the frequency into cell B3 e.g. 123
- Cell A4 "Nyquist Zone," type formula = ROUNDUP(B3/B2,0) into C4
- Cell A5 "Alias Frequency", type =IF(MOD(INT(B3/B2),2)=0,MOD(B3,B2), B2-MOD(B3,B2)) into B5

The example above should give answers of, Nyquist zone 3 and an aliased signal displayed on the panadapter at 23 MHz.

## Receiving in the 1st Nyquist zone

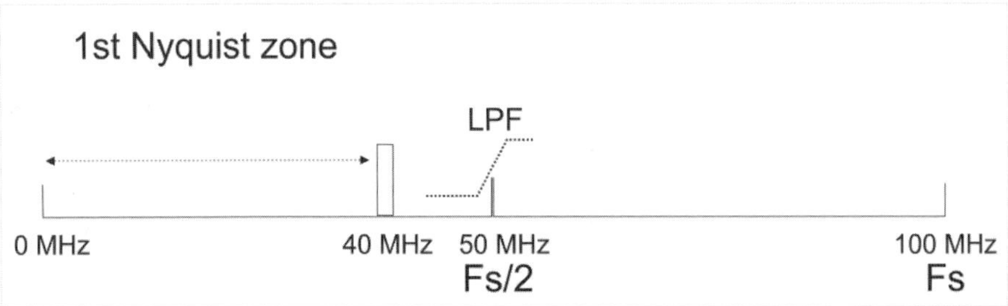


Figure 42: Receiving in the first Nyquist zone

This is the standard configuration for a direct sampling HF receiver. Signals in the 1st Nyquist zone are sampled by the ADC and displayed on the panadapter. Signals from above the 1st Nyquist zone are not sampled because they are attenuated by the anti-alias low pass filter centred on the Fs/2 frequency.

## Receiving in the 2nd Nyquist zone

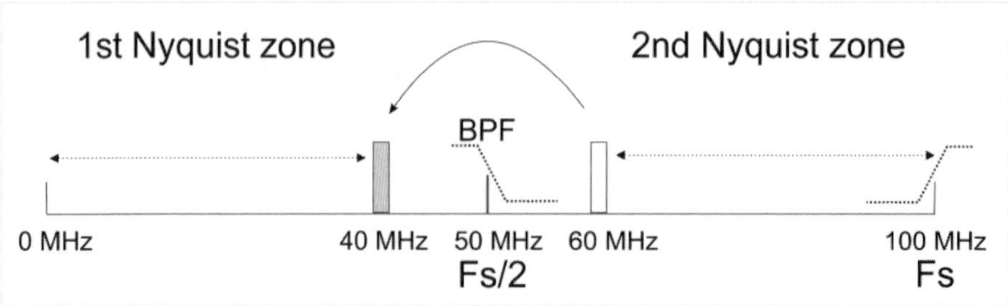


Figure 43: Receiving in the second Nyquist zone

If we want to receive signals in the 2nd Nyquist zone we replace the low pass filter with a bandpass filter that covers the zone from 50 MHz to 100 MHz. The 60 MHz signal is reflected and it appears on the panadapter at the 40 MHz position. In a real SDR receiver, the panadapter would be flipped to show 50 to 100 MHz and the inverted sideband would be corrected as well.

## Receiving in the 3rd Nyquist zone

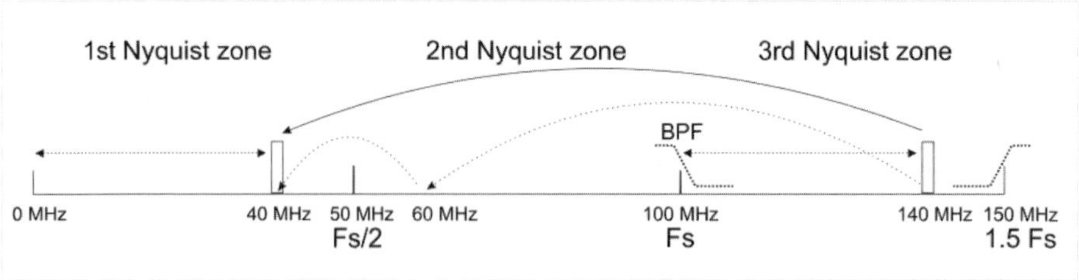


Figure 44: Receiving in the third Nyquist zone

To receive signals in the 3rd Nyquist zone we move the bandpass pass filter so that it covers the zone from 100 MHz to 150 MHz. The 140 MHz signal is folded back. It appears on the panadapter at the 40 MHz position.

It is not inverted so no correction to the sidebands or panadapter is required, although the panadapter frequency labels will need to be changed. You can think of this as a double hop i.e. inverted twice if it makes it more understandable, but really it is just the way that the ADC under-samples the input signal.

The filters are essential to prevent signals from multiple Nyquist zones from appearing on top of each other on the panadapter. Switching out the anti-alias filter on an HF SDR typically results in lots of FM broadcast stations appearing on the wideband scope.

## Another way to describe Nyquist zones

The Nyquist zones are rather like a semi-transparent fan-folded paper map. If you look through page one, the second page folds over and is reversed but the third page folds over the right way around.

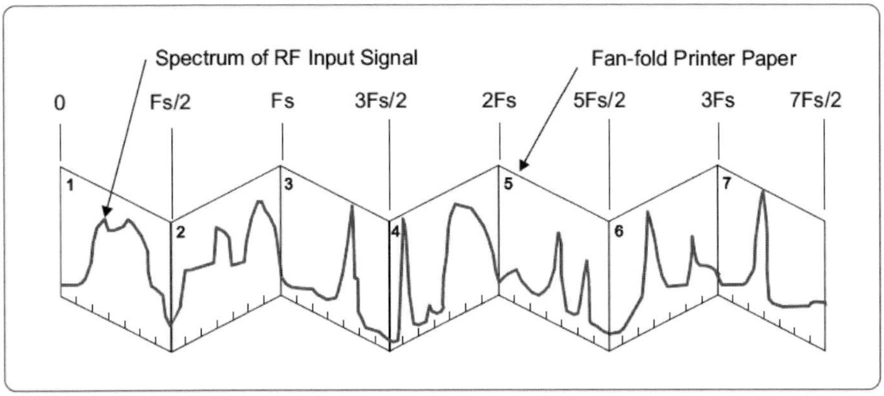


Figure 45: Nyquist zone folding

(Drawing from Pentek document, 'Putting under-sampling to work')

This drawing illustrates the way that 'even' numbered Nyquist zones are reflected or folded back in an inverted way where the panadapter would appear to be reversed. The 'odd' numbered zones are folded the same way as zone one so the panadapter and sidebands are not inverted.

If you folded all the pages together and looked through the folded paper stack with a bright light behind, you would see the signals from the higher Nyquist zones overlaying the signals on the first Nyquist zone. That is the situation that arises if you don't use an anti-alias or bandpass filter to make sure that the ADC is only sampling signals from a single Nyquist zone. It is a horrible mess and once the alias signals have been sampled there is no way to separate them from the signals in the wanted zone.

Note that the sideband reversal does not matter for double sideband modes like AM or FM because the sidebands are identical mirror images of each other. They carry the same information. This includes the digital voice modes used on amateur radio, including D-Star, P25, C4M, and DMR, because they are all FM transmissions.

SSB voice, digital text modes carried as audio on a single sideband transmission, ISB, and some phase shift modes need to be corrected because the sideband inversion causes USB to become LSB. PSK31 will decode OK because the information is carried by the phase changes, but the operating frequency would be offset below rather than above the nominal 'carrier' frequency. RTTY will be affected as it uses high and low tones, which would become reversed. Modes like FT8 will be affected because the tone sequence will be reversed.

## OVERSAMPLING AND UNDERSAMPLING

This is a subject that confuses a lot of people. I have even seen product brochures from leading SDR manufacturers that get this wrong. So let's sort it out!

When you use an ADC to sample frequencies above the Nyquist frequency, you are using under-sampling. As shown in the drawings above, if the sampling frequency Fs is 100 MHz, the 1st Nyquist zone is 0 to 50 MHz. When you deliberately receive a signal at 60 MHz, from inside the 2nd Nyquist zone. The signal has been sampled accurately using a sampling frequency that is under (less than) what would be required if you were direct sampling and not using the aliasing mode.

Over-sampling is completely different. It is the practice of using a higher sampling frequency than necessary. Take the example used above. If you increased the sampling rate from 100 Msps to 200 Msps, the signals in the wanted range of 0 to 50 MHz would be sampled four times per input RF cycle rather than the two times required by the Nyquist theorem. Why would you want to do that?

Oversampling by a factor of two can reduce the amount of quantisation noise by up to 3 dB, which is equivalent to adding 0.5 bits to the published or calculated ENOB. The 'real world' improvement resulting from a 3 dB increase in the dynamic range is pretty marginal considering that the noise floor of the receiver is usually much lower than the typical received noise level with the antenna connected. It is "nice to have" but arguably not worth the additional cost.

This technique is used by FlexRadio Systems in their FLEX-6600 and FLEX-8600 radios which use an ADC sample rate of 245.76 Msps instead of the 122.88 Msps sampling rate used in the FLEX-6400 and FLEX-8400. The oversampling clock rate is usually an integer multiple of the normal Nyquist sample rate. It is typically two times Fs, but it could be three or four times the Fs rate.

Another way to reduce quantisation noise is to use two ADCs sampling the same range of frequencies. Again the maximum improvement would be a 3 dB improvement in the dynamic range, (0.5-bit increase in the ENOB).

The 'classical' textbook approach is to employ an ADC that uses more bits to describe each sample. This is difficult to do if you are building a direct sampling HF radio. The LTC2208 used in many direct sampling radios is a 16-bit 130 Msps ADC. I can't find any 100+ Msps ADCs that have a 24 or 32-bit output. If such a device exists at all it would be fantastically expensive. As discussed before, upgrading a radio with a 14-bit ADC to a 16-bit ADC has almost no effect on the noise performance because they have similar ENOB values.

- Undersampling is the practice of using an ADC to sample signals above the Nyquist frequency.
- Oversampling is the practice of sampling the signals at a higher rate to improve the receiver's dynamic range by reducing quantisation noise.

# THE FPGA

Many direct sampling receivers use an FPGA (field programmable gate array) to manage the data coming from the ADC. Some also include a microprocessor or microcomputer to manage controls or display functions. Sometimes the microcomputer is a 'soft processor,' which is a software emulation of a microprocessor running on the FPGA fabric. Some radios include dedicated DSP processors or radio tuner chips.

If the radio has real knobs, buttons, and a display, the microprocessor manages those functions. It probably also handles filter and antenna switching, PTT, CW keying, band switching, and the sending and receiving of control and signal data between the radio and any attached PC.

An FPGA acts as an interface between the data coming from the analogue to digital converter and the data being sent to the DSP stage of the radio. It is used to create the I and Q data streams, digital down conversion (DDC), and for decimation and filtering. In a transmitter, it performs the digital up-conversion (DUC) using interpolation filters.

## RECEIVER CHAINS

The software code 'blocks' in the FPGA are joined together to form a 'receiver chain' between the ADC and the DSP software or chip. There is one receiver chain for each panadapter the radio can create.

The FPGA in the Apache Labs ANAN G2 creates ten receiver chains which are assignable to either ADC. It can provide data over the Ethernet cable for up to ten independent panadapters. The actual number of panadapters that can be displayed is dependent on the SDR software you are using. Thetis can display two panadapters with two slice receivers on the top panadapter and one on the bottom. PiHPSDR supports two panadapters with one slice receiver on each panadapter. SDR Console v3 supports up to eight slice receivers on a single panadapter.

The FlexRadio 6000 series radios have onboard DSP so they are not as constrained by the bandwidth of the Ethernet interface to the PC. They can display two 7 MHz panadapters and two slice receivers, or four 14 MHz panadapters and four slice receivers depending on the model.

There is a very large amount of digital information coming from the analogue to digital converter. In the case of an Apache Labs ANAN G2 radio, there are 122,880,000 sixteen-bit, samples every second. This amount of data is difficult for the DSP stage to manage and it would be difficult to transport it all to the PC over the Ethernet connection. An FPGA is used to break the data into more manageable chunks. The process is called decimation and filtering.

FPGA chips are very good at parallel processing. In the ANAN transceiver, the signal data is copied into ten streams, which are all decimated and filtered at the same time, creating up to ten panadapters. You can split the incoming data signal inside the FPGA into ten data streams without causing any loss of level or introducing any noise or distortion. This is different from splitting an analogue signal with an RF coupler, where you get a 3 dB loss of signal every time you split the signal into two paths. Actually, the data is not really split in the FPGA. It is copied into ten input buffers, one for each parallel processing stage. This is much the same as using your PC to make a copy of a document or a photo except the process is continuous while the radio is running.

Different radio models use varying amounts of signal processing inside the FPGA. Some radios don't even have an FPGA. They use a tuner chip or some other form of computer. I will describe the FPGA firmware used on a Hermes board as an example of how an FPGA can be used.

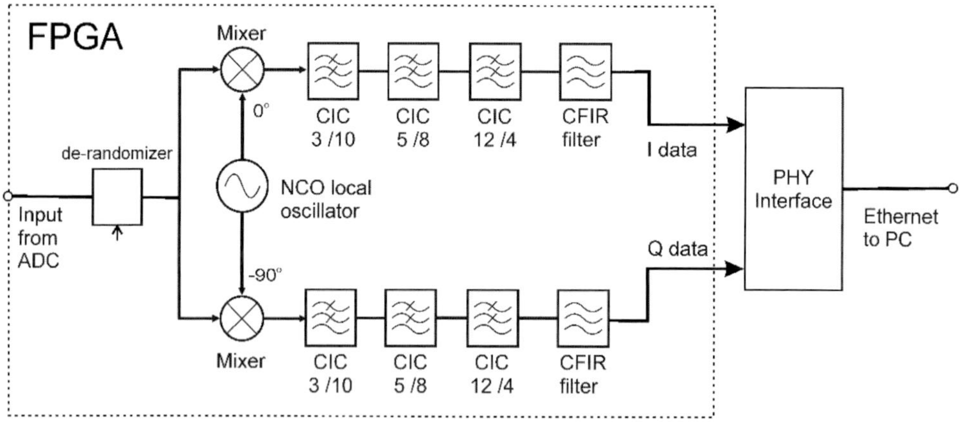


Figure 46: Receiver signal code blocks in a Hermes FPGA

The picture above shows the FPGA code blocks that are used in the Hermes / ANAN-100 receiver. Everything inside the dotted line is done using software. The blocks after the de-randomizer are repeated for each of the ten receiver chains. The picture is an expansion of the schematic on page 190 showing the three CIC (cascaded integrating comb filters) and one CFIR (compensating finite impulse response filter) in the I and Q data paths. Each CIR filter has a number showing the number of filter stages in it, and the amount that it reduces the data rate.

The de-randomizer code block is used to remove the effect of the ADC random function if it has been turned on. It uses an 'exclusive OR' function on bit 0 to remove the randomisation of the data signal.

### NCO and mixer

The first part of the chain consists of a numerically controlled oscillator (NCO) and two mixers.

You should recognize the circuit, it is a single conversion receiver the same as the Tayloe detector. But it is implemented by software working with numbers, not hardware working with analogue signals. The software used to implement the NCO in the FPGA is called a 'CORDIC' algorithm in the OpenHPSDR documents.

*CORDIC (coordinate rotation digital computer), "Volder's algorithm," is a simple and efficient algorithm to calculate trigonometric functions.* [In this case, a sine wave and a cosine wave at a settable frequency]. *CORDIC was conceived in 1956] by Jack E. Volder at the Aero-electronics department of Convair to replace the analogue resolver in the B-58 bomber's navigation computer with a more accurate and faster real-time digital solution.* [Wikipedia]

*Technical TIP: You might ask, "Won't the single conversion stage create image signals on the IQ streams?" Yes, it will and just like the QSD receiver, they will be removed by using the phasing method in the DSP section of the radio. When we use the Tayloe QSD method, imperfections in the phase and level relationship between the, I and Q signals make it impossible to completely eliminate the image signals right across the band. But inside the FPGA, the IQ phase and level relationship is perfect so the image signals can be completely eliminated.*

The numerically controlled oscillator creates two streams of 16-bit numbers, which represent a sine wave and a cosine wave. The frequency can be set by the software from about 10 ksps up to the incoming data rate of 122.88 Msps. These signals are 'mixed' with the incoming data, which shifts the received spectrum, downwards by the difference between the NCO and the incoming frequency. It is all done with mathematics but the result is the same as using a hardware mixer and then converting the data to a digital format. The I and Q streams end up with frequencies from above the NCO frequency and reflected 'image' frequencies from below the NCO frequency. Eventually the phasing method will be used to create a panadapter centred on the NCO frequency. But first the data rate has to be reduced, so the two data streams passed on to the first of the CIC filters for filtering and decimation.

### Decimation

The idea of decimation stems from the Romans who killed one in ten soldiers in a mutinous Roman legion as a punishment. The word comes from Latin 'decimat' which means, "taken as a tenth."

Decimation in the ANAN FPGA is the process where the data rate is slowed down in stages from the ADC data rate of 122.88 Msps to the rate sent to the computer for the panadapter. That is normally around 192 ksps but it can be as high as 1.536 Msps. The 'CORDIC' NCO and mixers shift the frequency range to the data we want to display and then the CIC filters literally throw away a lot of the data bits, to slow the data rate down.

## CIC filters

The following example is based on the Hermes SDR board in my ANAN-100 with the panadapter set to the 192 kHz bandwidth. As we pick up the story, the signal has been mixed down in frequency by the CORDIC stage.

After the CORDIC stage, the data is still at the 122.88 Msps rate, but it has been shifted by the mixer. As the CORDIC is a mixer the IQ data output will have numbers representing signals both above and below the NCO frequency we selected. We don't need such a high data rate to describe a 192 kHz bandwidth, so we throw a lot away. This is done in four stages using a process known as decimation and filtering. In this case decimation means dividing the sample rate by an integer amount, not necessarily 10.

To get from a sample rate of 122.88 Msps to a sample rate of 192 ksps, we need a decimation ratio of 122880 / 192 = 640 so we are effectively throwing away 639 out of every 640 samples. One of the advantages of the CIC filters is that the decimation can be built into the same code block as the filter.

*CIC filters are 'cascaded integrating comb filters.' They are multi-stage filters with the advantage that the mathematics is easy to code into software. The drawbacks are a certain amount of latency (delay) and they are not completely flat. They have a roll-off at the top end. The more CIC stages you use, the more roll off occurs and that is why the last filter in each of the two receiver chains are CFIR filters. CFIR stands for compensating finite impulse response. They are trickier to implement in the FPGA because they use more FPGA elements and resources.*

Reducing the sample rate causes a problem for Mr. Nyquist. The low pass filters before the ADC removed all of the signals above 61.44 MHz, but each stage of the decimation process causes new alias signal problems. *ANAN-100 uses a 55 MHz filter.*

There are three CIC filters in each of the, I and the Q signal paths. They are relatively easy to implement in the FPGA since they do not require a multiplier, just shift registers and addition. To avoid unwanted alias signals, we need to eliminate any signals from outside of the resulting Nyquist bandwidth before each decimation process takes place. The CIC stages filter then decimate.

The final result is a 192 kHz wide panadapter spectrum display, centred on the NCO local oscillator frequency, with all of the signals above the 192 kHz bandwidth filtered out.

- **First CIC filter**

The first CIC is a 3 stage filter with a decimation ratio of 10. It reduces the sample rate from 122.88 Msps to 12.288 Msps. The Nyquist bandwidth of a 12.288 Msps data stream is 6.144 MHz. That means that any signals above 6.144 MHz are unwanted and they could be aliased back into the new receiver bandwidth. To prevent that from happening, the CIC filter works as a low pass filter at a little above 6.144 MHz.

Note that the maximum frequency being received at this stage is the NCO frequency, shifted to 0 Hz by the CORDIC, plus the 6.144 MHz filter frequency.

A side effect of the decimation process is an improvement in the receiver's dynamic range resulting from process gain. The output of the first CIC filter is at 12.288 Msps, one-tenth of the original rate. Reducing the sample rate by a factor of 10 gives us 10 dB more dynamic range.

$$\text{Process Gain} = 10 * \log_{10}(122.88/12.288) = 10 \text{ dB.}$$

We need more bits to describe the additional dynamic range so the number of bits per sample is increased from 16 bits to 24 bits. This is enough for all further stages, so it remains at 24 bits throughout the rest of the process.

- **The second and third CIC filters**

The decimation ratio of the second and third CIC filters can be changed by commands from the Thetis PC software in order to select the wanted output bandwidth; 48 kHz, 96 kHz, 192 kHz, 384 kHz, 768 kHz, or 1536 kHz.

For a 192 kHz panadapter, the second (5 stage) CIC filter will decimate by 8, down to 1.536 Msps and the third (12 stage) CIC filter will decimate by 4, down to 384 ksps.

At the end of this block, there have been a total of 20 stages of CIC filtering which is enough to reduce aliased signals from above the wanted panadapter by more than 100 dB. One more decimate by two stage will bring us down to the wanted 192 ksps sample rate and bandwidth. The final stage uses a different kind of filter.

### The CFIR filter

The CIC filters are easy to code and they don't use too many FPGA elements but their output passband is not flat. After the 20 CIC filter stages, the output will have a roll off, of about 8 dB at the high-frequency end. This is not desired, as it would show on the panadapter as a roll off in the noise floor and receiver sensitivity at both edges of the spectrum display. The final filter and decimation by two, down to 192 ksps in our example, is a CFIR (compensating finite impulse response) filter.

A standard FIR filter has a flat response and a sharp roll off above the wanted passband. The compensating 'CFIR' filter is designed to have a shape, which rises at the higher frequencies and then has a sharp roll off. Combined with the CIC filters the output is flat across the wanted receiver panadapter with at least 100 dB rejection of signals from above the cut off frequency.

### How are the filters designed?

All of these filters were designed using software like MATLAB and the resulting Verilog code is loaded into the FPGA as a firmware file. They operate mathematically on the data bits representing the signals, which were sampled by the ADC, but their performance can be measured and displayed like hardware filters.

Verilog is a 'hardware description language' (HDL) with a syntax similar to the C programming language.

## ETHERNET COMMUNICATIONS

For the ANAN receiver, IQ signal creation, frequency translation, decimation, and filtering is all that happens inside the radio. After the CFIR stage, the IQ signals for up to ten panadapters are combined into data frames and sent over the Ethernet connection to the DSP software running on the PC. The data also contains control signals, microphone audio, and data for the wideband spectrum display across the whole HF spectrum.

Data flowing from the PC to the radio contains the audio signal from the demodulators, which is sent to amplified speakers plugged into the radio, various control signals, and the transmitter baseband signal.

## OTHER FPGA FUNCTIONS

The FPGA manages signals including, microphone audio, received signal audio for speakers plugged into the radio, CW keying, keying waveforms, and sidetone. CW is not generated in the PC software. Control signals such as PTT, antenna switching, attenuator and preamplifier switching, switching the receiver bandpass filters, and the transmitter low pass filters, are all processed inside the FPGA. The phase locked loop controlling the frequency of the 128.88 MHz clock for the ADC and DAC is also implemented in the FPGA.

## TRANSMITTER CHAIN

The code blocks for the transmit side are much simpler than the receiver software. Again, this example is based on The Hermes / ANAN design.

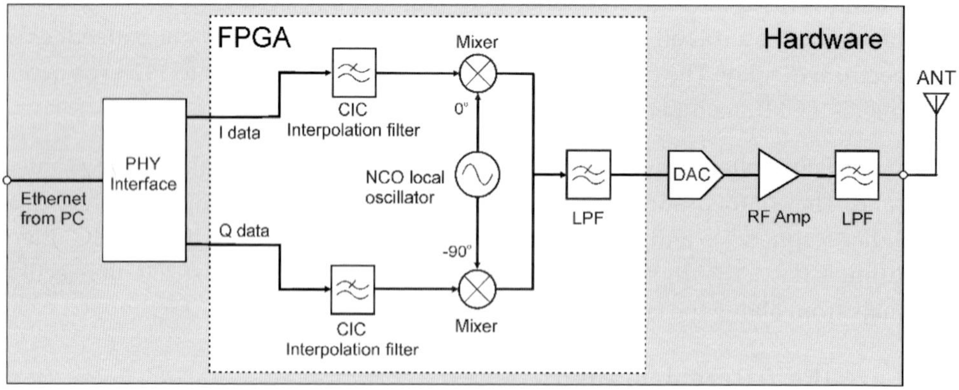


Figure 47: Transmitter signal code blocks in a Hermes FPGA

It is relatively easy for the DSP software and the Ethernet connection to manage the transmitter signal because it is only a narrow-band data stream carrying the modulation signal.

The IQ data arriving from the PC is at a much lower speed than the rate required for the DAC. It only has to be at least twice the highest modulating frequency. In the Hermes radio, the transmit data is at 48 ksps so the transmitter has a maximum bandwidth of 24 kHz. This is more than enough for amateur radio signals, which are typically less than 3 kHz wide. The 24 kHz wide transmit bandwidth carrying the modulation signal is referred to as the 'transmit baseband' signal.

The 48 ksps IQ streams received from the PC software are decoded from the Ethernet packets in the PHY interface. Interpolation CIC filters increase the data rate up to 122.88 Msps. The full speed signal is mixed using a CORDIC NCO and mixer stage so that the DAC will generate an RF signal on the wanted RF frequency. The transmitter CORDIC uses the same algorithm as the receiver CORDIC but it is separate code as the radio works in full duplex. The ten receiver chains are all working at the same time as the transmitter.

After the FPGA, the DAC converts the 122.88 Msps bit stream to an analogue signal at the RF frequency, which is amplified and filtered before being sent to the antenna.

## FPGA – TECHNICAL DESCRIPTION

It's not necessary to know how an FPGA works, so feel free to skip this section if you are not interested.

An FPGA (Field Programmable Gate Array) is an integrated circuit containing thousands of configurable logic blocks. They can be programmed to connect together the logic elements to perform the logic functions that you want. Some of them can be configured to emulate a microprocessor or 'soft processor' while still having heaps of space for the developer's project. This is especially useful for SDR developers since they can use the emulated microprocessor to drive the screen, read the buttons and knobs, and communicate with an external computer. A big strength of the FPGA is that once you have designed a process, for example an SDR receiver chain with a CORDIC, CIC filters, and a CFIR filter, you can easily duplicate that to create multiple panadapters of different sizes, working at different frequencies. Unlike a microprocessor or microcontroller which must process instructions in sequence, the FPGA can process all ten receiver chains and the transmitter process, in parallel at the same time. An FPGA has no fixed memory. The SDR software code must be loaded into it every time the radio is started. The ability to change that code with a firmware update makes the ANAN radio a true software defined radio.

# SERVER / CLIENT ARCHITECTURE

There is a strong trend towards using a server/client architecture for software defined radios. With this new approach, the traditional SDR program is split into two new programs with different functions. The server program is often running on a microprocessor, an embedded Windows or Linux computer board, or a single board computer (SBC) like the Raspberry Pi compute module, inside the radio. Or, it could be running on an external computer that is connected via a USB, WiFi, or Ethernet connection to the radio hardware. The KiwiSDR is an example of this. The server in the radio is available to clients which connect to it over the Internet.

The server program usually has no direct user access. Its job is to act as a channel between the radio and the client software.

The SDR client software may be running on a computer device (PC, tablet, iPad, phone) using the same computer network, connected via an Ethernet cable, USB, or WiFi. Or it might be on a computer at a remote location connected to the server via the Internet. It is possible to have more than one client application connected to a server application at the same time.

The SDR client usually does the display and control 'GUI' functions and the server does the DSP and signal routing. An API (page 143) will describe the data requirements and controls between the client and the server. You could use different SDR client programs to talk to the same SDR radio via the same server application. For example, you might use SmartSDR on your PC and Dogpark on a Mac computer at the same time. You might listen for 6m band openings on your phone while talking to a ham radio station on 20m SSB in your shack, while your children are listening to the BBC on a notebook PC in the lounge.

OK maybe you do not want to do all of that, but the client/server architecture does have several advantages. Firstly, the server can do much of the computer-intensive digital signal processing. There will be a large continuous amount of data flowing between the radio hardware and the server program. The data flow between the server application and the client is be much reduced. It only has to carry the graphics for the panadapters being displayed, audio in both directions, information data like the S meter, frequency readouts, and the control commands (PTT, volume, filter selection, etc.)

Developers will be able to provide a range of different client software applications designed to work with the server. This should mean shorter development times, as they would only need to create a GUI. The difficult DSP and FFT code resides in the server program. You could have a choice of user interface styles based around using the same server program and radio. Digital mode or specialist programs could interface directly with the radio.

FlexRadio Systems has already followed this route for their Signature 6000 and 8000 range of transceivers. The developers wanted to create a radio that was capable of being used across a computer network. They were aware that transferring a huge amount of data would be a severe limitation especially for wireless and Internet networking, so they moved to a server/client design. The FlexRadios have a powerful ARM processor and DSP chips inside the radio. Firmware running on these chips acts as the server. The SmartSDR software running on the PC or a Maestro unit is the client. The data requirement between the radio box and the PC is much reduced compared to a traditional 'DSP on the PC' approach and this allows for flexible networking options and the 14 MHz wide panadapters.

## THICK CLIENT

This is not a slur on the operator, it is a term used to describe a type of data connection between the radio and the PC. It does not refer to the actual electrical connection, usually Ethernet or USB, but rather to the amount of data traffic that is required. The thick client model is the situation where the DSP functions are performed by DSP software running on the PC. The computer also manages the control functions and displays the panadapters. This model of operation requires much more data to be transferred between the radio and the PC than the thin client model used by FlexRadio Systems. It also requires that the data flow is continuous, it can't be in bursts, or the latency will be too variable, possibly resulting in interrupted audio or disruptions in the spectrum or waterfall images. The thick client model tends to have a faster development cycle since PC software is easier to write than embedded software, especially FPGA software.

The thick client model is neither better nor worse than the thin client model, although the manufacturers and supporters of various radios may have a differing opinion. Both methods have advantages and disadvantages.

## THIN CLIENT

No, this does not mean that you have to be a thin person to use the radio. That would rule out a lot of hams, including myself. A thin client is the flip side of the thick client model. The term is used to describe a data connection, which requires a low amount of continuous data. The FlexRadio 6000 and 8000 radios work on the 'thin client' model where the DSP functions are performed by DSP and computer chips inside the radio. The PC is only required for control functions and to display the panadapters. This model requires less data to be transferred between the radio and the PC than the thick client model used by Apache Labs and almost all other SDRs. Thin client connections allow the use of less powerful client computers, including Netbooks, tablets, and phones. It also makes remote operation over the Internet easier to achieve. The downside is that performing the DSP inside the radio makes the radio more complex and hence more expensive.

# FFT MAGIC

Arthur C Clark wrote that "*Any sufficiently advanced technology is indistinguishable from magic.*" He meant that if somebody demonstrates a device to you and you are completely unaware about how it works, such as showing a caveman a car, or your neighbour your ham radio transceiver, then "it works by magic" is as good an explanation as any other.

As far as I am concerned, Fast Fourier Transformation (FFT) is mathematical magic. It uses very complex mathematics including 'real' and 'imaginary' numbers to convert signals in the 'time domain' into signals in the 'frequency domain' or back the other way. This is essential for the creation of the panadapter spectrum display. What the heck are time and frequency domains? I hear you ask.

All software defined radio receivers use an analogue to digital converter to sample the incoming RF, I.F., or audio signals and represent them as a series of digital numbers. The analogue to digital converter samples the level of the incoming wideband received signal at instants in time and each sample is represented by a digital number appearing on the ADC output pins. On the next clock cycle, another sample is read and its level becomes the next number on the output pins. The parallel data signals representing these numbers are often converted to a serial stream of digital data bits. Since the data coming from the analogue to digital converter represents the signal at specific 'sample' times, it is in the 'time domain.' This serial data is ok for decoding the modulation and for DSP functions like filtering and noise reduction, but it is no good for creating a spectrum and waterfall display. We want to see a display representing signals at their operating frequencies across the band, not signals at different times. The FFT converts the data into a form that can be displayed on the panadapter.

*FFT creates a series of numbers representing the signal level at different frequencies instead of a series of numbers representing the signal level at different times.*

After the FFT process, each digital number in the output buffer represents the signal level of a narrow section of the frequency spectrum called a bin. As well as being able to display the spectrum on your panadapter, having the signal represented in terms of frequency is useful for some DSP functions. For example, to create something like a low pass filter, you could progressively reduce the level of the bins above the cut off frequency. By altering the level in individual bins, you can make fine adjustments to the shape, or the high and low cut off frequencies of a filter. Frequency translation can be done simply by moving the position of bins in the software buffer. Amplification or equalisation can be achieved by multiplying the level in each bin by a calculated amount. Or, you could create a notch filter by simply reducing the level in a few bins. The spectrum display is made by plotting the level of each bin across the bandwidth of the panadapter.

Because the bins are only a few Hertz wide, they have a low noise floor and a large dynamic range due to process gain. Stacking a thousand or more bins side by side gives you a wide display with a low noise floor. The same technique is used in modern spectrum analysers. Without FFT, a wide panadapter display would have less process gain and therefore a higher noise floor than the noise level within the receiver bandwidth. Using FFT, the panadapter can display a lower noise floor than the noise within the receiver bandwidth.

The mathematics used in the FFT process is outside the scope of this book. It is for serious 'Boffins' only. Even though they are very smart, it is a bit too hard for some of the SDR software developers. Luckily, there are open source FFT algorithms available in the form of already published open source .dll files. Software developers can write a SDR program that makes calls to the FFT.dll file and gets the converted data back in a data array or buffer. They send the FFT routine time-related data and it returns frequency-related data, or they send frequency-related data and it returns time-related data.

I believe that most amateur radio SDR software uses a .dll library called, FFTW, 'Fastest Fourier Transformation in the West.'

*Technical TIP: A .dll file contains reusable software subroutines that can be used by other programs. This allows software developers to avoid having to reinvent the wheel. Under license conditions, they can incorporate code that another developer has written, into in their software applications.*

## TIME DOMAIN

An oscilloscope displays signals in the 'time domain.' It shows the signal voltage on the vertical Y-axis and time on the horizontal X-axis. If we sample the analogue signal using an analogue to digital converter, the 8, 14, or 16-bit samples at the output of the ADC represent the received signal voltage at sequential times. The digital signal is still in the 'time domain.' It represents the original analogue signal in, 'samples per second.' A digital to analogue converter can recover the original analogue signal.

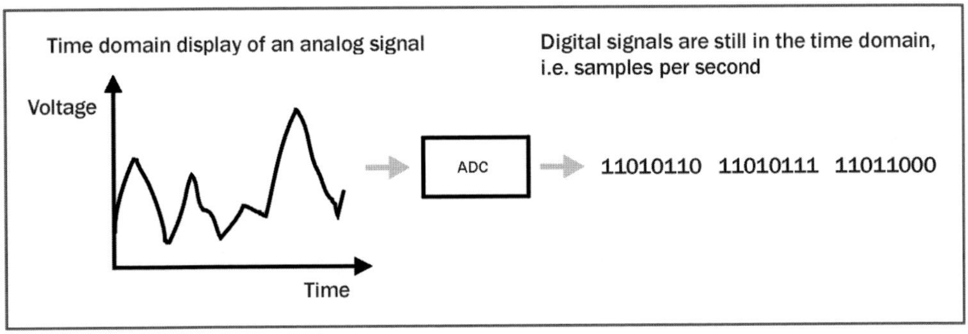


Figure 48: Analogue and digital signals in the 'time domain'

## FREQUENCY DOMAIN

The panadapter spectrum display on your favourite SDR program shows voltage on the vertical Y-axis and frequencies on the horizontal X-axis, essentially the same as a spectrum analyser display. Because the display shows signals at different frequencies instead of at different times, the digital signal and the display are said to be in the 'frequency domain.' Before the FFT stage, the analogue and digital signals are in the time domain. After the FFT stage, the digital bins are in the frequency domain.

To build the panadapter, the FFT stage outputs numbers representing the level of

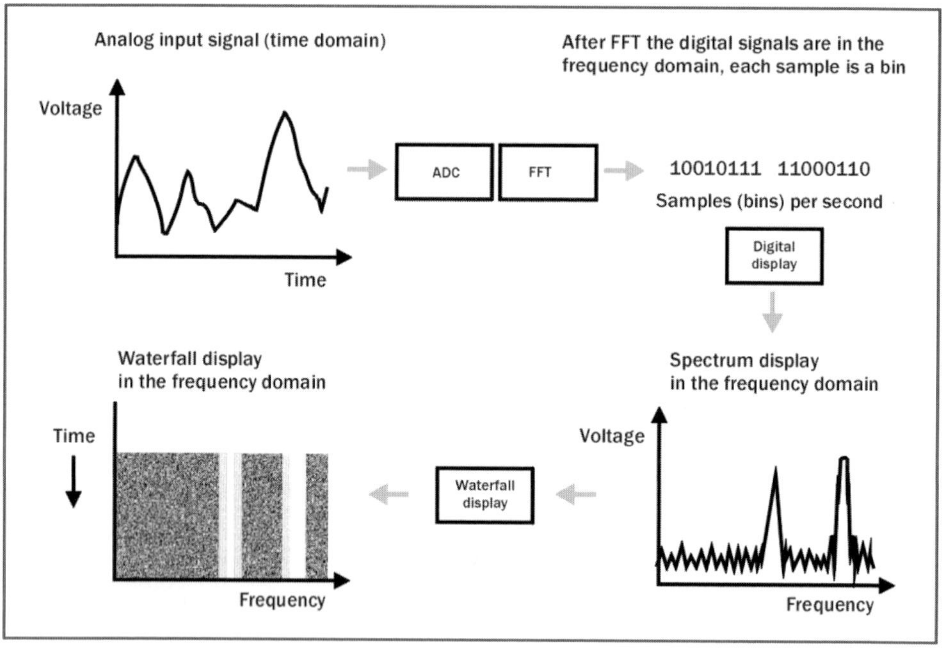


Figure 49: Analogue and digital signals in the 'frequency domain'

each bin into a software buffer. As the buffer is read, the level of each bin is displayed as a dot on the panadapter. The higher the level, the higher the dot is placed on the screen. Then the FFT program looks at the next input sample and calculates the frequency spectrum relating to that. This allows the spectrum display to be updated at a nominated number of frames per second. The spectrum line is drawn left to right across the panadapter and overlaid when the spectrum relating to the next ADC sample is drawn. The level of the bins is also used to calculate the brightness and colour of each waterfall pixel. A whole line is calculated, then the entire waterfall picture moves one line down the screen and the new line is inserted at the top.

The waterfall display is very useful because it indicates higher signal levels by using more brightness and colour, in the time domain down the screen, and in the frequency domain across the screen, at the same time.

## ANOTHER WAY OF LOOKING AT FREQUENCY AND TIME

I know this is a bit of a repeat of the above, but I think that this diagram illustrates the way you can imagine yourself looking through the "time window" on the left, or through the "frequency window" on the right.

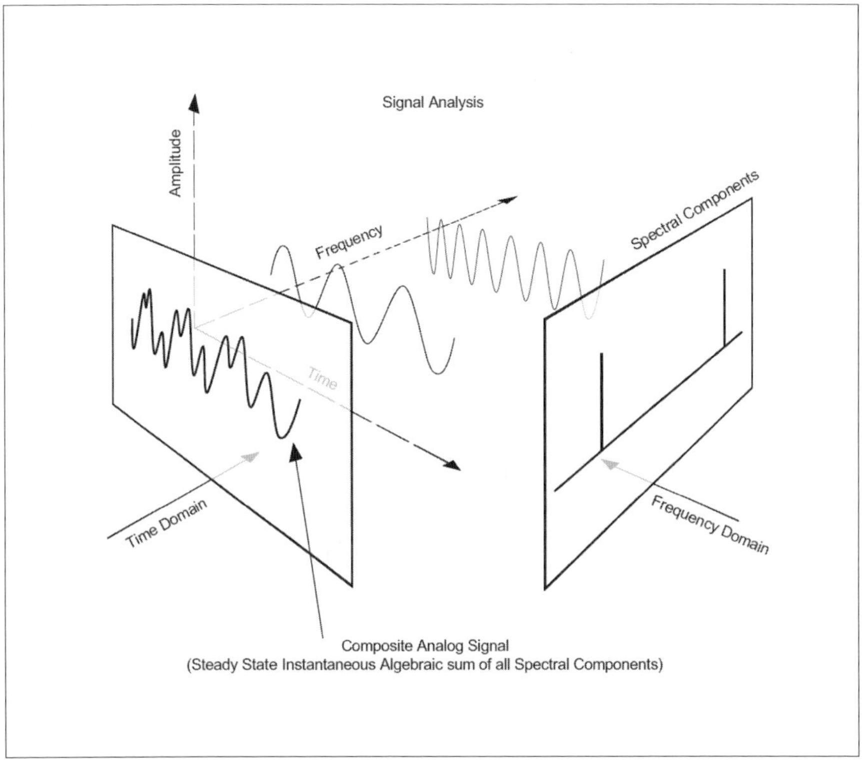


Figure 50: Time and frequency domains

The image above comes from 'Understanding Data Converters Frequency Domain Specifications,' Datel application note AN-4 by Bob Leonard. It shows how you can look at the same signal in two different ways. The picture on the left is the time domain signal, which shows the signal as you would see it on an Oscilloscope. The picture on the right is the frequency domain display of the same signal as viewed on a Spectrum Analyser or a panadapter.

If we consider radio signals being input to an ADC. When you look at the time domain signal on the left, you see the composite vector sum of all of the frequencies being sampled. When you look at the frequency domain, panadapter spectrum display, you see the amplitude of each individual signal. The frequency domain (panadapter) image is a much more useful display.

# SPECTRUM AND WATERFALL DISPLAYS

This chapter provides some more detail about how the spectrum and waterfall displays are created. The spectrum display looks very complex with noise at the bottom and a lot of signals across the band. It is always changing. In fact, it changes so fast that it is usually better to select an averaging function to make the rate of change smoother and less flickery. The waterfall flows down the screen, showing a history of the spectrum display. But these two displays only show a small fraction of the information that is being converted to digital by the ADC and then processed by the DSP. There is too much data arriving much too fast. The panadapter screen does not have enough pixels across it to display all of the bins that are being created by the FFT process and the sample rate is far faster than the rate that we need the spectrum and waterfall images to be refreshed.

## THE SPECTRUM DISPLAY

The IQ data contains the level of the signals at the sample times. However, this is no good for display on a panadapter spectrum. It must be converted to a series of samples or 'bins' that contain the level of the signal at different frequencies. The panadapter spectrum display shows these levels as dots on the screen. The higher the level, the further up the panadapter the dot is placed. Each dot is called a pixel.

As discussed in the last chapter, the conversion from time domain data samples, to frequency domain samples or bins is done using FFT. To keep things going without losing any of the information coming from the radio, the FFT processes the samples at the same rate as the input data. It takes a chunk of the incoming data, for example, a 'frame' of 32770 or 65529 samples, converts each frame to bins and then goes back for the next frame. FFTs with larger buffers create more bins with a narrower bin width but they take more time to calculate and use more computer resources. Narrowing the bin width lowers the noise floor on the panadapter spectrum display.

*TIP: You might not see a difference if you change the FFT size when your antenna is connected because the received noise will be much higher than the receivers natural noise floor. This is especially noticeable on the HF bands as they are nosier than the VHF or UHF bands.*

If we use the example of a 192 kHz wide panadapter, the sample rate being managed by the digital signal processing software is at 192 ksps. The FFT is calculating the spectrum bins at 192,000 samples per second. Each 16-bit time domain sample that is sent to the FFT becomes a 16-bit frequency domain 'bin' in the FFT output buffer.

There is a difference between the FFT bin width which determines the noise floor performance of the panadapter display and the resolution of the panadapter display. For example the FFT might be set to create 65529 bins at the 192 Ksps rate. That makes a bin width of 192000 / 65529 = 2.93 Hz. The real bin width will be a little larger due to the windowing process which reduces sudden transitions.

The panadapter display might be able to display 1024 pixels across the screen. If the panadapter is 192 kHz wide, the display resolution is 192000 / 1024 = 187.5 Hz per pixel. Each displayed pixel is showing the peak or average (selectable) of 64 bins. 64 * 2.93 Hz per bin = 187.32 Hz per pixel.

If the FFT and the sample rate are left at the same settings and the display is zoomed in so that the panadapter is only showing 96 kHz across the screen, each display pixel is now showing the peak or average of 32 bins * 2.93 Hz per bin = 93.76 Hz per pixel. The displayed noise floor stays the same because the FFT bin width is the same.

*TIP: Most SDR software lets you change the sample rate into the FFT, which sets the maximum (not zoomed) panadapter width. Thetis also lets you change the number of FFT bins that the software creates.*

The spectrum display does not update at the same rate as the data stream. If it did, the spectrum information would be updated at about 187 lines per second. That is much too fast. Depending on the refresh rate that you choose, the software may only display every 10th or 20th frame. The full data rate is not required for the panadapter, but it is necessary for the DSP and demodulator.

A 192 kHz wide panadapter displayed on a screen with 1024 pixels shows a pixel for every 187.5 Hz of the input spectrum. A 14 MHz wide panadapter on the same screen only displays a pixel dot every 13.675 kHz across the band. As you zoom in, the resolution improves. This is called the display resolution.

The maximum range of frequencies that can be displayed on the panadapter is the 'panadapter bandwidth.' It could be 192 or 1536 kHz on an ANAN radio or as much as 14 MHz on a FlexRadio. The range of frequencies that you can actually see on the screen when you zoom in, or out, is called the 'span.'

This is not the whole story! The panadapter displays the amplitude of the received signals using a logarithmic dBm scale, not the linear scale that comes out of the FFT. The numbers are adjusted mathematically before the pixels are displayed on the panadapter.

The number of frames displayed per second is usually adjustable. It is most obvious when you look at the speed that the waterfall is flowing down the screen. On Thetis you can set the frame rate for the spectrum display on the **Setup > Display > General** tab. You can make the waterfall update slower than the spectrum display on the **Setup > Display > RX1 > Update every x frames.** For example if you set the frame rate at 50 frames per second and the waterfall update to 2, the spectrum will display a new line every 20 ms and the waterfall will get a new line every 40 ms.

## Spectrum display settings

Most SDR software lets you change the colour of the spectrum line and whether you want the space below the line filled or unfilled. I prefer it unfilled.

There may be a peak hold option. I don't like it because that is what the waterfall is there for, however you may prefer to leave it turned on. Some software includes the ability to change the background image or display the callsigns of received stations.

You can usually change the level of the spectrum display so that you can set the noise "grass" level so that it is just visible at the bottom of the spectrum view. At that point the display can display any received signal that can be heard on a slice receiver, because the panadapter's noise floor is lower than the receiver's noise floor. At that point the display has the maximum dynamic range between the level indicated at the top line of the spectrum display and the noise level at the bottom. Usually the waterfall colours will be set so they look good when the spectrum level is set correctly. Most panadapters do not let you adjust the gain of the spectrum display. They are fixed at 10 dB per division and display a dynamic range of 80 to 100 dB. Thetis and SmartSDR do let you change the gain and the level of the display. Often you can set an averaging number to slow down the spectrum refresh rate so that it is not flickery and distracting.

You can usually select the sampling rate that SDR software sends to the FFT. In most cases other than Thetis you cannot change the number of bins that the FFT calculates. The sample rate affects the width and the noise floor of the panadapter. If you want to zoom in a lot and make the panadapter only a few kHz wide, you should increase the display resolution by selecting a slower sampling rate. You cannot usually change the sampling rate on radios 'with knobs.' The bandscope 'Centre' display lets you change the display span in steps up to the maximum that the radio supports. Typically ±500 kHz on a Yaesu or Icom radio. The panadapter 'Fix' display lets you set a range of frequencies up to the maximum that the radio supports. Typically a 1 MHz bandwidth on a Yaesu or Icom radio.

## THE WATERFALL

The waterfall display can be updated at the same rate as the spectrum display. Some programs let you select a slower rate so that it scrolls down the screen at a half or a quarter the rate that the spectrum display is being updated.

The waterfall is created from the same data samples as the spectrum display. If we continue to assume a 1024 pixel screen, the value of each of the 1024 displayed spectrum pixels is recorded as a dot across the top line of the waterfall display. Large value numbers (big signals) are displayed as bright colours and small values (weak signals) are displayed as dark colours. Every time a full line of spectrum information has been displayed, a complete waterfall line is added to the top of the waterfall display. In PowerSDR, the waterfall is stored as an image file that has 1024 pixels by 256 lines. (It may be different in Thetis).

In SmartSDR, the waterfall is stored in a matrix that is much larger than the number of lines that are displayed on the screen. You can drag the whole waterfall display up to see what the signals looked like as far as 10 minutes in the past.

If you change the panadapter bandwidth or centre frequency, the whole displayed waterfall is recalculated so the waterfall lines stay straight. In PowerSDR and Thetis, the lines once written to the image are not recalculated, so the vertical traces indicating traces will appear to bend if you change the panadapter centre frequency or bandwidth.

## RBW - TECHNICAL DESCRIPTION

The sample rate sets the maximum span of frequencies that can be displayed on the panadapter, with no zoom applied. If you set a sample rate of 192000 Hz the panadapter could display 192 kHz of the band. It actually displays less than that to avoid the user seeing roll off at the edges. Setting the size of the FFT buffer changes the bandwidth of each bin. That sets the noise floor of the panadapter display, with no antenna connected. Note that if you set the 1 Hz/BW checkbox in Thetis the panadapter will be offset to show the noise level in a 1 Hz bandwidth, rather than the noise level in the actual bin size. This is a feature used on spectrum analysers to make the instrument noise floor the same irrespective of the selected FFT size. Select the 1 Hz/BW checkbox if you want to use the radio as a spectrum analyser. The windowing feature is used to prevent level errors or artifacts caused by signals appearing in two bins at the same time. It has the effect of making the bin bandwidth slightly wider and increasing the noise floor very slightly. About 2 dB for a Hann Window. Considering the received background noise at my house is about 20 dB above the receiver noise floor, this minor increase is completely irrelevant.

If we assume that the panadapter is able to display 1024 pixels across the screen, the spectrum and waterfall will display 1 dot for every 187.5 Hz across a 192 kHz panadapter. An FFT calculating a buffer of 32770 16-bit data samples per second, will create 32770 bins that represent the level in a 5.859 Hz bandwidth. The Hann window degrades that to 8.8 Hz per bin. That works out at about 21 bins per pixel.

The 21 bin levels are averaged, or the peak is taken (selectable), and a dot is placed on the spectrum scope. If you zoom in, so that there is only 96 kHz being displayed on the panadapter, around 11 bin levels are averaged for each dot on the spectrum. Eventually you will zoom in to the point where two or more dots are drawn from the level of a single bin. With the 192 kHz sample rate and the Hann filter that will happen when the display span is less than 9011 Hz wide. Zooming in after that point the display loses resolution and looks lumpy rather than spikey.

Lower sample rates have a smaller bin width, so if you select a lower sampling rate you can zoom in to a greater level before the display is degraded. SmartSDR, automatically reduces the sample rate as you zoom in, to avoid this issue affecting the SmartSDR panadapter.

*TIP: If you want to view a very narrow panadapter reduce the sampling rate to 48 ksps and/or increase the FFT buffer size.*

# DSP

Digital signal processing (DSP) is the manipulation of analogue signals using digital circuits, usually programmable digital circuits. To use digital signal processing the analogue signal must first be converted into a digital data stream using an analogue to digital converter. The signal is normally converted back to analogue using a digital to analogue converter at the output of the DSP stage.

In a receiver, the input to the DSP stage is at RF or a low intermediate frequency and the output is at audio frequencies. The primary advantage of DSP is the ability to filter, modulate, demodulate, amplify, and attenuate the signal without adding any significant noise or distortion. Other advantages are the ability to make extremely sharp filters without adding ringing distortion and being able to create various noise and notch filters. The receiver filters can have variable bandwidth and shift options rather than a selection of fixed I.F. filters. DSP filters are usually better, or at least more versatile, than using hardware filters, and high-quality hardware filters are more expensive to make.

The digital signal processing is normally done using a software code block or .dll running within our SDR software, or by a dedicated DSP chip. You can also use an embedded microcomputer, or a softcore computer running on an FPGA. The hardware or softcore computer arrangement is similar to using a dedicated DSP chip, in that the DSP is being performed inside the radio box.

In a typical SDR, the ADC creates a series of digital words that represent the level of the sampled signal as a continuous stream of numbers. If it is a 16 bit ADC, each sample is represented by a 16-bit number. If the ADC samples at 100,000,000 samples per second (100 Msps), the DSP stage must handle 100,000,000, sixteen-bit numbers every second. A parallel data rate of 100 Msps, or serial data rate of 1.6 Gbits.

Each DSP feature can be expressed as a set of mathematical operations applied to the numbers representing the signal. Some of the mathematics is very complex, particularly the FFT code. On the other hand, some of the DSP code is surprisingly simple, such as the single line of code that is used to demodulate an AM signal. The line of code calculates the magnitude or 'amplitude' of each IQ sample,

AM = the square root of $I^2+Q^2$.

Another simple DSP operation is making a signal louder. You just multiply all of the 16-bit samples by the same number. For example, multiplying each sample number by two will double the level of the data stream. To multiply a binary number by two all you have to do is move all of the bits one place to the left and add a zero into the least significant bit position. E.g. 0111 x2 = 1110 or (7 x 2 = 14). You do have to be careful to avoid losing the most significant bit if it is a one.

## DTTSP

DttSP is the DSP code module that was used in early versions of PowerSDR, GHPSDR, and cuSDR. Versions of PowerSDR mRX later than V3.2.7 use WDSP instead. It runs as a program under Linux or as a .dll file under Windows. PowerSDR and other Windows SDR programs call the DSP code blocks in the .dll file. DttSP, in turn, uses the FFTW .dll to do the fast Fourier transformation mathematics.

*DttSP is an open-source project started by Dr Frank Brickle and Dr Robert McGwier of the DTTS Microwave Society to provide code to be used in various DSP projects with an emphasis on Software Defined and Cognitive Radio. DttSP implements the basic modulation, demodulation, signal conditioning, and synchronisation processes required to operate a high-performance transceiver using DSP as the detection and synthesis stages. While the development is done primarily on Linux, the code is also available for use as a Visual Studio 6 or Visual Studio 2003 project for Microsoft Windows®. It uses FFTW for much of the heavy lifting. http://dttsp.sourceforge.net/*

FFTW = Fastest Fourier Transformation in the West – a software library used by many SDR programs. FFTW calculates the data that is used for the panadapter 'frequency domain' display. Some DSP filters are also applied to the frequency domain signal, rather than the standard 'time domain' samples.

## WDSP

Wdsp is a DSP library written by Warren Pratt, NR0V. It is used in Thetis and it replaced DttSP in the later versions of PowerSDR mRX PS.

## SHARPDSP

SharpDSP was written by Phil Covington one of the pioneers of SDR for hams. It is the DSP code used in the Kiss Konsole SDR application.

*TIP: Not to be confused with SDR# (SDRsharp) which was also named because it was written in the C# (C sharp) programming language.*

## HARDWARE DSP

Hardware DSP can be done inside the radio using code running on a microcomputer or a softcore in an FPGA. However, the most common choice is to use a dedicated DSP chip. They are the same as the ones used in conventional I.F. DSP radios. The DSP chips are firmware upgradable while still installed in the radio. The FlexRadio 6000 radios use a Texas Instruments TMS320C6A8167 DaVinci chip for DSP. It incorporates an ARM Cortex-A8 microprocessor, which is used for control functions. The Texas DaVinci series are described as 'digital media processors.' They are actually designed for video signal processing, but they are commonly used for digital signal processing inside amateur radio transceivers. The Elecraft KX3 uses the Analogue Devices ADSP21479 SHARC chip. The Icom IC-7300 uses a Texas Instruments TMS320C6745 DSP chip.

# RADIO PERFORMANCE TESTING

Much has been said about how well software defined radios work and sound. In most cases, the transmitter performance is about the same as a radio using a conventional architecture. The excellent receiver performance that you can expect from an SDR receiver is mostly due to the elimination of the sections in the receiver that cause noise and distortion to be added to the signal that you want to listen to. Software defined radios replace the mixers and oscillators that cause noise and intermodulation with digital algorithms that don't.

Online comments and reviews can indicate whether a radio performs exceptionally badly, or is a low-quality product, but much of the time, the reviews are very subjective. You only get people's opinions and that makes comparisons difficult. While there is, no substitute for actually using the radio 'on the air' it is valuable to be able to compare radios in an objective and repeatable way. The best way to do that is to compare the test results gathered by a lab. The ARRL lab performs tests on many transceivers and receivers and the results of these tests are published in QST magazine. Radcom, Sherwood Engineering, and Adam Farson (SK) also publish top quality lab reports and reviews.

When I started this book, one of the questions I asked myself, was; "Do SDRs perform better than conventional Superhet architecture radios?" There is no clear answer to my question because there is a wide variance in the performance of both conventional and software defined radios. Some of it is related to the price you pay, the internal architecture of the radio, quality of components, and how recent it is. Many of the recent offerings from the 'big three' equipment suppliers have much better performance statistics than the radios they released ten years ago. Black box, no knobs, SDR performance is affected by the PC and the SDR software as well as the radio hardware.

Based on the numbers published online and in QST lab tests, one or two conventional superheterodyne receivers do outperform even the best SDR receivers. But if you compare radios at similar prices, the software defined radio is likely to beat the conventional radio in some of the key tests.

In the real world with many signals on the band, a software defined radio will usually have superior intermodulation distortion (IMD) performance because, the IMD performance of an SDR does not degrade in a linear 3:1 ratio like it does in a superheterodyne radio (or hybrid SDR). It may be worse in some situations and better in others, but so long as the IMD remains below the noise level when you have your antenna connected, that doesn't matter. The signals you receive will almost certainly sound better on a software defined radio, especially when the bands are full, such as during an amateur radio contest. Possibly not if you are using a "cheapo" 8-bit RTL dongle or similar 8-bit SDR board. You do have to make some commitment.

A big problem is that several of the tests that are traditionally used to compare radios on league tables like the Sherwood Engineering list and as the key measurements in Radcom, and QST reports are not as relevant to SDRs because of fundamental differences in technology.

For example, the Sherwood Engineering list is sorted on the results of the '3rd order dynamic range - narrow spaced', two tone IMD performance test. This is because that test indicates how well a receiver will perform when the band is busy such as during a contest. The problem in using this approach when comparing software defined receivers against traditional superheterodyne receivers is that the causes and effects of intermodulation distortion are completely different. In an SDR, the intermodulation distortion is not caused by oscillators and mixers but in preamplifiers and filters and as a product of the analogue to digital conversion. As a result, software defined radio receivers often have very good 3rd order dynamic range and they often rank very high on the Sherwood list and in reviews. Unlike superheterodyne receivers, the IMD performance is usually the same at a close 2 kHz offset as it is at a 20 kHz offset. See http://www.sherweng.com/table.html.

## TRANSMITTER PERFORMANCE TESTS

All of the common transmitter tests such as RF power, carrier and unwanted sideband suppression, two tone IMD performance, transmitter harmonics, and composite noise, are just as important in SDR transmitters as they are in conventional ones. In broad terms, you can expect that your SDR transmitter will work just as well as the transmitter in a conventional radio. The only real exception is the exceptionally good transmitter IMD performance from the few software defined radios that can perform transmitter pre-distortion.

## MINIMUM DISCERNIBLE SIGNAL (MDS)

Minimum discernible signal (MDS) is a measurement of the weakest signal, which can be heard from the receiver. Having a good MDS figure is important for receiving weak signals when the band is quiet, or on 160m when using a low gain antenna like a Beverage. It is also important for receivers working in the less noisy VHF, UHF, and SHF bands.

The noise level on the spectrum display should rise when you connect your antenna to the radio, indicating that the background noise being received at your location is higher than the noise being generated inside the receiver. As long as this is the case, you will be able to hear any signal that is above the received noise level.

*TIP: If the noise floor does not rise a few dB when you attach an antenna, you either have a very poor antenna or a very poor receiver.*

## RECEIVER INTERMODULATION DISTORTION

Intermodulation is the mixing of two or more signals resulting in unwanted distortion products. In a receiver, this can occur when two signals that are close together in frequency are received at the same time, or when an incoming signal mixes with the signal from an oscillator within the radio. Intermodulation distortion can happen in any active stage of a receiver or transmitter and even within filters. It is very often generated within receiver mixer stages and in RF amplifiers. Bandpass filters and pin diode switches in the receiver front end can also cause problems. When a band is busy, the reception of many strong signals can result in so much intermodulation distortion that wanted signals are badly affected. Old-timers who complain that contest weekends fill the band with unintelligible noise are likely to be hearing intermodulation distortion generated inside their own receivers.

For this reason modern 'contest grade' receivers have excellent 'two tone IMD dynamic range' and 'RMDR' test results. This applies to the top quality conventional superheterodyne receivers like the Kenwood TS-890 or TS990, hybrid transceivers like the Yaesu FTdx101, and direct sampling SDRs like the Elecraft K4, Icom IC-7300 and Icom IC-7610. To test the 'many signals being received at once' problem we apply two unmodulated RF signals to the receiver at fixed offsets from the receiver frequency and measure the chaos that ensues. It is called a 'two tone IMD test.'

Receiver IMD testing is often performed at 2 kHz, 5 kHz, and 20 kHz offsets from the nominal receiver frequency. The spacing between the two incoming signals is usually the same as the offset from the receiver frequency. If the receiver was tuned to 14.000 MHz USB, the test signals will be inserted at 14.002 MHz and 14.004 MHz, i.e. 2 kHz offset plus 2 kHz spacing. Superheterodyne and hybrid receivers will usually exhibit worse receiver IMD performance at the 2 kHz spacing than at the 20 kHz spacing. Direct sampling SDR receivers usually exhibit identical IMD performance at any spacing.

As well as the two tone test, we also apply a single signal, to test for intermodulation between the incoming signal and the oscillators within the radio. The tests are called 'blocking dynamic range' and 'RMDR.' In this case, a single test signal is applied at either the 2 kHz or 20 kHz offset from the receiver frequency.

## TRANSMITTER INTERMODULATION DISTORTION

Intermodulation can also be generated within your transmitter and linear amplifier. The most common source of transmitter intermodulation distortion is non-linearity in the RF amplifier. In a perfect amplifier, the ratio of the output power to the input signal would be constant at all input levels. i.e. the relationship is linear. Unfortunately, this never happens. The power amplifier in your transceiver may have a little less gain when the input signal is small and it might not put out as much power as it should when the input signal is large.

This is usually due to the power supply voltage drooping under a high current load. Components heat up under high current conditions. This can cause them to have more resistance or to change their capacitance or inductance, which might de-tune the transmitter slightly. All of these factors combine to introduce two kinds of nonlinearity. Amplitude nonlinearity where the gain of the amplifier does not remain the same for all input levels and phase nonlinearity is when the phase of the output signal is being affected by the input level or frequency.

IMD can also be caused by signal overload in components especially the Toroid cores in the low pass filters. If the signal is over modulated or clipped, then severe intermodulation distortion will occur. You can see stations with this kind of distortion on the receiver panadapter. The signals are far wider than they should be and the shape looks more like a triangle than the rectangular shape we desire.

To test a transmitter for IMD we measure the effect on the spectral purity of the transmitted signal when the transmitter is modulated with two tones. It is called the 'transmitter two tone IMD test.' The intermodulation causes harmonics of the spacing between the tones. On the transmitted RF spectrum, they appear both below and above the two test tones. The first set are called 3rd order IMD products, the next set are 5th order. It is common to record the products out to the 9th order but the benchmark test is the 3rd order IMD. A perfect transmitter would have no harmonics just the two test tones.

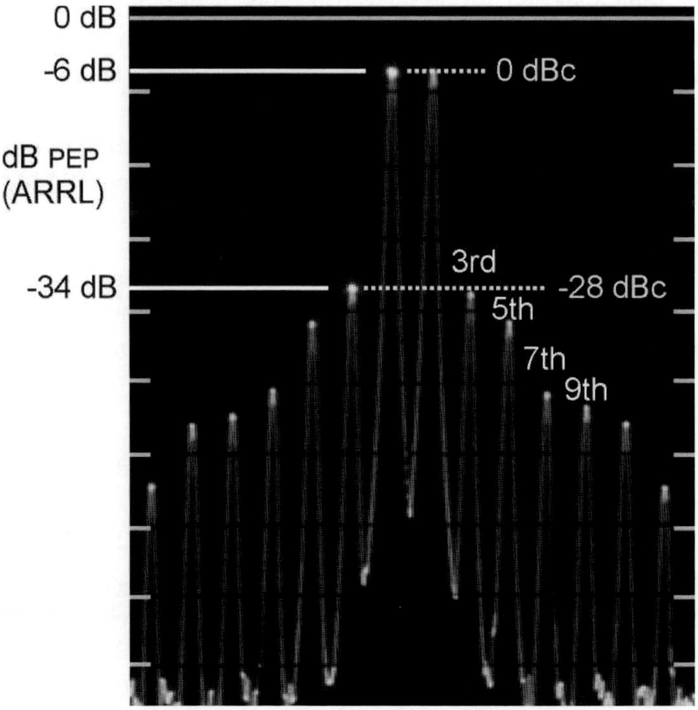


Figure 51: Transmitter 'two tone IMD' test

Image from https://www.dj0ip.de/gallery/tx-spectral-purity-imd/

In the image we can see the two audio test tones translated up to the transmitter frequency. In this case 14.158 MHz at 100 watts. The vertical display is set to a reference level of 0 dB and a scale of 10 dB per division. We can see that the radio has generated 3rd, 5th, 7th and out to the 13th order intermodulation products on both sides of the test tones. The most important are the 3rd order IMD products because they are the largest. Although arguably the products that extend outside the nominal 3 kHz SSB passband are the most likely to cause a problem to another station.

This test result is not very good. It is easy to tell that this transmitter is not using adaptive pre-distortion.

Unfortunately, there are two ways of recording the 2 tone transmitter IMD test results. The ARRL method and the European ITU method. The test is the same but the ARRL results look 6 dB better.

You perform the test by sending a single test tone to the SSB transmitter and adjusting the tone level so that the transmitter outputs full power. That is the PEP (peak envelope power) level. The spectrum analyser is adjusted until the single test tone is at the top 0 dB reference line. Some images show it one line down from the top. Then the second test tone is applied at exactly the same level as the first. It is important that the tones are not harmonically related i.e. the higher frequency one should not be a multiple of the lower one. When the second tone is applied the level of the two tones displayed on the spectrum analyser will drop by 6 dB. You will also see the intermodulation products created by the two tones mixing, primarily due to non-linearity in the power amplifier. The ARRL measures the 3rd order IMD product in reference to the PEP level at the top of the screen. The other method measures the 3rd order IMD product in reference to the level of the two tones. This is a dBc measurement. Neither method is better, but you need to be aware of the 6 dB difference so that you compare "apples with apples."

The IMD 3rd order and other products may not be exactly the same level on each side of the test signals. You can either record the worst result or average the two results.

## 3RD ORDER DYNAMIC RANGE (DR3)

This test of the receiver is radically different when testing a software defined radio. In a superheterodyne receiver, the non-linear process of mixing causes intermodulation distortion (IMD) products. Large signals near the receive frequency can cause interference on your receiver frequency. This is a big problem on contest weekends when you want to hear weak signals operating on frequencies close to stronger stations. Direct sampling SDRs do not have hardware mixers so they don't suffer from the problem of mixers causing IMD, however, the front end components and the ADC do cause some intermodulation distortion, so the test is still useful if it is measured correctly. Preferably with the results presented on a graph.

The traditional practice of measuring the input level at the point that the intermodulation products become equal to the MDS level does not apply to SDR receivers. It will produce unreliable and unpredictable results. Use the IFSS test instead.

*In a 'normal' superheterodyne receiver or a hybrid, the IMD gets worse as the level of the two tone test signal is increased. In a direct sampling software defined radio, it does not.*

In the traditional method of testing the 3rd order dynamic range (DR3), you record the input level of the two tone test signal when intermodulation products become noticeable on the receive frequency, i.e. when you can just hear or measure the intermodulation level above the receiver's noise floor. The distortion is measured in the same way as the wanted signal in the MDS test. This is the level where the dynamic range between the IMD products and the test signal is greatest. It is a 'best case' scenario because as the two tone level is increased further, the IMD dynamic range continues to degrade linearly. The test is easy to measure. You simply inject two RF signals that are 2 kHz, 5 kHz, or 20 kHz apart and increase their level until the resulting intermodulation distortion is 1 dB above the receiver noise floor.

In a software defined radio, the IMD dynamic range does not degrade linearly. It stays the about same when the two tone test level is increased and at some input levels, it actually improves. The IMD performance crashes when the two tone test signal level reaches the receiver clipping level. In the real world, this means that a conventional receiver may suffer worse intermodulation distortion in the presence of several medium-size signals like a contest going on, than a software defined radio will.

*Technical TIP: Because the 3rd order dynamic range of a conventional radio is usually worse when the interfering signals are close to the tuned receiver frequency, it is traditionally tested at a 'narrow' 2 kHz 2 tone spacing and also at wider spacings of 5 kHz and 20 kHz. Software defined radios usually have the same 3rd order dynamic range at any spacing.*

The 3rd order IMD dynamic range of an SDR should not be measured at the MDS level, but at the level where there is the greatest difference between the two tone input level and the IMD product level. This is at a high level, usually just before clipping occurs.

*To quote Leif Åsbrink, SM5BSZ in QEX Nov/Dec 2006, "Rather than measuring what level is required for getting IM3 equal to the noise floor, one should measure the largest difference (in dB) between the test tones and the intermodulation product. It will be close to saturation of the A/D converter on the SDR-14, while it will be at the noise floor for an analog receiver."*

The dynamic range of a superheterodyne receiver stays constant until the intermodulation products start to become higher than the noise floor (MDS level) of the receiver.

As the input signal is increased, the intermodulation distortion increases at three times the rate of the RF input signal. The highest dynamic range is the point where the IMD is just measurable above the noise floor.

The dynamic range of an SDR is affected earlier, but the IMD products are usually well below the background noise level when you have an antenna connected, so they are not a problem. The IMD products of a superheterodyne receiver usually cross the receiver noise background noise threshold about 3 or 4 dB before the SDR, but that varies from radio to radio.

## 3RD ORDER INTERCEPT POINT (IP3)

The often quoted 3rd order IMD intercept point, or IP3 is never measured directly. It is usually calculated from the 3rd order IMD dynamic range.

The calculated IP3 = (1.5 * DR3) – MDS (dBm).

With a conventional receiver, a plot on a chart of the RF input level against the receiver's audio output level is a straight line. The slope of the line is dependent on the gain of the receiver. Then add a plot of the 3rd order intermodulation relative to the level of the two RF input signals, (two tones). The 3rd order IMD is also a straight line which increases at three times the rate of the fundamental signal. If there was no AGC action, the two lines would eventually cross on the chart. That point is the 3rd order IMD intercept point. However, the IP3 point is never reached because the AGC reduces the receiver's gain as the input signals get larger. This means that the IP3 cannot usually be measured, it is either calculated, or the lines on the chart are extrapolated to the point where they would cross if the receiver had no AGC.

IP3 can't be measured for an SDR receiver because the relationship between the intermodulation products and the input tone levels is not linear. Plotted on a chart of RF input level vs receiver output level, the two lines diverge rather than crossing. There is no IP3 point. However, this does not stop many SDR manufacturers from quoting an IP3 using a calculation based on the 3rd order dynamic range. I guess you can use IP3 as a point of comparison between receivers, but IP3 is actually meaningless for a software defined radio.

## IFSS TEST

The IFSS (interference-free signal strength) test was proposed by Adam Farson (SK) to replace the traditional 3rd order IMD dynamic range test which is not good for SDR receivers. It simply involves plotting the level of IMD distortion against the input level of the two test signals. You can see from the chart on the next page that the IMD response of the SDR receiver is nowhere near linear. The flat diagonal line is the IMD response for a typical superheterodyne receiver. In a real test, it would probably not be as flat as that line due to receiver variations.

The dynamic range is DR3 = Input 2 tone level – measured IMD level.

Points to note.

1. The SDR creates noticeable IMD distortion products at much lower input signal levels than the superheterodyne receiver. But they are not a problem because they are too low to be heard over the typical -109 dBm noise level in a quiet rural location when an antenna is connected.

2. The SDR outperforms the superheterodyne receiver in the range where the IMD products are higher than the typical -109 dBm noise level in a quiet rural location when an antenna is connected.

3. The best dynamic range for this IC-7300 SDR is 94 dB when the IP3 control is turned off and 103 dB when it is turned on. This occurs at an input level of -25 dB. (-25)-(-119) = 94 dB. This is a high input level, near the clipping level, not at the point where the IMD level is just noticeable above the noise floor.

4. The dynamic range when the IMD is at the noise floor on the IC-7300 (with IP+ turned off) is (-55)-(-129) = 74 dB. Testing the dynamic range using the old test is unfair since, unlike the superheterodyne receiver, the dynamic range is not at its highest at that point.

5. The best dynamic range for the superheterodyne receiver is 105 dB. This occurs when the IMD is at the noise floor level. (-25)-(-130) = 105 dB. It degrades to 85 dB at the -15 dBm input level.

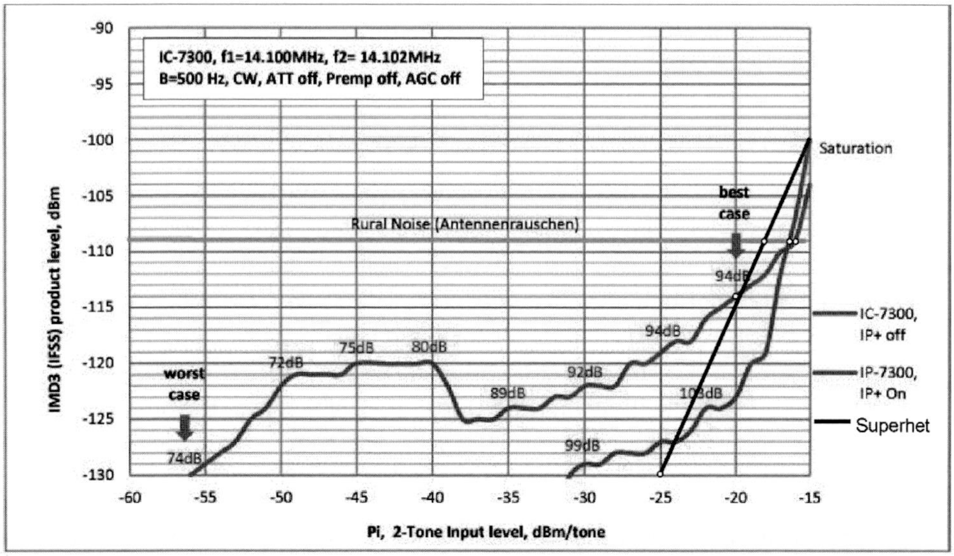


Figure 52: IFFS 3rd order IMD test by Werner Schnorrenberg DC4KU

6. With the IP+ on, the IC-7300 SDR outperforms the superheterodyne receiver at input levels above -24 dBm. With the IP+ control turned off, the IC-7300 SDR outperforms the superheterodyne receiver at input levels above -20 dBm. But this does not matter until the IMD distortion becomes bad enough to hear over the received noise level when an antenna is connected.
7. Hybrid SDRs tend to have an IFSS chart that is halfway between a superheterodyne straight line and a direct sampling SDR. A wiggly line with roughly a 3:1 IMD slope.

It is important to note the horizontal line at -109 dBm. The accepted local noise level for an urban environment is -103 dBm and the accepted local noise level for a rural environment is -109 dBm. If you have less local noise than that, you are lucky. I don't think that plotting the chart is necessary although it does demonstrate the difference between an SDR and a conventional superheterodyne receiver, which shows a flatter diagonal line.

The key points are where the IMD products exceed the urban and rural noise thresholds. It does not matter what level the intermodulation products are if they are less than the noise level at your place when the antenna is connected. Therefore the critical numbers are the input level at which the intermodulation distortion products are higher than the noise floor, with your antenna connected. In this receiver, the input signals need to reach at least -17 dBm before the intermodulation could even be seen at the bottom of the panadapter. It's a pretty good receiver! The second wiggly line (to the right) shows the IMD with the IC-7300 IP+ control turned on. Since this circuit is specifically designed to improve this result. It is not surprising that it is better.

## RECIPROCAL MIXING DYNAMIC RANGE (RMDR)

The RMDR test measures how well a receiver can cope with a single large signal just outside the receiver's passband. I call it a "Ham next door test" because it demonstrates how well your receiver works when the ham down the street transmits near your operating frequency. In a conventional receiver, reciprocal mixing noise is caused when noise from the LO (local oscillator) mixes with strong adjacent signals, generating noise and distortion at the output of the mixer. The generated noise can degrade a receiver's sensitivity and it is most noticeable when the offending signal is close to your receiver frequency, which is why RMDR is usually measured at a close offset of 2 kHz and also at a wider offset of 20 kHz.

In a direct sampling SDR, reciprocal mixing noise is caused when phase noise from the analogue to digital converter mixes with the offending signal. So, in a software defined radio, the RMDR measurement is an indicator of the spectral purity of the ADC clock. The RMDR is usually completely independent of the offset from the receiver frequency. It is normally the same at the 2 kHz and 20 kHz offset.

## NOISE POWER RATIO (NPR)

Noise Power Ratio, sometimes called 'white noise testing' can be measured in either type of receiver. It is a very good test of SDRs because they are wideband receivers. The idea of the NPR test is to load up the receiver with lots of signals instead of just the two test frequencies used in the two tone IMD test. This is simulated by connecting a 4 or 8 MHz wide 'white' noise source to the receiver input. You increase the noise level until the signal is slightly under the clipping level. Then you cut a slot in the noise exactly 3 kHz wide using a very good band stop filter and measure the amount of noise caused by intermodulation products generated in the receiver that fall into the narrow quiet spot.

Typically, you use a 2.4 kHz wide receiver bandwidth in the bottom of the slot since this is a typical bandwidth for SSB and it fits neatly inside the 3 kHz slot. The NPR is the ratio of the noise power inside the slot to the level of noise power in a channel with the same bandwidth outside the slot. Since the SDR receiver has a spectrum display, the result can be read straight off the screen. A 4 MHz wide noise signal, applied to the receiver input at just under clipping level is equivalent to more than 1200 SSB signals at S9 +30 dB. This is the ultimate test of how well the SDR will perform in a busy band.

## RECEIVER DYNAMIC RANGE

The 'receiver dynamic range measurement,' tests a receiver's ability to handle a range of received signal strengths from very weak signals near the receiver's internal noise floor to very strong signals like S9 +60 dB. The number of bits the ADC uses and the decimation ratio both affect the dynamic range of SDR receivers. Receiving very large signals can shrink the dynamic range due to ADC gain compression. The dynamic range of software defined radios is quite large compared to conventional receivers. You can expect about 120 dB of dynamic range in a 2.4 kHz bandwidth. Conventional superheterodyne receivers use AGC to extend their dynamic range, but this is not necessary with software defined radios. The AGC in SDRs is used to limit the receiver's audio output level rather than managing the receiver's gain.

There is a great deal of misinformation on the Internet about the dynamic range of SDR receivers. It is due to a lack of understanding about the process gain arising from the ratio between the ADC sampling bandwidth and the actual receiver bandwidth. Superheterodyne receivers have process gain too, but it is less because the ratio between the RF bandwidth and the I.F. bandwidth is much smaller than the ratio of ADC bandwidth to receiver bandwidth in an SDR.

An SDR receiver's dynamic range is the difference between the MDS which is the weakest signal that the radio can 'hear' and the clipping level which is the largest signal that the ADC can sample. We normally back off the maximum input level to 1 dB below the clipping point. So the usable dynamic range is from the MDS level up to just under the clipping level.

One downside of SDR receivers is that the dynamic range of the ADC is affected by the sum power of all of the signals in the filtered range of frequencies that are being presented to it. If the ADC is sampling very large signals, the dynamic range shrinks. This can be demonstrated using the Noise Power test set which simulates a very high loading of signals. In a test performed by Adam Farson, a single tone injected into a FLEX-6700 demonstrated a clipping level in excess of +13 dBm, but a worst-case white noise signal 8 MHz wide from the NPR test set reduced the clipping level to -2 dBm. This represents a 15 dB decrease in dynamic range, so it is lucky the receiver has a very large dynamic range, to begin with. Of course, this level of reduction could never happen in the real world because the loading could never approach the white noise level. Eight-bit SDRs, which have a limited dynamic range and don't have filters to reject the AM broadcast band, definitely suffer degraded dynamic range due to this effect. This is one reason why the RTL dongles and many of the experimenter's boards are not designed to work below 24 MHz.

## BLOCKING DYNAMIC RANGE (BDR)

In the Blocking Dynamic Range test, a wanted signal on the receiver frequency and an unwanted signal at either a 2 kHz or 20 kHz offset are input to the receiver. The unwanted signal is increased until the wanted signal measured at the receiver output decreases in level by 1 dB. The input level at which this occurs is the blocking level. The blocking dynamic range is the difference between the blocking level and the noise floor or MDS level. Some labs refer to this test as 'blocking gain compression, dynamic range.'

SDRs don't normally suffer from this blocking de-sensitivity. The 'unwanted' test signal normally reaches the clipping level with no effect on the wanted signal. So although BDR can be measured for a software defined radio it may not mean much since you are just recording the ADC clipping level.

BDR is commonly measured differently, using only the unwanted offset signal with no signal present on the receiver frequency. In this case, the BDR is stated as the input level when the noise floor in a 2.4 kHz bandwidth at the wanted frequency rises by 1 dB. Instead of measuring a decrease in the wanted signal level, you measure an increase in noise within the 2.4 kHz bandwidth. The assumption is that a 1 dB increase in the noise level is equivalent to a 1 dB decrease in the signal to noise level of a wanted signal if it was present. Blocking dynamic range tested this way can be used as a measurement of the ADC dynamic range compression noted above and the method is easier to perform since you only need one test tone.

## IMAGE REJECTION

In a conventional superheterodyne receiver, the image signal is related to the first I.F. frequency. If the local oscillator frequency is greater than the receiver frequency, the image frequency is the receiver frequency plus twice the intermediate frequency.

For example, if the receiver is tuned to 14.020 MHz and the receiver uses a 9 MHz I.F. then a test signal will be inserted at the image frequency of 32.020 MHz. It is increased until a signal is just noticed on the receiver frequency i.e. at the MDS level. The image rejection is the difference in dB between the input level of the image signal and the receiver's MDS level.

If the local oscillator frequency is less than the receiver frequency, the image frequency is the receiver frequency minus twice the intermediate frequency.

Image rejection is not very relevant to direct sampling HF software defined radios because they don't have hardware mixers. Hybrid SDR receivers do have at least one down-conversion mixer and oscillator stage, so you can measure their image rejection.

Direct sampling SDRs are direct conversion receivers so the I.F. frequency is zero. This means that the image frequencies are spaced at the same offset from the local oscillator and panadapter centre as the inserted signal. For example, if the panadapter is centred on 28.3 MHz and you insert a signal at 27.3 MHz (Fo – 1 MHz) the image will show on the panadapter at 29.3 MHz (Fo + 1 MHz). The image rejection can be read directly off the panadapter. It is the difference in dB between the level of the inserted signal and the level of the image signal at an equal frequency offset on the other side of the panadapter. If the IQ balance and phase relationship is good, the image signal will be small. Hopefully, you will not be able to see the image frequencies at all.

QSD receivers don't usually have perfect image cancellation across the whole panadapter. Especially if one audio channel is at a slightly different level. Some QSD receivers use an offset frequency of around 12 kHz to shift the whole panadapter. This is effectively creating a non-zero I.F. frequency. You still get images but they are no longer equidistant on the other side of the centre of the panadapter. They get the same offset shift as the wanted signal.

## ALIAS SIGNAL REJECTION

A better 'rejection' test for digital down conversion SDRs is to check the filtering used to prevent 'aliased' signals appearing in the receiver passband. The biggest problem for HF receivers is usually interference from FM broadcast stations transmitting on frequencies in the 2$^{nd}$ Nyquist zone.

For the second Nyquist zone, a signal that is 1 MHz above the Fs/2 frequency will appear on the panadapter at 1 MHz below Fs/2. A signal that is 5 MHz above the Fs/2 frequency will appear on the panadapter at 5 MHz below Fs/2. The order of the reflected frequencies is backward and the sidebands are inverted. You can check spot frequencies but I have proposed a sweep test that can draw the alias response onto your panadapter.

Aliasing is not a fault. It is integral to the way an SDR works. However inadequate filtering can cause severe interference from large signals above the Nyquist Fs/2 frequency.

One radio that I tested suffered very bad interference from FM broadcast stations. It used a 160 Msps sample rate, so the FM broadcast band was close to the bottom of the 2nd Nyquist zone. The choice of ADC sampling frequency is important and so is the cut-off frequency of the anti-alias filter. Some manufacturers fail to realize that the sharpness of the alias filter roll-off can be critical to the performance of the radio.

If the alias filter is at the same frequency as the Fs/2 frequency, there is almost no alias rejection of signals just above Fs/2. The rejection gets better the further above Fs/2 that you go. The amount of alias rejection is directly proportional to the attenuation provided by the filter.

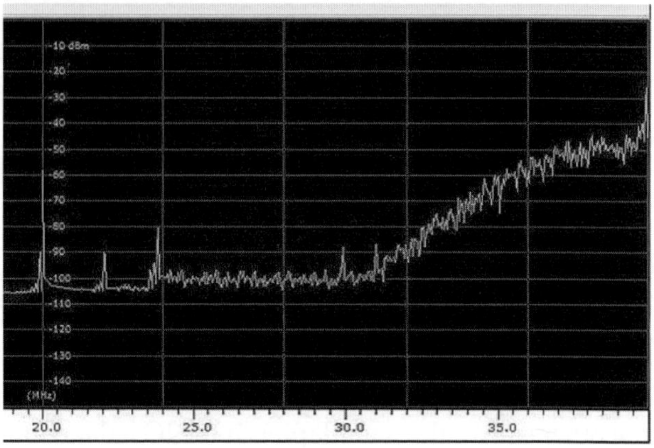


If you apply a strong sweep signal from Fs/2 up to Fs/2 plus your panadapter bandwidth, you will see a line of signal curve down from a maximum at the top right of the panadapter. The curve that is drawn is an exact representation of the roll off shape of the anti-alias filter and any other front-end filters.

This sweep result was performed on an SDR that has a sample rate of 80 Msps therefore an Fs/2 of 40 MHz. Unfortunately in an effort to maximize the receiver's frequency range the anti-alias filter is also at 40 MHz. As you can see, there is little protection from signals between 40 and 50 MHz being aliased into the wanted passband. The input sweep was from 40 to 58 MHz.

The effect of alias signals can be almost completely eliminated if the filter frequency is well below the Fs/2 frequency because any interfering alias signals will be well attenuated. Hopefully to a level below the noise floor. The Apache Labs ANAN radios have an Fs/2 of 61.44 MHz and a filter cutoff at 55 MHz. This limits the receiver to working below 55 MHz, but it also eliminates any alias image problems. I was not able to see any alias signals above the noise floor. You will not see a sweep result like this one.

## MORE READING

You may be interested in my "Testing 123" book if you want to know more about radio performance testing and how you can do these tests at home using cheap test equipment.

# CATALOGUE OF SOFTWARE DEFINED RADIOS

This chapter is a sample of some popular software defined receivers and transceivers used by listeners, hackers, experimenters, and amateur radio operators. It doesn't include very expensive, commercial, or specialist radios used by military and commercial users. It is not possible to list all of the software defined radios that are available. New models are appearing all the time and there are many variations of the same technology, particularly with small box receivers. The radios are listed approximately in order of the type of radio; 100 watt transceivers, 100 watt hybrids, QRP transceivers, and receivers including dongles, small box receivers, experimenter's cards, and a few QSD radios.

*The information provided is as accurate as my online research allows and I apologize in advance if any errors have been made. Placement, inclusion, or exclusion from this list in no way indicates any endorsement or criticism of any of the products mentioned or omitted. The listing is not ranked in any preferential order.*

If you are buying an SDR radio, it is worth considering a few things:

- Make sure the radio suits your needs, particularly regarding the frequency range that it covers.
- If you buy a QRP transceiver but you want to run higher power, you need to factor in the additional cost of a linear amplifier. It is often cheaper to buy a 100 watt radio.
- Don't expect a $30 RTL dongle to perform as well as a high end conventional or software defined radio. They are cheap and fun to play with, but they will not replace a real radio.
- SoftRock, dongle, or small box receivers can be used as bandscopes when connected to the IF output of a conventional receiver.
- In most cases, direct sampling HF receivers that use a 14 bit ADC will perform just as well as those using a 16 bit ADC. Although sometimes they cover less spectrum as they use a lower ADC sample rate.
- Receivers with band specific front-end bandpass filters or pre-selectors offer more protection from overload and ADC gain compression.
- Receivers with a maximum panadapter bandwidth of 192 kHz or less are almost always QSD based receivers. Receivers with multiple, or wider, panadapters usually use direct digital sampling or hybrid designs.
- A 192 kHz wide panadapter is more than adequate for use on the HF amateur radio band and listening to shortwave stations. At VHF and UHF frequencies where the bands are bigger and wanted channels are further apart, wider panadapters are better.

## 100 WATT DIRECT SAMPLING SDR TRANSCEIVERS

### FLEX-6600 / FLEX-6600M (FlexRadio)

The FLEX-6600 and FLEX-6600M ceased production in May 2024. They have been replaced with the FLEX-8600 and FLEX-8600M. The M model has the Maestro front panel knobs and a very high resolution 1920x1200, 8″ touchscreen display. The front panel has an identical layout to the Maestro external controller.

- Frequency range: 30 kHz – 54 MHz
- Bands: LF, MF, HF +6m
- Two ADCs, diversity reception. 16 Bits at 245.76 Msps
- Frequency stability: ±0.5ppm TCXO, GPSDO option.
- Four slice receivers on up to four 14 MHz panadapters
- Class: direct sampling 100 W transceiver
- Two models. One with no knobs, one with Maestro front panel
- SDR type: SDR with embedded computer, dedicated DSP chip
- Software architecture: client/server
- Interface: Ethernet 100 or 1000 Mbit
- Receive filters: 7$^{th}$ order on contest bands, 3$^{rd}$ order other bands
- Transmitter power: 100 W HF +6m
- Low latency CW generated in the radio FPGA firmware
- Built-in antenna tuner
- DAX audio connectivity: 4x audio plus 4x IQ
- Software: SmartSDR (supplied), SmartSDR for Mac, SmartSDR for iOS, dogpark SDR (for Mac).
- M model or external Maestro
    - 8″ touchscreen, 1920x1200 pixels
    - Can display two 14 MHz panadapters
    - HDMI 1920x1200 video output
    - Spectrum dynamic range: >155 dB
    - Spectrum RBW (min): 1.6 Hz/px
- SO2R operation with one radio
- Dimensions (Imperial) 6.74″ H X 14″ W X 13.25″ D
- Dimensions (Metric) 17.1 H x 35.6 W x 33.7 D cm

*TIP: you may have noticed that the ADC sample rate for the FLEX-6600 and FLEX8600 transceiver is 245.76 Msps. Twice the sample rate of the FLEX-6400 and FLEX8400. This 'oversampling' averages the quantisation errors reducing the quantisation noise. Less noise improves the dynamic range of the receiver.*

## FLEX-8600 / FLEX-8600M (FlexRadio)

The FLEX-8600 and FLEX-8600M have the same specifications listed above. The radios look identical to the 6000 series, other than the model number. There are several key differences internally.

The FPGA is much larger in the new radios which will allow FlexRadio the capacity to develop new features for the radios. The microprocessor is also much faster. New features include a built-in GNSS (GPS) receiver for very accurate frequencies. There is still the option to install a GPSDO (GPS disciplined oscillator) if you want super accuracy. You would only need that if you were using a microwave frequency transverter. The front-end preselector filters are now solid state switched. A sampler has been installed to allow for adaptive transmitter pre-distortion. I assume that this feature will be added to SmartSDR. There will also be improved spectrum display features and a new option to display a wideband view of several bands at once. Future developments include the ability to transmit and receive modes such as FT8 over a 40 kHz range without moving the VFO frequency. DX stations will be able to transmit multiple FT8 signals at the same time.

## FLEX-6400 / FLEX-6400M (FlexRadio)

The FLEX-6600 and FLEX-6600M ceased production in May 2024. They have been replaced with the FLEX-8600 and FLEX-8600M. The M model has the Maestro front panel knobs and a very high resolution 1920x1200, 8″ touchscreen display. The front panel has an identical layout to the Maestro external controller.

- Frequency range: 30 kHz – 54 MHz
- Bands: LF, MF, HF +6m
- Single ADC: 16 Bits at 122.88 Msps
- Frequency stability: ±0.5ppm TCXO, GPSDO option.
- Two slice receivers on up to two 7 MHz panadapters
- Class: direct sampling 100 W transceiver
- Two models. One with no knobs, one with Maestro front panel
- SDR type: SDR with embedded computer, dedicated DSP chip
- Software architecture: client/server
- Interface: Ethernet 100 or 1000 Mbit
- Receive filters: 3rd order on all bands
- Transmitter power: 100 W HF +6m
- Low latency CW generated in the radio FPGA firmware
- DAX audio connectivity: 2x audio plus 2x IQ
- Software: SmartSDR (supplied), SmartSDR for Mac, SmartSDR for iOS, dogpark SDR (for Mac).
- Antenna tuner option

- M model or external Maestro
    - 8" touchscreen, 1920x1200 pixels
    - Can display two 7 MHz panadapters
    - HDMI 1920x1200 video output
    - Spectrum dynamic range: >145 dB
    - Spectrum RBW (min): 5.85 Hz/px
- Dimensions (Imperial) 6.74" H X 14" W X 13.25" D
- Dimensions (Metric) 17.1 H x 35.6 W x 33.7 D cm

### FLEX-8400 / FLEX-8400M (FlexRadio)

The FLEX-8400 and FLEX-8400M have the same specifications as the FLEX6400 and FLEX-6400M listed above. Other than the model number, the radios look identical to the 6000 series. They have the same improvements as the FLEX-8600 models. These include a bigger FPGA, faster microprocessor, a coupler for adaptive transmitter pre-distortion, GPS frequency control, and solid-state pre-selector switching.

### FLEX-6700 (FlexRadio)

The FLEX-6700 transceiver is 'out-of-stock' and only available as a special order. It is very expensive and I expect that in spite of having slightly better specifications that it has been superseded by the new 8000 series radios.

- Frequency range: 30 kHz – 77 MHz and 135 – 165 MHz
- Bands: LF, MF, HF +6m, and 2m at a very low power
- Two receiver ADCs (called Spectral Capture Units), diversity reception
- ADC resolution: 16 Bits, 245.76 Msps,
- Class: direct sampling 100 W transceiver
- Models: one model, no knobs
- SDR type: SDR with embedded computer, dedicated DSP chip
- Software architecture: client/server
- Interface: Ethernet 100 or 1000 Mbit
- Receive filters / pre-selector: yes (except 60m band)
- Transmitter power: 100 W HF +6m (+7 dBm transverter output on 2m)
- Low latency CW generated in the radio FPGA firmware
- Built-in antenna tuner, eight-channel DAX audio connectivity
- Eight slice receivers on up to eight 14 MHz panadapters
- Software: SmartSDR (supplied), SmartSDR for Mac, SmartSDR for iOS, dogpark SDR (for Mac).

## K4, K4D, K4HD (Elecraft)

The K4 series are direct sampling 100 watt HF transceivers, with impressive connectivity and very good specifications. There are two models with a third model in development.

The K4 has a single ADC. The K4D has two ADCs with two completely independent receivers each having its own set of bandpass filters. It is capable of diversity reception with two antennas.

The K4HD is not yet available. It has all the features of the K4D but also includes a 'dual superhet receive section' making the radio an optional hybrid with an 8 MHz I.F. before the SDR stage. I can't see very much advantage in this idea other than perhaps improving the receiver specification a little.

- Frequency range: 100 kHz – 54 MHz
- Bands: MF, HF +6m
- K4 - Single 16-bit ADC. Wideband 'dual watch.'
- K4D - Two 16-bit ADCs, two receivers, diversity reception
- K4HD - Two 16-bit ADCs, diversity reception. Plus two hybrid-style superheterodyne modules with an 8 MHz I.F.
- Frequency stability: ±0.3 ppm TCXO
- Reference clock input
- Two panadapter/bandscopes
- Class: direct sampling 100 W transceiver
- Three models K4, K4D, and K4HD (not released yet)
- SDR type: SDR transceiver with dedicated DSP chip
- Transmitter power: 100 W HF +6m
- Supports controlled envelope single sideband compression
- Silent CW QSK using pin diode switching
- Antenna tuner option: KAT4 ATU 10:1 range with three antenna jacks.
- Text modes: CW, PSK31, PSK63, RTTY decoders and keyers
- Power supply: DC 11-15 volts
- Software: No external software. Can be remotely controlled.
- User manual available on-screen
- Display: 7" touchscreen
- HDMI video output
- Dimensions (Imperial) 5.2" x 14.0" x 12.5"
- Dimensions (Metric) 13.2 x 35.5 x 31.7 cm
- A microphone is not included

### ANAN-G2 and ANAN-G2 'Ultra' (Apache Labs)

The ANAN-G2 uses the Saturn SDR board, designed by the OpenHPSDR group (high level design, firmware & software) and Apache Labs (pcb and hardware). It is marketed as, "the new generation ultra high performance SDR ultra high performance SDR." The receiver performance specifications are very good with an RMDR of 116 dB at a 2 kHz offset, and image rejection of 90 dB. Using Pure Signal transmitter pre-distortion the transmitter IMD performance is amazing. The G2 comes in three formats.

The G2 Ultra version has the same Saturn SDR board and specifications as the older G2. It has a larger 8" touchscreen and more controls on the front panel. The front panel controls are managed by an Arduino, taking some of the load off the Raspberry Pi compute module. The front panel runs piHPSDR on the embedded Raspberry Pi4 compute module. You can use an external keyboard, mouse and monitor if you want to. Or you can use Thetis running on your PC. The controls on the radio can control Thetis while it is running on your PC.

There is a black box, no knobs, option that can be used with piHPSDR using the embedded Raspberry Pi4 compute module in the radio with and external keyboard, mouse and monitor. You can also use Thetis running on your PC.

There is a black box, no knobs, version that includes a 1000 watt power amplifier. Making it possibly the first 1000 watt SDR transceiver. This unit has a screen but it is only used to control the linear power amplifier. The company is currently not taking orders for this model due to a backlog.

- Frequency range: 9 kHz – 60 MHz
- Bands: LF, MF, HF +6m
- Two 16-bit ADCs at 122.88 Msps, diversity reception
- Frequency stability: ±0.1ppm TCXO
- Clock phase noise -149 dB at 10 kHz offset
- Reference clock input
- Two panadapters and two slice receivers on piHPSDR. Two panadapters and three slice receivers on Thetis. The hardware is capable of ten panadapters. Panadapter bandwidth up to 1.536 MHz depending on computer power and interface.
- Class: direct sampling 100 W transceiver
- Four models G2, G2 black front, G2-1000, G2 Ultra (not released yet)
- SDR type: SDR with embedded computer. DSP in FPGA and software
- Transmitter power: 100 W HF +6m
- Supports controlled-envelope single sideband CESSB
- Low latency CW generated in the radio FPGA firmware

- Pure Signal transmit pre-distortion – best transmit IMD in class
- Power supply: DC 13.8 volts
- Software: piHPSDR (internal or external), Thetis, SDR Console V3.
- Display: 7″ touchscreen on G2 model, 8″ on G2 Ultra MKII
- 2x HDMI 4K video output
- Dimensions (Metric) 314mm (W) x 138mm (H) x 368mm (D) (Not including extrusions).
- Microphone is not included

https://www.youtube.com/watch?v=su8GbKmX28A

https://www.youtube.com/watch?v=-ZNW3m_DmA4&t=441s

Primary members of the Saturn development team:

Laurence Barker – concept, FPGA design, test software

Rick Koch – Software development

Abhi Prakash – Schematic, PCB design manufacture, and sales.

Figure 53: Apache Labs ANAN G2 Ultra

## MB1 (Expert Electronics)

The MB1 is a direct digital sampling HF/VHF SDR transceiver with an Intel Core i7 based PC inside the radio box. The computer runs on either Linux or Windows. It covers the 2m VHF band or the HF band but not both at the same time. The radio supports two panadapters on the same or different bands, with two receivers on each panadapter. Four receivers in total. Output power is 100 watts on HF and 50 watts on the 6m and 2m bands. The radio has an impressive list of specifications, with a quoted RMDR of 110 dB, and BDR on HF of 130 dB. The main oscillator has a 0.5 ppm accuracy.

The ExpertSDR2 software runs on the built-in touchscreen, or an external Windows, macOS, or Linux computer.

*Note: Expert Electronics is a Russian company. Due to the current political situation, it seems that this radio is currently impossible to buy outside of Russia. NSI Communications in the USA reports no stock. The European dealers Pileup and WiMo say that the radio is no longer available.*

### IC-7300 (Icom)

The IC-7300 was the first direct digital sampling HF transceiver from one of, "the big three," (Icom, Yaesu, & Kenwood). It was truly groundbreaking providing many features that had been previously unavailable, or only available on the most expensive radios. I believe that it is currently the best selling HF amateur radio in the world. Icom recently announced they have sold 100,000 units.

Firmware revision v1.40 added the FT8 (and other digital mode) presets, and spectrum scrolling modes.

- Frequency range: 30 kHz – 70.5 MHz
- Bands: HF +6m
- One ADC, probably 16-bit
- Digital down conversion to 36 ksps in FPGA
- Frequency stability: ±0.5ppm TCXO
- One bandscope (panadapter) with one receiver
- Split uses the second VFO frequency. Use A/B to hear the TX frequency
- Maximum panadapter bandwidth 1 MHz. Bandscope span ±500 kHz
- Class: direct sampling 100 W transceiver
- SDR type: SDR with dedicated Texas Instruments TMS320C6745 DSP chip
- Transmitter power: 100 W HF +6m
- Antenna tuner 3:1 (16.7 Ohms to 150 Ohms)
- Power supply: DC 13.8 volts ±15%
- Software: no external SDR software
- Display: 4.3" touchscreen
- No video output
- Dimensions (Metric) 240(W) × 94(H) × 238(D) mm
- Dimensions (Imperial) 9.4"(W) × 3.7"(H) × 9.4"(D)
- A microphone is included
- Voice, CW and RTTY keyer, RTTY decoder
- Recording mode which lets you record signals off air and optionally your transmissions as well.
- FT8 (and other digital modes) preset

### IC-7610 (Icom)

The IC-7610 was the second direct digital sampling HF transceiver from one of "the big three," (Icom, Yaesu, & Kenwood). It added a second receiver and other features including a bigger display, a PSK keyer and decoder, and 'Digisel' digital preselectors. It has similar controls and a similar menu structure to the IC-7300.

- Frequency range: 30 kHz – 54 MHz
- Bands: HF +6m
- Two ADCs, LTC2208 16-bits @ 130 Msps
- Digital down conversion to 12 ksps in FPGA
- Frequency stability: ±0.5ppm TCXO
- Two bandscopes (panadapters) with two completely independent receivers
- Maximum panadapter bandwidth 1 MHz. Bandscope span ±500 kHz
- Class: direct sampling 100 W transceiver
- SDR type: SDR with dedicated Texas Instruments TMS320C6745 DSP chip
- Transmitter power: 100 W HF +6m
- Antenna tuner 3:1 (16.7 Ohms to 150 Ohms)
- Power supply: DC 13.8 volts ±15%
- Software: no dedicated external SDR software. IQ out for HDSDR extio.dll.
- Display: 7" 800x480 touchscreen
- DVI-D video output
- Dimensions (Metric) 340(W) × 118(H) × 277(D) mm
- Dimensions (Imperial) 13.4"(W) × 4.67"(H) × 10.9"(D)
- A microphone is included
- Voice, CW, PSK and RTTY keyer, PSK and RTTY decoder
- APF (audio peak filter for CW), TPF (twin peak filter for RTTY)
- 'Digisel' digital preselectors on both receivers to improve receiver selectivity

### FT-710 (Yaesu)

The FT-710 is Yaesu's first direct sampling HF transceiver. The received radio spectrum is sampled by two ADCs located after the selectable attenuator, bandpass filters, and RF preamplifiers. An FPGA (field programmable gate array) carries out filtering and frequency translation before the signal is passed to a DSP (digital signal processing) stage. The transmitter is also an SDR design. It is interesting that the FTdx10 hybrid transceiver is currently number three on the Sherwood list and the FT-710 is at number four, with the same narrow spaced two tone receiver IMD result. Proving that direct sampling transceivers can compete head-to-head with conventional and hybrid HF transceivers.

The quoted performance figures are excellent. Especially for an 'entry level' radio.

RMDR 113 dB at 2 kHz offset, BDR 127 dB, IMDR 102 dB, TX phase noise -143 dBc/Hz at a 2 kHz offset.

The use of a dual channel 14-bit ADCs for a single receiver is unusual. The Yaesu brochure states that "the digital signal is combined in the FPGA to reduce overload and overflow of the A/D converters and improve blocking characteristics." The TI ADS62P45 dual channel ADC contains a DSP section which can perform offset and gain correction between the two inputs, and filtering and decimation by 2, 4, or 8. I don't know if this is used in the FT-710.

- Frequency range: 30 kHz – 75 MHz (presumably using undersampling for the frequencies above 54 MHz)
- Bands: HF +6m, plus 70-70.5 MHz in the U.K.
- ADS62P45 dual channel ADC 14-bits @ 125 Msps
- Digital down conversion to 24 ksps in FPGA
- Frequency stability: ±0.5ppm TCXO
- One bandscope (panadapter) with one receiver
- Maximum panadapter bandwidth 1 MHz. Bandscope span ±500 kHz
- Class: direct sampling 100 W transceiver
- SDR type: SDR with dedicated NXP DSP chip
- Transmitter power: 100 W HF +6m +4m
- Antenna tuner 3:1 (16.7 Ohms to 150 Ohms)
- Power supply: DC 13.8 volts ±15%
- Software: no dedicated external SDR software.
- Display: 4.3" touchscreen
- DVI-D video output
- Dimensions (Metric) 239(W) × 80(H) × 247(D) mm
- Dimensions (Imperial) 9.4"(W) × 3.1"(H) × 9.7"(D)
- A microphone is included
- Voice and CW keyers, no decoders
- APF (audio peak filter for CW)
- Some packages include the 'AESS' acoustic enhanced speaker system audio

## 100 WATT HYBRID SDR TRANSCEIVERS

### FTdx101D and FTdx101MP (Yaesu)

The FTdx101D has superb technical specifications. It currently holds the top position on the highly regarded Sherwood Engineering transceiver performance table. The quoted performance figures are excellent. RMDR 123 dB at 2 kHz offset, BDR 150 dB, IMDR 110 dB, TX phase noise -150 dBc/Hz at 2 kHz offset.

The two radios share the same architecture as the FTdx10 except they are much larger, with two independent receivers. In accordance with previous models the D model has a 100 watts output. The MP variant has 200 watts of transmitter power and an external mains power supply and speaker. The D model has a 'VC Tune' variable preselector on the main receiver, the MP has one on each receiver. It also has a 300 Hz CW filter which is an option on the FTdx101D. They both have a 7" touchscreen. Most of the menu functions and features are the same as the FTdx10.

The internal layout especially the front-end of the receiver is 'very similar' to the FTdx5000. The main difference is that the 2nd I.F. stage has been replaced by an ADC and FPGA. A second ADC, sampling before the first mixer is used to create the 1 MHz bandscope display.

- Frequency range: 30 kHz – 70 MHz
- Bands: HF +6m, plus 70-70.5 MHz in the U.K.
- Two ADCs
- Digital down conversion to 24 ksps in FPGA
- Frequency stability: ±0.1ppm TCXO
- Two bandscopes (panadapters) with two completely independent receivers
- Maximum panadapter bandwidth 1 MHz. Bandscope span ±500 kHz
- Class: Hybrid 100 W transceiver
- Receiver type: Superheterodyne conversion to a 9 MHz (approx.) I.F., ADC, and FPGA as the 2nd I.F. With digital down conversion to 24 kHz for the dedicated Texas Instruments TMS320C6746 DSP chip
- Transmitter power: D version 100 W, MP version 200 W
- Antenna tuner 3:1 (16.7 Ohms to 150 Ohms)
- Power supply: D version DC 13.8 volts ±15%, MP mains power
- Software: no dedicated external SDR software.
- Display: 7" 800x480 touchscreen
- DVI-D video output
- Dimensions (Metric) 420(W) × 130(H) × 322(D) mm
- Dimensions (Imperial) 16.6"(W) × 5.17"(H) × 12.7"(D)
- A microphone is included
- Voice, CW, PSK, and RTTY keyers and decoders
- APF (audio peak filter for CW)
- 'VC-Tune' analogue preselectors to improve receiver selectivity. Main RX only on the D version
- 3DSS '3d spectrum display' presents a history of the spectrum as a 3d tapering image. More of a gimmick than a useful tool, as the waterfall also presents a history of the received spectrum

### FTdx10 (Yaesu)

The FTDX10 has similar styling to the FTDX101, but it is a lot smaller. Under the skin, the radio employs the same hybrid SDR architecture as the FTDX101. The 2nd I.F. stage has been replaced by an ADC and FPGA. A second ADC sampling the RF signals before the first mixer is used to create the 1 MHz bandscope display.

The quoted performance figures are excellent, but not as good as the FTdx101. RMDR 116 dB at 2 kHz offset, BDR 141 dB, IMDR 109 dB, TX phase noise -145 dBc/Hz at 2 kHz offset.

- Frequency range: 30 kHz – 70 MHz
- Bands: HF +6m, plus 70-70.5 MHz in the U.K.
- One ADC, one receiver
- Digital down conversion to 24 ksps in FPGA
- Frequency stability: ±0.5ppm TCXO
- One bandscope (panadapter) and one receiver
- Maximum panadapter bandwidth 1 MHz. Bandscope span ±500 kHz
- Power supply: DC 13.8 volts ±15%
- Software: no dedicated external SDR software
- Receiver type: Superheterodyne conversion to a 9.005 MHz I.F., ADC, and FPGA as the 2nd I.F. With digital down conversion to 24 kHz for the dedicated Texas Instruments TMS320C6746 DSP chip
- Class: Hybrid 100 W transceiver
- Transmitter power: 100 W on all bands
- Antenna tuner 3:1 (16.7 Ohms to 150 Ohms)
- Display: 5″ 800x480 touchscreen
- DVI-D video output
- Dimensions (Metric) 266(W) × 91(H) × 263(D) mm
- Dimensions (Imperial) 10.5″(W) × 3.6″(H) × 10.4″(D)
- A microphone is included
- Voice, CW, PSK, and RTTY keyers and decoders
- APF (audio peak filter for CW)
- 3DSS '3d spectrum display' presents a history of the spectrum as a 3d tapering image. More of a gimmick than a useful tool, as the waterfall also presents a history of the received spectrum
- FT8 (and other digital modes) preset

### IC-9700 (Icom)

The IC-9700 is the first full-power, multi-mode, VHF/UHF amateur radio transceiver to be based on SDR technology. It is also the first VHF and UHF transceiver to feature a panadapter spectrum and waterfall display.

The receiver uses direct digital sampling for the 2 m and 70 cm bands and a single down-conversion stage followed by direct digital sampling for the 23 cm receiver.

The transmitter uses direct digital up-conversion for the 2 m and 70 cm bands and a heterodyne up-conversion stage for 23 cm.

The radio has two receivers, so you can receive signals on two bands at the same time, and it is capable of full-duplex crossband operation, being able to receive signals on one band while transmitting on another band. There is provision for the connection of a 3rd party GPS receiver for use with the D-Star and digital voice gateway modes and connection of an external 10 MHz frequency reference for superior oscillator stability.

- Frequency range:
    - 144 MHz – 148 MHz (146 MHz in EU region)
    - 430 MHz – 450 MHz (440 MHz in EU region)
    - 1240 MHz – 1300 MHz (all regions)
- Bands: 2m, 70cm, 23cm. No wideband receiver
- Hybrid I.F. output frequency on 23cm band: 311-371 MHz
- Dual channel ADC, LTC2156 14-bits @ 196.608 Msps
- 2m is in the 2nd Nyquist zone, and 70cm is in the 4th Nyquist zone. The 23cm 311-371 MHz I.F. is in the 4th Nyquist zone.
- Frequency stability: ±0.5ppm TCXO. Can connect external 10 MHz reference
- Two bandscopes (panadapters) with two receivers (any two bands)
- Full duplex satellite mode
- Maximum panadapter bandwidth 1 MHz. Bandscope span ±500 kHz
- Class: direct sampling 100 W transceiver on 2m and 70cm, hybrid on 23cm
- SDR type: SDR with dedicated Texas Instruments TMS320C5517 DSP chip. There is a second DSP chip for D-Star digital voice
- Transmitter power: 100 W on 2m, 75 W on 70cm, 10 W on 23cm
- Power supply: DC 13.8 volts ±15%
- Software: no dedicated external SDR software
- Display: 4.3″ 800x480 touchscreen. No video output
- Same form factor as the IC-7300
- Dimensions (Metric) 240(W) × 94(H) × 238(D) mm
- Dimensions (Imperial) 9.4″(W) × 3.7″(H) × 9.4″(D)

- A microphone is included
- Voice, CW, PSK and RTTY keyer, RTTY decoder
- TPF (twin peak filter for RTTY)
- D-Star

## QRP (LOW POWER) SDR TRANSCEIVERS

QRP (low power) transceivers can be loads of fun. However, my advice is to only buy a QRP radio if you want to operate QRP. Be aware that the cost of buying one of these radios and adding a 100 wat linear amplifier later, is likely to be more expensive and less versatile than simply buying a 100 watt radio.

### IC-705 (Icom)

The IC-705 is a 5-watt (on battery) or 10-watt (on DC) 'QRP' radio with the HF bands, the 6m and 2m VHF bands, and the 70cm UHF band. This radio takes ham radio back into the wilderness, urging you to "be active in the field." I am sure this radio will become very popular for those of us who enjoy the great outdoors. You may be activating mountain tops for SOTA, camping near a stream, hiking in a forest, operating as a cycle or pedestrian mobile, or just walking the dog.

- Frequency range: 30 kHz to 200 MHz and 400 MHz to 470 MHz.
- Bands: HF, 6m, 2m, and 70cm
- One ADC
- Frequency stability: ±0.5ppm TCXO
- One bandscope (panadapter) with one receiver
- Maximum panadapter bandwidth 1 MHz. Bandscope span ±500 kHz
- Class: portable direct sampling QRP transceiver
- SDR type: SDR with dedicated DSP chip. There is a second DSP chip for D-Star digital voice.
- Transmitter power: 5 watts on battery power, 10 watts on external DC
- Power supply: Internal battery or external DC 13.8 volts ±15%
- Software: no dedicated external SDR software
- Display: 4.3" 800x480 touchscreen
- No video output
- Same form factor as the IC-905 head unit
- Dimensions (Metric) 200(W) × 83.5(H) × 82(D) mm
- Dimensions (Imperial) 7.9"(W) × 3.3"(H) × 3.2"(D)
- A microphone is included
- Voice, CW, and RTTY keyers, RTTY decoder
- TPF (twin peak filter for RTTY)

- D-Star
- Internal GPS receiver for location information

### IC-905 (Icom)

The IC-905 consists of a head unit that is similar to the IC-705, and a remote-mounted RF unit. This makes operating on the microwave bands possible, as the loss in traditional feeder cables would be excessive at frequencies above 2 GHz. The radio works on the amateur bands only. It covers the 2m, 70cm, 1.2 GHz, 2.4 GHz, and 5.6 GHz amateur radio bands. The optional CX-10G transverter module the 10 GHz band. Plans for a 24 GHz transverter have been announced. It provides full coverage of the entire band, unlike some transverters that can only work on a single frequency.

- Frequency range:
    - 144 MHz – 148 MHz (146 MHz in EU region)
    - 430 MHz – 450 MHz (440 MHz in EU region)
    - 1240 MHz – 1300 MHz (all regions)
    - 2390 MHz – 2450 MHz (2300 MHz – 2450 MHz EU region)
    - 5650 MHz – 5925 MHz (5650 MHz – 5850 MHz EU region)
    - 10000 MHz - 10500 MHz with CX-10G transverter
- Bands: 2m, 70cm, 23cm, 12cm, 5cm. No wideband receiver
- Hybrid I.F. output frequency on 23cm band: 311-371 MHz. Double conversion I.F. on 12 and 5cm band: 1st 915 MHz band, 2nd 346 MHz band.
- One ADC, LTC2156 14-bits @ 196.608 Msps
- Frequency stability: Internal GPS disciplined oscillator(s) ±66 ppb. 10 MHz reference output for transverters.
- Maximum panadapter bandwidth 50 MHz. Bandscope span ±25 MHz. The usable display depends on the size of the amateur band.
- One bandscope (panadapter) with one receiver
- Class: direct sampling transceiver on 2m and 70cm, hybrid on the three higher bands
- SDR type: hybrid SDR with dedicated Analogue Devises SHARC DSP chip. There is a second DSP chip for D-Star digital voice.
- Transmitter power: 2m, 70cm, and 23cm 10 watts. 12cm and 5cm 2 watts
- Power supply: DC 13.8 volts ±15%
- Software: no dedicated external SDR software
- Display: 4.3" 800x480 touchscreen
- Video input and output for the video mode only, not a panadapter display
- The control head has the same form factor as the IC-705
- One Ethernet cable between the control head and the RF unit

- Dimensions Controller (Metric) 200(W) × 83.5(H) × 82(D) mm
- Dimensions Controller (Imperial) 7.9″(W) × 3.3″(H) × 3.2″(D)
- Dimensions RF unit (Metric) 172(W) × 87(H) × 210(D) mm, weight 3.2 kg
- Dimensions RF unit (Imperial) 6.8″(W) × 3.4″(H) × 8.3″(D), weight 7.1 lb
- A microphone is included
- Voice, CW, and RTTY keyer and an RTTY decoder
- TPF (twin peak filter for RTTY)
- D-Star and analogue video

### Hermes Lite 2 (Steve Haynal KF7O / Makerfabs)

The Hermes-Lite is a low-cost direct down and up conversion software-defined amateur radio HF transceiver based on a broadband modem chip and the Hermes SDR project. It is an entirely open source and open hardware. It is not a kit, but you do have to install the pcb into the case if you elect to buy a case. If you wish to transmit you should buy the optional N2ADR filter board. There are two options for I/O boards. The Hermes Lite 2 IO board supports connection to an external power amplifier. http://www.hermeslite.com/ and https://www.makerfabs.com/hermes-lite-2.html

The Hermes Lite 2+ board from https://www.hermeslite2plus.com/ provides jacks for microphone input, phones output (or amplified speakers), and a key. Plus connection to an external power amplifier.

Hermes Lite 2 will work with OpenHPSDR Compliant SDR software including, Thetis, piHPSDR, SDR Console v3, LinHPSDR, Quisk, SparkSDR, and PowerSDR. There is a fork of Thetis especially created for the HL2.

- Frequency range: 100 kHz to 38 MHz
- Bands: HF
- One ADC, one receiver
- Frequency stability: unknown
- Two panadapters, one receiver on each panadapter (Thetis)
- Maximum panadapter bandwidth 345 kHz. (Thetis sampling at 384 ksps). This radio uses OpenHPSDR Protocol 1 data transfer so both panadapters must use the same sample rate and have the same maximum bandwidth.
- Power supply: DC 11-16 volts
- Software: Usually Thetis, see the list above
- Receiver type: direct digital conversion SDR. Uses an Analogue Devices AD9866 broadband modem with an integrated 80 Msps 12-bit ADC. Most DSP is performed in the Cyclone IV FPGA.
- Class: direct sampling QRP 5 W HF transceiver

- Transmitter power: 5 watts
- No antenna tuner
- No display – black box SDR
- Dimensions (Metric) 105(W) × 40(H) × 150(D) mm
- A microphone is not included
- Open-source hardware, manufactured and sold by Makerfabs

**X6100 (Xiegu)**

There are a number of Chinese manufactured SDR transceivers available at a range of prices and configurations. I included this one in the list because I have owned a Xiegu X6100 and found it to be a good quality radio. Xiegu also sells the X6200 and several other models. Be careful when reading the specifications and shop around for a good deal.

My first impression was that the radio is well constructed and quite heavy at 880g including the built-in battery. It has an excellent 4" (100mm) display which is very crisp and bright, although sadly it is not a touchscreen. I was impressed that the radio supports filter bandwidth control, RIT (receiver incremental tuning), a noise blanker (NB), noise reduction (NR), and a digital notch filter (DNF). And it includes an antenna tuner and VOX (voice-operated PTT switch) operation. There is even a built-in microphone and PTT (press to talk) button in case you don't want to use the supplied microphone. It was a big surprise to discover that not only does the radio include a voice keyer and a CW message keyer. It also has built-in CW, PSK, and RTTY decoders just like the IC-7300. The CW messages can be used for PSK and RTTY when the decoders are active.

An STM-32 microprocessor takes care of the SDR (software defined radio) and DSP (digital signal processing) functions. The radio's control and display functions are performed on a separate Allwinner R16 ARM (advanced RISC machine) quad-core processor running Linux.

- Frequency range: 500 kHz to 54 MHz
- Bands: HF +6m
- Frequency stability: ±1.5ppm TCXO
- One bandscope (panadapter) with one receiver
- Maximum panadapter bandwidth 1 MHz. Bandscope span ±500 kHz
- Class: portable QSD/QSE QRP transceiver
- SDR type: QSD SDR with STM-32 microprocessor DSP and an Allwinner R16 ARM chip for control and display functions.
- Transmitter power: 5 watts on battery power, 10 watts on external DC
- Power supply: Internal battery or external DC 13.8 volts ±15%
- Software: no dedicated external SDR software. Works with HDSDR

- WiFi and Bluetooth
- Display: 4" (100mm). Not a touchscreen!
- No video output
- Dimensions (Metric) 180(W) × 86(H) × 49(D) mm, weight 880g
- A microphone is included
- Voice, CW, PSK, and RTTY keyers and a CW, PSK, and RTTY decoder

### mcHF (designed by Chris Atanassov, M0NKA)

The mcHF is a very interesting SDR transceiver project. It is available partly assembled or as a kit, which includes the bare pc boards and components, excluding the RF power transistors. All of the construction information and software is available online, for free. There is a design file for a '3D' printed case as well. The radio is a bit of an odd mix because it is a standalone transceiver with built-in DSP and a very nice 2.8 inch LCD screen display, which includes a spectrum scope panadapter. But the performance is limited by the processing power that is available.

The panadapter can only display a 48 kHz bandwidth. The mcHF transceiver is a generation three SDR with quite a sophisticated interface and display. Like many other generation three transceivers, the radio uses an SI570 programmable oscillator as its VFO.

- Frequency range: 3 MHz – 30 MHz
- Bands: HF
- ADC resolution: the Codec samples at audio frequencies
- Panadapter bandwidth: 48 kHz (on built-in display)
- Class: standalone QRP transceiver (QSD / QSE)
- Interface: USB for control & programming
- Receive filters / pre-selector: four bandpass filters and four low pass filters, used for both receive and transmit.

### Recent RS-918 / RS-928

A Russian company called Recent produced two slightly different copies of the mcHF transceiver. They are complete transceivers, not kits. They have similar specifications to the mcHF and can apparently use the same firmware files.

### Note

There seems to be many Chinese clones of the mcHF and Recent RS-918 and RS-928 radios from the usual online sources. I think that the Xiegu radios are a better bet since they have a much wider frequency coverage and a wider bandscope.

## KX3 (Elecraft)

The Elecraft KX3 is a standalone QSD/QSE QRP transceiver, available as a 'no solder kit' or completely assembled. I recommend buying the kit as it is very easy to assemble. All of the boards are complete and there is no soldering required. Nevertheless, it is immensely satisfying to know that you 'built it yourself.' It has one of the best receivers available in any amateur radio, at any price. Yes, it really is that good! Check out the QST review and benchmarking websites.

The KX3 has several options that can be fitted inside the radio; these include; an antenna tuner, 2m transverter, roofing filters (at digital audio), batteries, NiCad charger, and a CW paddle (mounted on the radio). There is also a companion 100 W linear amplifier that can include a 100 W antenna tuner and the Elecraft PX3 external panadapter.

The transceiver has some neat features, including CAT control, DSP filters (I.F. width and shift, or high and low cut), a voice keyer, a CW memory keyer, and a BPSK / RTTY / CW decoder. Weirdly you can use the CW paddle or an external key to send RTTY or BPSK. The radio converts Morse code sent via the paddle to the digital modes. This allows you to send and receive BPSK or RTTY from a 'mountaintop' or holiday location with just the radio and an antenna. The radio's front panel is quite similar to the Elecraft K3 transceiver, but the internal architecture is completely different.

The radio is a generation three SDR with a stereo ADC and DSP chip on board. There is an analogue audio IQ output for connection to external SDR 'soundcard' software.

The KX3 was designed for portable 'SOTA' type operation, so it is unfair to criticize its capabilities as a home station. But I wish it had a USB interface for the audio and control interface to the PC. It would mean that there would only be one cable instead of three and it would eliminate the earth loop, and hum problems that many people experience with the audio cables.

- Frequency range: 31.8 kHz – 50 MHz
- Bands: HF +6m, (+2m with optional internal transverter)
- ADC resolution: the ADC samples at audio frequencies
- Panadapter bandwidth: 48-192 kHz (depending on PC soundcard) 200 kHz on PX3 panadapter
- Class: standalone QRP 10 W transceiver (QSD / QSE)
- Onboard DSP and MCU
- Interface: audio and a serial to USB adapter for CAT control
- Receive filters / pre-selector: bandpass filters and low pass filters are used for both receive and transmit.
- SDR software: NaP3, HDSDR, Win4K3, SDR#, and others.

FDM-DUO (ELAD)

ELAD is an Italian company that makes rather stylish SDR receivers and one SDR transceiver. They are rather unusual because they include front panel controls and a small display but no panadapter spectrum and waterfall display. You can use the radios as a stand-alone radio, or connect them to a computer running the FDM-SW2 Software to display four simultaneous receivers in 'single channel mode' or eight simultaneous receivers in 'double channel mode.'

- Frequency range: 10 kHz – 54 MHz
- Bands 160m to 6m
- Direct sampling receiver operating @122.88MHz, ADC: LTC2165 16bit
- DDC (Digital Down Converter) made with FPGA Xilinx Spartan 6
- DSP using ARM STM32F4 microcontroller with floating point unit
- Demodulation: CW, CWR, LSB, USB, AM, FM narrow
- Transmitter power: 5 watts
- 10 MHz reference input
- CW decoder
- Dimensions (Metric) 180(W) × 70(H) × 155(D) mm, weight 1.2 kg
- Dimensions (Imperial) 7.0″(W) × 2.75″(H) × 6.1″(D), weight 2.4 lb
- IQ output bandwidth 192 ksps (stand-alone), Remote: 384ksps, 768ksps, 1536ksps, 3072ksps, 6144ksps in single channel mode and 384ksps in double channel mode.
- Optional 100 watt linear amplifier available

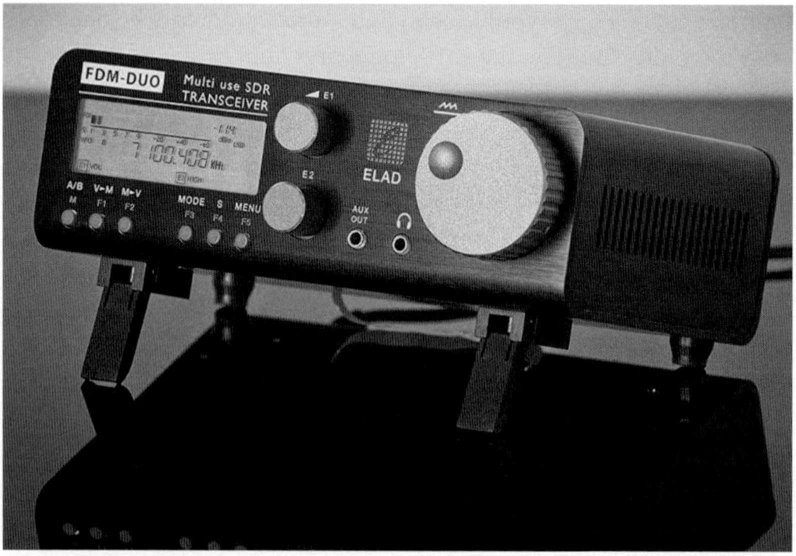


Figure 54: FDM-DUO QRP transceiver

## SUNSDR2 pro & DX (Expert Electronics)

There are three SunSDR2 'no knobs' models. The SunSDR2 Pro is QRP SDR transceiver covering HF and the VHF 6m and 2m bands. The SunSDR2 DX is a 100 watt version of the same radio.

The free Expert SDR2 software supports CAT (remote transceiver control) and VAC (virtual audio cable) operation. The higher ADC sampling speed means better protection from FM broadcast interference (except on 6m).

*Note: Expert Electronics is a Russian company. It seems that their radios are currently impossible to buy outside of Russia. Probably due to the current political situation. NSI Communications in the USA reports no stock. The European dealers Pileup and WiMo say that the radio is no longer available. The Russian website is still up and you may be able to order direct from them. There may be import restrictions in the USA and other countries.*

- Frequency range: 90 kHz to 65 MHz and 95 MHz to 155 MHz
- Bands: LF, MF, HF, VHF (6m + 2m)
- ADC: LTC2209, 16 bits, 160 Msps
- Frequency stability: ±0.5ppm TCXO
- Panadapter bandwidth: 2x 312 kHz, wideband display to 80 MHz
- Main and sub-receiver on each panadapter (4 in total)
- Class: direct sampling HF and VHF transceiver
- Interface: Ethernet
- Receive filters / pre-selector: bandpass filters
- SDR software: Expert SDR2 (supplied)
- Transmitter power:
    - SunSDR2 Pro 20 W HF, 15 W on 6m and, 8 W on 2m
    - SunSDR2 DX 100 W HF, 80 W on 6m and, 8 W on 2m
- Dimensions DX model (metric): 185 x 165 x 10.5 mm, weight 1.8 kg
- Dimensions DX model (imperial): 7.48" x 6.69" x 3.14", weight 3.9 lb
- Software: ExpertSDR2
- Display: none
- No video output
- A microphone is not included

## SUNSDR2 QRP (Expert Electronics)

The SunSDR2 QRP is a 5 watt 'no knobs' radio that covers the HF and 6m amateur bands. It uses a different a ADC clock at a slower rate than the pro and dx models.

*Note: Expert Electronics is a Russian company. It seems that their radios are currently impossible to buy outside of Russia. Probably due to the current political situation.*

*NSI Communications in the USA reports no stock. The European dealers Pileup and WiMo say that the radio is no longer available. The Russian website is still up and you may be able to order direct from them. There may be import restrictions in the USA and other countries.*

- Frequency range: 100 kHz to 55 MHz and 70 MHz to 70.5 MHz
- Bands: LF, MF, HF, and 6m. [4m band receive only]
- ADC: LTC2165, 16 bits, 122.88 Msps
- Frequency stability: ±2.5ppm TCXO
- Panadapter bandwidth: 2x 312 kHz, wideband display to 80 MHz
- Main and sub-receiver on each panadapter (4 in total)
- Class: direct sampling HF and VHF transceiver
- Interface: Ethernet
- Receive filters / pre-selector: bandpass filters
- SDR software: Expert SDR2 (supplied)
- Transmitter power: 5 watts on HF and 6m
- Dimensions (metric): 185 x 106 x 45 mm, weight 0.8 kg
- Dimensions (imperial): 6.5″ x 3.9″ x 1.4″, weight 1.8 lb
- Software: ExpertSDR2
- Display: none
- A microphone is not included

### QDX, QMX+, QCX+, QCX Mini (QRP Labs)

The QDX and QMX+ QRP SDR transceivers from QRP Labs use a different modulation technique. They determine the frequency of the audio tones coming from the digital modes software on your PC and offset the transmitted frequency by that amount. Rather than using an SSB modulator, these radios are transmitting FSK (frequency shift keying). This works for any mode that transmits a single tone at a time. It does not work for PSK which uses phase shifts, or for modes that transmit more than one tone at a time, like Winlink.

Frequency shift keying is not new. We have been using it to transmit direct keyed RTTY for years. It is, or was, also used for commercial data traffic, particularly for maritime services. The difference here is that in normal FSK, the radio receives a digital 'keying' signal. In this radio, the software sends audio tones and the radio works out what FSK offset is required. As far as I know, this is a unique method.

The receiver in the QDX and QMX+, and I assume the other models, is a QSD (quadrature sampling detector) feeding a 24-bit stereo ADC at audio frequencies. The audio is sent to the WSJT-X or other digital mode software in the usual way. It has four bandpass filters on the input side of the balanced Tayloe QSD detector. The oscillator is a SiLabs SI5351A clocked by a 0.25ppm 25 MHz TCXO.

The QCX models are for CW transmission only. The QMX model is for CW and digital modes, and the QDX is for digital modes only.

The radios are available as a kit or fully assembled. There is a delay for assembled models as they are built on demand. The kits have the smd devices preinstalled, but you will have to wind toroids and solder components on to the board.

- Bands: most models: 80, 60, 40, 30, and 20m, or 60, 40, 30, 20, 17, and 15m, or 20, 17, 15, 12, 11 and 10m. The QMX+ model supports all amateur bands up to 10m
- Class: QSD/QSE small box transceiver
- No panadapter display. Some models have no display at all
- The kit is fun to build
- Emphasis on CW and digital modes
- 4 – 5 watts output (usually lower on the higher bands)
- DC. The kits can be built for a 12 volt (not 13.8v), or a 9-volt supply
- Optional metal case – strongly recommended

### SDR Cube (Midnight Design Solutions)

The SDR Cube is quite unique as it uses any of the SoftRock radios as an RF front end. The SDR Cube hardware adds onboard DSP and control circuitry to create a stand-alone receiver or transceiver, on whatever bands the SoftRock card can cover; (usually one or two ham bands). There is a receiver board available as well, which can cover 2 MHz - 30 MHz.

The SDR Cube can be provided either assembled and tested, or in kit form. You end up with a sophisticated interface to a basic radio front end, making the radio a generation three SDR. The performance will be much the same as any other QSD / QSE based SDR but you do not need a PC to use it.

A 20 W linear amp based on the OpenHPSDR Pennywhistle design is also available.

As far as I can tell, there is no facility to output IQ to an external panadapter and the internal display is too small to be of any real use, so this is an SDR for people who want a QRP radio but don't want a panadapter.

- Frequency range: 2 MHz – 30 MHz (depends on the front end used)
- Bands: HF
- ADC resolution: the ADC samples at audio frequencies
- Panadapter bandwidth: 8 kHz (on the tiny internal display)
- Class: standalone sub QRP transceiver (QSD / QSE)
- Interface: USB for control & programming
- Receive filters / pre-selector: yes, with the receiver board
- Transmit power: 700 mW to 1 W (depends on the SoftRock used)

### Pocket SDR / Portable SDR / PDSDR (Michael Colton)

The Pocket SDR, also known as the Portable SDR or PSDR, takes yet another unique approach. As I write this, it is still in development and there has been very little technical data published. The radio is designed to be a rugged, ultra-portable, standalone, sub QRP transceiver small enough to pop in your pocket. It has one knob, which also has a push to click function. This single control is used to change all of the adjustable parameters. There is a built-in paddle for CW operation and a tiny two-inch colour LCD display.

The screen can display a waterfall, as well as the mode, frequency, and filter settings. But wait there's more in this little miracle! The radio also includes a GPS receiver, and it can work as a vector network analyser (VNA) to check out your antenna.

It can work as a standalone radio as it has onboard DSP and an ARM processor to manage the control and display functions. I am very impressed with the innovation demonstrated in this little transceiver. The PSDR is an open-source project funded by a Kick Starter campaign. https://www.kickstarter.com/projects/1703258614/psdr-pocket-hf-sdr-transceiver-with-vna-and-gps

See the video at https://www.youtube.com/watch?v= iCIqv1zF-I

- Frequency range: up to 35 MHz
- Transmit power: modulated exciter only. It needs external low pass filters and a power amplifier – hopefully under development.
- Internal GPS receiver – could in future work as an emergency locator beacon on HF, or display your location on the display.
- VNA mode – antenna analyser
- Built-in CW paddle
- Receives AM and SSB, and transmits SSB (either USB or LSB)
- Standalone SDR for SOTA or ultra QRP operation

### HackRF One (Great Scott Gadgets)

HackRF One is an experimenter's radio. It may be of interest to radio amateurs operating on the 23cm, 12cm, 9cm, and 5cm microwave amateur bands. It has very wide frequency coverage and transmitting capability. HackRF One is an open source project initially backed by Kickstarter funding. The hardware was designed by Michael Ossmann. It is able to receive and send wideband signals like WFM and WiFi. You can synchronize several HackRF One radios so that they work as a diversity reception system or as a MIMO radio. The radio runs off power supplied by the USB connection. There are clone versions that include a 'portapack' display four buttons and a VFO knob. It has limited dynamic range due to its 8 bit ADC.

- Frequency range: 10 MHz to 6 GHz
- Bands: upper HF, VHF, UHF, SHF

- ADC resolution: 8 Bits, (2 - 20 Msps)
- Panadapter bandwidth: 20 MHz
- Class: hybrid architecture sub-QRP low power transceiver (half duplex)
- Interface: USB and USB powered
- Receive filters / pre-selectors: none
- Onboard Arm Cortex M4 microcontroller and a CPLD for DSP experimentation
- Transmitter power -10 dBm to +15 dBm (0.1 to 31 mW) depends on the frequency of operation
- SDR Software: GNU Radio, SDR#, SDR Radio.

### Blade RF 2.0 (Nuand)

The BladeRF 2.0 is an experimenter's board, which may be of interest to radio amateurs operating on the 23cm or 12cm microwave amateur bands. It has very wide frequency coverage and transmitting capability.

A transverter board can be added to enable HF and VHF coverage.

The radio uses USB 3.0 format for higher data speed, which allows for a very wide output 'panadapter' bandwidth. It can also work at lower bandwidth with a USB 2.0 port. It uses a LimeMicro LMS6002D RF transceiver chip.

- Frequency range: 47 MHz to 6 GHz transmit, 70 MHz to 6 GHz
- Bands: UHF, SHF
- ADC resolution: 12 Bits, 61.44 Msps
- Panadapter bandwidth: 56 MHz
- Class: Uncased PCB - sub QRP transceiver (Full Duplex)
- 2x2 MIMO (two receivers two transmitters)
- Separate receive and transmit SMA antenna connectors
- Interface: USB 3.0, USB or 5V DC powered
- Receive filters / pre-selectors: none
- Onboard 200 MHz ARM926EJ-S processor and a Cyclone 5 FPGA for DSP experimentation
- Transmitter power +8 dBm (6 mW) typical.

### USRP B200 (Ettus Research)

Ettus Research makes a large range of professional SDR transceivers and receivers, both cased and un-cased. Of these, the B200 and B210 are the most accessible to amateur operators and experimenters. The B200 is really an experimenter's board, but it may also be of interest to radio amateurs operating on the microwave amateur bands. It has very wide frequency coverage and sub-QRP transmitting capability.

The radio uses a USB 3.0 interface to carry the high data load required for its very wide output 'panadapter' bandwidth.

- Frequency range: 70 MHz to 6 GHz
- VHF, UHF, SHF
- ADC resolution: 12 Bits, 61.44 Msps
- Panadapter bandwidth: 56 MHz
- Class: Uncased PCB - low power transceiver (sub-QRP)
- Type: Transceiver (Full Duplex)
- Interface: USB 3.0 and USB-powered
- Receive filters / pre-selectors: none
- Onboard Xilinx Spartan 6 FPGA for DSP experimentation
- Transmitter power +10 dBm (10 mW) maximum
- SDR software: GNU Radio.

### USRP B210 (Ettus Research)

The B210 is really an experimenter's board, but it may also be of interest to radio amateurs operating on the microwave amateur bands. It is similar to the B200 but it has two receivers and two transmitters, allowing the radio to work in MIMO mode. The twin receivers could be used for diversity or beam steering applications. The radio uses a USB 3.0 interface to carry the high data load required for its very wide output 'panadapter' bandwidth.

- Frequency range: 70 MHz to 6 GHz
- VHF, UHF, SHF
- ADC resolution: 12 Bits, 61.44 Msps
- Panadapter bandwidth: 56 MHz (or 30.72 MHz, in the MIMO mode)
- Class: Uncased PCB - low power transceiver (sub-QRP)
- Type: Transceiver (single or 2x full duplex)
- Interface: USB 3.0, separate 6V DC powered
- Receive filters / pre-selectors: none
- Onboard Xilinx Spartan 6 FPGA for DSP experimentation
- Transmitter power +10 dBm (10 mW) maximum
- SDR software: GNU Radio
- USRP = Universal Software Radio Peripheral.

### USRP 2900 and 2901 (National Instruments)

The National Instruments USRP 2900 and 2901 are small box radios designed primarily for educational use and for experimenters. The 2901 has two receivers and transmitter front ends and the 2900 only has one.

National Instruments owns Ettus Research, which explains why these radios have essentially the same specifications as the Ettus radios.

- Frequency range: 70 MHz to 6 GHz
- VHF, UHF, SHF
- ADC and DAC resolution: 12 Bits, 61.44 Msps (or 30.72 MHz in the MIMO mode)
- Panadapter bandwidth: 56 MHz (max)
- Class: small box - low power transceiver (sub-QRP)
- Type: Transceiver (single or 2x full duplex)
- Interface: USB 3.0, separate 6V DC powered
- Receive filters / pre-selectors: none
- Transmitter power +20 dBm (100 mW) maximum
- USRP 2901 1-channel single input single output (SISO)
- USRP 2901 2-channel multiple input multiple output (MIMO)
- SDR software: LabVIEW, GNU Radio

### Red Pitaya (redpitaya.com)

The Red Pitaya SDRs are small cards with an onboard FPGA and a softcore processor that can run Linux. They have two RF receiver ports and two transmitter ports, as well as the usual digital i/o pins. They are development boards with RF functionality but could be used as the basis for a microwave frequency transverter.

The board is marketed for use as a two-channel 50 MHz digital oscilloscope. I expect that it could be used as a vector network analyser or an antenna analyser as well.

STEMlab 125-14

- Frequency range: DC – 60 MHz
- 2x2 MIMO (two receivers two transmitters)
- Separate receive and transmit SMA antenna connectors
- ADC resolution: 14 Bits at 125 Msps
- Class: uncased PCB – sub-QRP transceiver (1x or 2x full duplex)
- Interface: Ethernet, USB
- Onboard FPGA for DSP experimentation

SDRLAB 122-16

- Frequency range: 300 kHz – 550 MHz
- 2x2 MIMO (two receivers two transmitters)
- Separate receive and transmit SMA antenna connectors
- ADC resolution: 16 Bits at 122.88 Msps
- Class: uncased PCB – sub-QRP transceiver (1x or 2x full duplex)

- Interface: Ethernet, USB
- Onboard FPGA for DSP experimentation

### Pluto (ADALM)

Pluto SDRs are often used as the base for transverters used for the QO-100 amateur radio satellite. They are a developer's board housed in a plastic case.

- RF coverage from 325 MHz to 3.8 GHz
- Up to 20 MHz of instantaneous bandwidth
- Sample rate 61.44 Msps
- 20 MHz maximum panadapter bandwidth
- One transmitter and one receiver, half or full duplex
- MATLAB®, Simulink® support
- GNU Radio sink and source blocks
- Class: sub-QRP hybrid SDR transceiver development board
- USB interface
- Analogue Devices AD9363--Highly Integrated RF Agile Transceiver and Xilinx® Zynq Z-7010 FPGA. The AD9363 has an integrated ADC sampling at 61.44 Msps. It is a similar arrangement to the chip used in the Hermes Lite 2 except it has two 12-bit ADCs and two DACs. The received RF is mixed down with an analogue mixer and an external local oscillator then sampled by the ADCs.

### Pluto+ (Howard and Jun Su)

Pluto+ SDRs are often used as the base for transverters used for the QO-100 amateur radio satellite. The Pluto plus is not made by ADALM. It has a completely different circuit board. The radio claims compatibility with the original Pluto SDR. It has 2x receivers and 2x transmitters.

The supplier states, "The design draws on the well-known ADI ADALM-PLUTO and improves its shortcomings."

- RF coverage from 70 MHz to 6 GHz
- Analogue Devices AD9363--Highly Integrated RF Agile Transceiver and Xilinx® Zynq Z-7010 FPGA, (see above).
- Ethernet and USB interface
- 40 MHz TCXO with 0.5 ppm stability
- 2x receivers and 2x transmitters with separate antenna jacks
- Power supply 5V DC
- Metal case

https://github.com/plutoplus/plutoplus

# RECEIVERS

## IC-R8600 (Icom)

The IC-R8600 is rather expensive, but it has features that make the price worthwhile.

This receiver provides the "best of both worlds." It is a desktop receiver with buttons and knobs, and it can be connected to an SDR program running on your PC to give the same functionality as many of the 'black box' SDR receivers on the market. Very few desktop receivers offer a spectrum and waterfall display, although they are common on amateur radio transceivers. The IC-R8600 has many features that make a big difference. It has superb performance figures, covers a super-wide range of frequencies, has an excellent band scope, and can receive several digital voice modes as well as the usual SSB, AM, CW, and FM modes. There are 2000 memory slots and a very fast scanning mode.

The IC-R8600 is ranked at number five on the prestigious Sherwood Engineering website. http://www.sherweng.com/table.html.

*TIP: Some people moan that the radio cannot decode DMR. This is a valid criticism since DMR is the most popular amateur radio digital voice mode. However it is not surprising because Icom is dedicated to promoting the D-Star mode which it helped to develop.*

The radio has three connectors to suit different types of antennas. ANT 1 is a Type-N connector, which can be used for receiving the whole range from 10 kHz to 3 GHz. ANT 2 is an SO239 connector that takes the standard PL259 plug. It is used for the LF to HF bands, covering 10 kHz up to 30 MHz. ANT 3 is an RCA connector with a port impedance of 450 ohms. It is designed for a long wire, Beverage, or similar antenna. This input also covers the LF to HF bands, from 10 kHz up to 30 MHz.

The spectrum scope and waterfall span can be set between 5 kHz and 5 MHz. On most Icom and Yaesu transceivers the maximum display span is 1 MHz, although the new IC-905 has a 50 MHz maximum span.

- Frequency range 10 kHz to 3 GHz. Most regional models have small gaps in the coverage to suit local regulations
- ADC: 16-bits at 122.88 Msps
- Frequency stability: ±0.5 ppm
- 10 MHz reference input
- A dedicated DSP chip and a separate DSP chip for the digital voice modes
- Excellent specifications: L.O. noise -144 dBc/Hz, BDR 125 dB at 100 kHz offset, RMDR 117 dB at a 10 kHz offset.
- No internal or rear panel fans.
- 4.3" (93mm x 52mm) colour TFT LCD touchscreen

- One receiver on one panadapter. External software can support more receivers.
- Type: direct sampling SDR to 30 MHz, single conversion hybrid from 30 to 500 MHz, double conversion hybrid from 500 to 1500 MHz, triple conversion hybrid from 1500 to 3000 MHz
- Modes SSB, CW, FSK (RTTY), AM, FM, WFM (not stereo, no RDS decode), D-Star, NXDN, dPMR, P25 (tier 1). No digital broadcast or video modes.
- 3 antenna connectors
- IQ output for external SDR software, I.F. output for external receiver
- No Bias-T voltage output (a major disappointment)
- Fast scanning (no trunk scanning)
- 2000 channel memories in up to 100 groups, plus scanning channels. 200 auto write channels.
- No video output
- Dimensions: 220 × 90 × 230 mm, 8.7 × 3.5 × 9.1 in
- Weight: 4.3 kg, 9.5 lb

### Malahit DSP2

The DSP2 is a licensed Chinese clone of a Russian design. There is firmware specific to the clone version on the Russian Malahiteam website. This radio is exceptionally well made and it carries both America FCC and European CE, type approval.

The Malahit DSP2 is a compact receiver with great features and a huge frequency range, from 10 kHz to 380 MHz and 404 MHz to 2 GHz. Unlike most 'dongle,' and 'black box' SDR receivers, the DSP2 is completely self-contained. It has a colour LCD touchscreen, a built-in speaker, and an internal 5000 mAh 3.7-volt Lithium battery. The radio is portable making it ideal for plane spotters, and for use while walking or on holiday. I particularly like that you can charge the battery using the supplied Type-C USB cable and any USB power supply. It has the usual AM, SAM, DSB, FM, SSB, and CW modes, plus wideband FM stereo with RDS. There are decoders for CW, RTTY, and FT8.

- Frequency range: 10 kHz to 380 MHz and 404 MHz to 2 GHz
- Maximum panadapter bandwidth 48 kHz, 96 kHz, or 192 kHz
- Modes: AM, SSB, DSB, CW, NFM, WFM
- FM stereo with RDS (stereo to headphone jack only)
- 3.5" 480x320 LCD touchscreen
- One bandscope (panadapter) with one receiver
- Bias-T for powering a masthead preamplifier
- 5000 mAh 3.7-volt Lithium battery – charged via USB cable

- STM32H743 ARM CPU at 480MHz
- Mirics MSi001 multi-band, multi-mode tuner
- Class: Hybrid using a tuner chip
- SDR type: hybrid SDR with DSP performed by an STM32 ARM processor
- IQ output for HDSDR or SDR# software
- Built-in decoders for CW, RTTY, and FT8.
- FM Retro Scale for FM Broadcast Stations

*The original DSP2 was designed by Georgy 'George' Yatsuk (RX9CIM), Vladimir Gordienko (R6DAN), Vladimir Burlakov (R6DCY), and Igor Naumenko.*

### FDM-DUOr, FDM-S1, FDM-S2, FDM-S3 (ELAD)

ELAD is an Italian company that makes rather stylish SDR receivers and one SDR transceiver. They are rather unusual because they include front panel controls and a small display but no panadapter spectrum and waterfall display. You can use the radios as a stand-alone radio, or connect them to a computer running the FDM-SW2 Software to display four simultaneous receivers in 'single channel mode' or eight simultaneous receivers in 'double channel mode.'

DUOr

The DUOr is the receiver from the DUO transceiver, mounted in the same case. There are no specifications online, but I expect that they will be the same as the DUO.

- Frequency range: 10 kHz – 54 MHz
- Bands 160m to 6m
- Direct sampling receiver
- ADC: LTC2165 16bit operating at 122.88 MHz
- DDC (Digital Down Converter) made with FPGA Xilinx Spartan 6
- DSP using ARM STM32F4 microcontroller with floating point unit
- Demodulation: CW, CWR, LSB, USB, AM, NFM
- 10 MHz reference input
- CW decoder
- Dimensions (Metric) 180(W) × 70(H) × 155(D) mm, weight 1.2 kg
- Dimensions (Imperial) 7.0"(W) × 2.75"(H) × 6.1"(D), weight 2.4 lb
- IQ output bandwidth 192 ksps (stand-alone), Remote: 384ksps, 768ksps, 1536ksps, 3072ksps, 6144ksps in single channel mode and 384ksps in double channel mode.

FDM-S1

- 20 kHz – 30 MHz (Undersampling to 165 MHz)
- 14 bit 61.44 Msps ADC

- The FDM-S1 can decode DRM broadcast signals
- Dimensions (Metric) 90(W) × 27(H) × 108(D) mm, weight 360g
- External software: FDM-SW2

FDM_S2

The FDM-S2 can deliver 192 ksps, 384 ksps, 768 ksps, 1536 ksps, 3072 ksps, or 6144 ksps panadapters. It can receive HF +6m, the FM broadcast band, and VHF from 135-160 MHz using ADC under-sampling.

- HF: 10 kHz – 52 MHz, 74 MHz – 108 MHz FM, 135 MHz – 165 MHz
- Separate antennas for HF and VHF
- 16 bit, 122.88 Msps ADC
- USB 3 interface, USB powered
- MDS: -132 dBm @ 14 MHz
- FM: 74-108 MHz. Sensitivity <2uV 12 dB SINAD @ 98 MHz
- VHF: 135-160 MHz MDS -137 dBm @ 145 MHz
- Dimensions (Metric) 110(W) × 40(H) × 90(D) mm, weight 360g
- External software: FDM-SW2

FDM_S3

- 9 kHz – 108 MHz
- The FDM-S3 has 24 MHz of bandwidth and a GNSSDO (Global navigation satellite system Disciplined Oscillator) usually called a GPSDO.
- Two antennas for HF and two antennas for VHF
- DC 8-16 volts
- Dimensions (Metric) 230(W) × 60(H) × 155(D) mm, weight 2.15 kg
- Dimensions (Imperial) 9.0″(W) × 2.4″(H) × 6.1″(D), weight 4.74 lb
- External software: FDM-SW2

### RTL dongles

RTL dongles are a range of USB stick software defined radios. They are usually based on the RTL2832U and R820T / R820T2 chips. RTL dongles were designed to be cheap receivers for digital TV and FM broadcast signals, but they can be used as general-purpose receivers for the upper HF, VHF, and UHF bands. There are many models. NooElec sells RTL dongles and an up-converter plus dongle combo. It also sells excellent SAW filters for NOAA satellites on 137 MHz, L Band on 1542 MHz (Inmarsat), Iridium & Inmarsat on 1620 MHz, NOAA and GOESS on 1688 MHz, and 1420 MHz Hydrogen line astronomy. Plus band stop filters for protection from AM and FM broadcast stations, and also an LNA.

RTL-SDR.com https://www.rtl-sdr.com/about-rtl-sdr/ has a lot of information about RTL dongles.

They also sell two models. The RTL-SDR BLOG V3 and the RTL-SDR BLOG V4. These have been widely copied and the company warns against buying cheap copies. The website has specifications for their models, and photographs to help you identify the fakes.

RTL dongles are incredibly cheap and they are fun to play with. There are options to extend the range down to the HF bands by using an up-converter.

- Frequency range: varies with model. 24 MHz – 1766 MHz or 52 MHz to 2200 MHz are typical for the most common models
- Bands: VHF – UHF (and HF with modification or an upconverter)
- ADC resolution: 8 Bits
- Limited dynamic range and overload performance
- Panadapter bandwidth: 2.8 to 3.2 MHz typical
- Class: USB dongle hybrid SDR receiver (tuner/receiver chip)
- Interface: USB
- No band specific bandpass filters
- SDR software HDSDR, SDR#, and others
- There is software for maritime ADS that will only work with SDR dongles

### FlyCatcher (NooElec)

The FlyCatcher is a dual-channel UAT and ADS-B SDR receiver, configured as a Raspberry Pi HAT. It plugs into the Rpi 40-pin header, but it does not use any of the pins, so there is no need to mount it on the Rpi if you don't want to.

The unit consists of two RTL dongle receivers on one board. One includes a filter at 1090 MHz for the International ADS-B transmissions and the other has a 978 MHz filter for ADS-B UAT transmissions. This mode is used by aircraft operating under 18,000 feet mostly in US airspace. Above 18,000 feet large aircraft must transmit standard 1090ES ADS-B data. The unit includes low noise amplifiers which can be manually switched out if you are using an external masthead preamplifier. Each receiver is powered by a micro USB connector, so you can use one or both receivers as you wish.

The normal way to use the receiver is to plug a short USB cable between the receiver and a USB Type-A port on the Raspberry Pi, then run the server software on the Pi. The installation is easy, with only two lines to enter via SSH. You can run the web browser client on any computer connected to your network. The last line of the installation before the Linux command line, contains the URL to enter into your web browser. It is the Pi's IP address followed by /tar1090. (http://192.---.---.---/tar1090).

https://www.nooelec.com/store/flycatcher.html

https://github.com/wiedehopf/adsb-scripts/wiki/Automatic-installation-for-readsb

Thanks to Tech Minds for his video at,
https://www.youtube.com/watch?v=ov3JOIDLN90

Figure 55: FlyCatcher SDR hat by NooElec

**FUNcube Dongle pro+ (funcubedongle.com)**

The FCD pro+ is a USB stick 'dongle' style receiver designed by Howard Long G6LVB. It has automatically switched front-end filters to protect the receiver from overload, including excellent SAW filters for the 2m and 70cm amateur bands. The radio features a 'Bias-T' power injector to supply 5 Volts DC to a remote low noise amplifier (LNA) over the antenna feeder coax. You can select an internal low-noise pre-amplifier as well.

The FCD was developed specifically to create a cheap high-performance receiver that schools could use as a part of the educational goals of the FUNcube satellite program. One source states that the FCD pro+ uses the Mirics MSi001 tuner chip, which is also used in the SDRplay receiver.

- Frequency range: 150 kHz to 260 MHz and 410 MHz to 2.05 GHz
- Bands: LF to L band (LF, MF, HF, VHF, UHF)
- ADC resolution: 32 Bits (at audio frequencies). Decimated to 16 bits before being sent to the PC.
- Panadapter bandwidth: 192 kHz on pro+ (96 kHz on FCD Pro and FCD)
- Class: USB dongle receiver
- Interface: USB
- Receive filters / pre-selector: 11 filters, SAW filters on 2m and 70cm
- SDR software SDR#, HDSDR, and others.

### Airspy HF+ Discovery (Airspy.com)

The Airspy receiver was created by the same developer as the excellent SDR# software. It is a hybrid receiver, using a polyphase mixer followed by a pair of ADCs working at the I.F. There is also an Airspy HF+ dual port model with two antenna ports, which appears to have the same specifications as the Discovery model.

- Frequency range: 500 kHz – 31 MHz and 64 MHz - 260 MHz
- Bands: HF and 2m (not 6m)
- Very sensitive receiver -140 dBm MDS in 500 Hz on HF
- Frequency stability: 0.5 ppm high precision, low phase noise clock
- Class: Small box receiver, USB interface
- Software SDR#, GQRX (Linux and macOS), SDR Console v3, HDSDR and others with extio.dll, Krypto 500 Narrowband Signals Analysis and Decoding Suite, Krypto 1000 Wideband Signals Analysis and Decoding Suite
- Dimensions (metric): 45 x 60 x 10 mm
- Open source libairspyhf library can be used

### Ranger (Airspy.com)

- Frequency range: 500 kHz – 1.75 GHz
- One antenna port for VLF, LF, HF, and another for VHF and UHF
- 9 bandpass filters
- Uses an RF Soc (tuner chip). UHF bands are down-converted using a Rafael Micro R860 chip
- Metal case
- Internal Bias_T for powering a masthead amplifier
- Frequency stability: ± 0.5ppm TCXO
- Specified BDR 110 dB (HF) and RMDR 89 dB (HF)
- Software SDR#, GQRX (Linux and macOS), SDR Console v3, HDSDR and others with extio.dll, 'Krypto 500 Narrowband Signals Analysis and Decoding Suite,' 'Krypto 1000 Wideband Signals Analysis and Decoding Suite'.
- Open source libairspyhf library can be used.

### Airspy R2 (Airspy.com)

The Airspy R2 receiver uses the R8202T2 receiver chip and is similar to an RTL dongle, with similar frequency coverage. The 12-bit ADC is oversampled to improve the dynamic range. The maker claims that this makes it equivalent to a 16-bit sampling process and decimation.

- Frequency range: 24 – 1700 MHz, down to LF with the SpyVerter option
- 10 Msps IQ output
- Frequency stability: ± 0.5ppm TCXO
- Software SDR#, GQRX (Linux and macOS), SDR Console v3, HDSDR and others with extio.dll, Krypto 500 Narrowband Signals Analysis and Decoding Suite, Krypto 1000 Wideband Signals Analysis and Decoding.

### RSP1B, RSPdx, RSPdx-R2, RSPduo (SDRplay.com)

The SDRplay 'Radio Spectrum Processor' series are small box receivers, which cover a very wide frequency range. The receivers use a Mirics MSi001 RF tuner and MSi2500 'digital interface' chip. The Mirics RF tuner chip contains; five low noise amplifiers for several bands, the frequency synthesizer, and a quadrature mixer producing I and Q streams at an I.F. frequency. The digital interface chip performs the ADC and DSP functions. This is the same chipset as used in the Mirics MSi3101 USB TV dongle, but with the SDRplay you get front-end filters and established ham radio / general listening software compatibility.

The RSPduo is a dual tuner receiver that supports diversity reception when used with the SDR Connect or SDR Uno software. It has two SMA antenna connectors plus a high-impedance antenna connector for the HF bands.

- Frequency range: 1 kHz – 2 GHz
- Bands: LF, MF, HF, VHF, UHF
- ADC resolution: 14 Bits

  The ADC bits reduce at higher sampling rates. 14-bits to 10.66 Msps, 12-bits from 10.66-8.064 Msps, 10-bits from 8.064-9.261 Msps, and 8-bits between 9.261 and 10 Msps.
- Panadapter bandwidth: up to 10 MHz. (See bit rates above)
- Multiple receivers in the 10 MHz bandwidth
- Frequency stability: 0.5 ppm
- Class: USB small box receiver (tuner/receiver chip)
- Receive filters / pre-selector: eight switched filters
- USB for DC and PC connectivity
- Metal case
- Bias-T to power a masthead amplifier
- SDR software: SDR Connect, SDRuno, HDSDR, Cubic SDR, SDR Console, Soapy SDR, and others.

Figure 56: RSPduo receiver

## KiwiSDR (John Seamons ZL4VO/KF6VO)

http://kiwisdr.com/public/ and https://kiwisdr.nz/.

KiwiSDR is a web-based SDR network of KiwiSDR receivers. You can use any web browser to connect and listen to over 800 of these radios all over the world. They have been made available for public access by their owners. See http://kiwisdr.com/public/ and http://rx.linkfanel.net/.

The KiwiSDR receiver was designed and produced by John Seamons, a ham radio operator (ZL4VO/KF6VO) who lives in New Zealand. They are available from WiMo.com, Martin Lynch and Sons (hamradio.co.uk), and KiwiSDR.nz. You can add your receiver to the KiwiSDR network or simply use it over your LAN at home.

The radio is sold as an SDR 'cape' for the BeagleBone Black SBC. Or as a complete software defined HF radio (0-30MHz), complete with a BeagleBone Green, in an aluminum enclosure. This version includes a GPS antenna and self-test cable.

They tune from 0-30 MHz and can receive and decode many signals including:

- Amateur radio AM, FM, USB, and LSB voice signals around the world.
- Decode amateur radio FT4, FT8, and WSPR digital signals and CW (Morse).
- Decode and display weather FAX signals.
- Medium, Long Wave, and Very Low Frequency (VLF) stations.
- Shortwave international broadcast stations.
- DRM (Digital Radio Mondial) digital broadcast decoding.

### Titan / Titan Pro (Enablia)

This receiver has some rather unique features that are better suited to commercial users for surveillance and monitoring than amateur radio operators and shortwave listeners. It is designed to monitor up to 100 demodulated channels at the same time, which is not something your average ham needs to do. The radio displays a wideband panadapter between 9 kHz and 32 MHz wide, so you can display the whole of the HF spectrum. You then select between one and four wideband slots, which can be 312 to 2187 kHz wide. You can demodulate up to 100 'narrowband' signals within the four wideband slots. It has sixteen pre-selector bandpass filters, but they are probably not very useful unless you want to monitor many channels very close together.

- Frequency range:
    - Standard HF v2 model 9 kHz – 32 MHz.
    - VLF model 3 kHz to 2 MHz.
    - AB (air band) model 118 MHz to 137 MHz.
- Wide bandwidth: 9.66 MHz HF v2, or 19 MHz on the AB model
- 40 slice receivers on HF v1, and 100 slice receivers on HF v2.
- Class: direct sampling (see note below) small box receiver
- Interface: USB (the Titan Pro has a LAN port for the audio outputs)
- Receive filters / pre-selector: 16 switchable bandpass filters
- Useful for monitoring, interference investigations, propagation analysis, beacon monitoring
- SDR Software: TitanSDR (includes special recording features). Tight integration with Krypto500 (by Comint Consulting) and CODE300-32 (by Hoka Electronic)

### Colibri DDC (Expert Electronics)

The ColibriDDC is a small box, direct sampling receiver that uses an Ethernet connection to the PC. This enables the receiver to be controlled over a LAN, or it can be connected directly to your PC.

*Note: Expert Electronics is a Russian company. Due to the current political situation, their radios are currently impossible to buy outside of Russia. The Russian website is still up and you may be able to order direct from them. There may be import restrictions in the USA and other countries.*

- Frequency range: 10 kHz – 55 MHz (works up to at least 61 MHz)
- Bands: LF, MH, HF +6m
- ADC resolution: 14 bits, 125 Msps
- Panadapter bandwidth: two 312 kHz panadapters (4 slice receivers)

- Wideband scope to 62 MHz
- Class: small box direct sampling receiver
- Interface: Ethernet
- Receive filters / pre-selector: no, just a 55 MHz anti-alias filter
- Switch out 55 MHz to under-sample zones up to more than 900 MHz
- SDR Software: ExpertSDR2, ExpertSDR3 (Beta)
- 10 MHz SMA connector for optional external clock input
- SMA antenna connector
- TXCO +- 0.5 ppm
- DC 5V plug pack supplied

### ColibriNANO (Expert Electronics)

The ColibriNANO is a direct sampling SDR dongle receiver.

- Frequency range: 10 kHz – 55 MHz (undersampling to 500 MHz with appropriate external filters)
- ADS4145 ADC resolution: 14 bits, 122.88 Msps
- Main and sub-receiver on a single panadapter
- Class: USB dongle – direct sampling receiver
- Interface: USB
- SDR Software: ExpertSDR2, ExpertSDR3 (Beta)
- SMA antenna connector
- TXCO +- 0.5 ppm
- DC 5V plug pack supplied

### RX888 MKII

The RX888 MKII is a direct sampling receiver with a 16-bit ADC and tuning range from 1 kHz up to 1.7 GHz. The design is based on the RX-666 which is turn was based on Oscar Steila's (IK1XPV), BBRF103 open source design. The product is designed and manufactured in China.

It has an amazing bandwidth of up to 64 MHz between 1 kHz to 64 MHz, which means that you can display the entire HF spectrum at the same time. This works well with SDR Console because it has a setting to display all of the HF amateur bands at once. Between 64MHz - 1700MHz the receiver has a 10 MHz bandwidth.

You **must** use a USB 3 connection to your PC to provide a fast enough data speed for the large bandwidth and enough DC power.

- Frequency range: 1 kHz – 64 MHz with 64 MHz max BW, and 64 MHz -1.7 GHz with 10 MHz BW
- ADC resolution: 16 bits, 130 Msps

- Class: small box – direct sampling receiver
- Interface: USB 3 Type B to Type A
- SDR Software: SDR Console, HDSDR, SDR#, SDR++
- Two antenna connectors (low and high bands)
- LTC2208 16bit ADC at 130 MSPS
- Frequency stability: ±0.5ppm VCXO
- ATT adjustment range -32dB to 0dB
- VGA adjustment range -11dB to +34dB
- External 27Mhz reference clock support
- 3.3v software switched Bias-Tee HF/VHF independent control
- ADC random and dither enabled under software control
- Will probably require the installation of a Cyprus FX3 USB driver

TechMinds video https://www.youtube.com/watch?v=3rapmewZA20

There were problems with this radio initially. Critics started that it had poor performance and very badly written firmware code. However that seems to have been overcome now. *"A lot of very good work has gone on with the RX888 firmware over the last two years – I've just been catching up. Do not ignore the RX888!"* Simon Brown author of SDR Console. I have no experience with the radio and am neither endorsing it or criticising it.

### Perseus (Microtelecom)

Perseus is one of the earliest direct sampling SDR receivers. It comes with its own SDR software. The 14 bit ADC used in this radio offers very similar performance to a 16 bit ADC (almost the same ENOB), but the lower sampling rate means that the Perseus receiver does not include coverage of the 6m band. A separate (not in the same box) 'FM+' down converter can be purchased to add coverage of the FM broadcast band, 87.5 to 108 MHz. But it is rather expensive when compared to the cost of an average FM broadcast receiver, so it would only appeal to listeners looking for FM broadcast DX. The design has been static with no major changes or updates since 1988. However, a new model (described below) has recently been announced.

- Frequency range: 10 kHz – 30 MHz
- Bands: HF
- ADC resolution: 14-bit, 80 Msps
- Frequency stability: ±1ppm TCXO
- Panadapter bandwidth: 125 kHz to 1.6 MHz
- Class: direct sampling small box HF receiver
- Interface: USB

- FPGA: Xilinx Spartan IIIE
- Receive filters / pre-selector: yes 10 switched bandpass filters
- Software: proprietary Perseus software, LinRad, HDSDR, and others.
- BDR 124dB CW, 117dB SSB

### Perseus22 (Microtelecom)

The Perseus22 is the first new radio from Microtelecom since 1998. It is a four-channel direct sampling receiver with a continuous frequency coverage from 10 kHz to 225 MHz and a typical image rejection larger than 70 dB. Two channels are for VLF to HF coverage and the other two are for VHF coverage.

All channels are synchronously sampled by a high SNR, 14-bit ADC and processed by a software defined digital down converter, implemented on an FPGA. The outputs are routed to the host PC by a USB 3.0 controller, allowing wide-bandwidth IF applications. The dual channels in each frequency band can be used for diversity reception.

- Frequency range: 10 kHz - 225 MHz
- Bands: HF and VHF
- ADC resolution: 14-bit, 80 Msps
- Panadapter bandwidth: 125 kHz to 1.6 MHz
- Dynamic range: >100dB SSB and >105 dB CW (500Hz)
- Two receivers for VLF-HF (diversity reception) 10 kHz –70 MHz
- Two receivers for VHF (diversity reception) 70 MHz - 225 MHz
- USB 3 interface
- DC power adapter

### BB60C (Signal Hound )

The Signal Hound BB60C SDR is marketed as a 'real time' spectrum analyser and long-term recorder for industrial and scientific applications, rather than as a general use receiver. But it can be used as a receiver. It covers a very wide frequency range.

- Frequency range: 9 kHz – 6 GHz
- Bands: LF, MF, HF, VHF, UHF, SHF
- ADC resolution: 14 Bits, 312.5 ksps - 40 Msps
- Panadapter bandwidth (maximum): 27 MHz
- Class: USB small box receiver, USB-powered
- Interface: USB 3.0
- Streams at 140 Mbits to the PC
- Wide dynamic range -158 dBm to +10 dBm
- Receive filters / pre-selector: switched filters

### SA124B (Signal Hound )

The Signal hound SA124B SDR is marketed as a spectrum analyser for industrial and scientific applications. It has the highest maximum frequency and the widest range of all of the radios listed in this chapter. It is cheaper than the BB60C model and it can be used as a general purpose receiver.

- Frequency range: 100 kHz – 12.4 GHz
- Bands: LF, MF, HF, VHF, UHF, SHF
- Class: USB small box receiver, USB-powered
- Interface: USB 2.0
- IQ output up to 240 kHz bandwidth
- Wide dynamic range -151 dBm to +10 dBm
- Internal demodulation of AM/FM/SSB/CW
- Free API for software development.

### SA44B (Signal Hound )

The Signal hound SA44B SDR is marketed as a spectrum analyser for industrial and scientific applications. It is considerably cheaper than the SA124B model and it can be used as a general-purpose receiver.

- Frequency range: 1 Hz – 4.4 GHz
- Bands: LF, MF, HF, VHF, UHF, SHF
- Class: USB small box receiver, USB-powered
- Interface: USB 2.0
- Wide dynamic range -151 dBm to +10 dBm
- IQ output up to 240 kHz bandwidth
- Internal demodulation of AM/FM/SSB/CW
- Free API for software development.

### AR-2300 / AR-5001D (AOR)

The AOR-2300 is a black box no knobs version of the AOR AR-5001D receiver. They are quite complex receivers employing direct sampling technology to receive frequencies below 25 MHz and conventional receiver architecture above 25 MHz. They very have very limited SDR capability about the same as a generation two SDR, using I and Q audio signals at a 12 kHz I.F. from the DSP stage. You can also buy an optional 'digital board', which provides a digital IQ output when the receiver is tuned to frequencies between 25 MHz and 3.15 GHz.

- Frequency range: 40 kHz – 3.15 GHz
  
  40 kHz – 25 MHz direct conversion SDR
  
  25 MHz – 220 MHz double conversion superheterodyne
  
  220 MHz – 360 MHz triple conversion superheterodyne

360 MHz - 3.15 GHz double conversion superheterodyne

- Bands: LF, MF, HF, VHF, UHF, SHF
- ADC resolution: Audio IQ at 12 kHz I.F. or 14 bits at 65 Msps from the optional digital board (25 MHz to 3.15 GHz only)
- Panadapter bandwidth: 24 kHz or 1 MHz with the digital board
- Class: dual/triple conversion superheterodyne receiver with SDR direct conversion below 25 MHz only. (Hybrid SDR)
- Interface: USB
- Receive filters / pre-selectors: none below 25 MHz
- SDR software: proprietary AR-IQ software supplied with the digital card, or typical soundcard SDR software for the audio IQ output.

## WR-G305e/i (WiNRADiO by RADIXON)

WiNRADiO makes a large range of 'no knob' conventional architecture receivers. Some are designed to be installed as add-on PCI cards inside a PC. Other models like the WR-G305e/i are small box, no knobs, radios. The WR305e/i radio has very limited SDR capability about the same as a generation two SDR, using IQ audio signals out of the DSP stage.

- Frequency range: 9 kHz – 1.8 GHz
- Bands: LF, MF, HF, VHF, UHF
- ADC resolution: not applicable, (soundcard radio)
- Panadapter bandwidth: soundcard dependent, 48, 96, or 192 kHz
- Class: dual-conversion superheterodyne receiver with a software-defined last IF stage and demodulator. (Hybrid SDR)
- Interface: USB
- Receive filters / pre-selector: tracking pre-selector
- Expandable to 3500 MHz using an optional down converter
- SDR software: proprietary G305 software.

## WR-G31DDC 'EXCALIBUR' (WiNRADiO by RADIXON)

Unlike the WR-G305 model, the Excalibur is a direct sampling SDR. But it is LF to HF only and it does not include coverage of the 6m amateur band. There is a PCI version for installation inside a desktop PC and also a 'professional' version. The software has the ability to schedule recording of either demodulated audio or the entire 2 MHz wide IQ spectrum to the PC hard drive. Within the 2 MHz wide panadapter, the radio supports the reception and demodulation of three narrowband receiver channels at the same time.

- Frequency range: 9 kHz – 49.99 MHz
- Bands: LF, MF, HF

- ADC resolution: 16 bits, 100 Msps
- Panadapter bandwidth: 20 kHz – 2 MHz plus a low-resolution wideband display of up to 50 MHz
- Class: direct sampling HF receiver
- Three demodulated channels
- Interface: USB
- Receive filters / pre-selector: no, apart from a filter for the MF broadcast band.
- SDR software: proprietary G31DDC software,

### SoftRock (Tony Parks KB9YIG)

In many ways, the SoftRock is the design that started SDR for radio amateurs.* The SoftRock kits have been the first introduction to software defined radio for thousands of people, including me. The kits were originally designed and sold by Tony Parks KB9YIG. These little kits have surprisingly good performance and they were excellent value for money. But, these days the equally cheap and more versatile RTL dongles are more popular. SoftRock kits are still useful if you want to create a bandscope for an older ham transceiver or receiver, or if you want to "build it yourself."

The SoftRock kits are based on the Tayloe QSD design (Dan Tayloe N7VE).

**Note: Another early pioneer of SDR for radio hams is Gerald Youngblood AC5OG, the founder of FlexRadio, who brought the first commercial SDR transceivers to market. He wrote the early code for PowerSDR and the excellent, 'A software-defined radio for the masses' series of articles for QEX.*

There is information about the past and currently available SoftRock SDRs at https://www.wb5rvz.com/sdr/

Some kits are available from https://www.softrock40.com/category/all-products

While these kits are an important part of SDR history, they are out of date now. If you want a cheap SDR, buy one of the black box radios. If you want an interesting kit to build, look at the QRP Labs radios.

### Fobos SDR (Rig Expert)

The Fobos receiver has two coherent direct sampling Rx channels for the HF bands up to 25 MHz, and double conversion from 25 MHz to 6 GHz. The IQ output sample rate can be set between 4 MHz and 50 MHz. This implies a maximum bandwidth of the full 25 MHz on the HF bands and 50 MHz on the higher bands.

- Frequency range: 100 kHz – 25 MHz direct sampling, 25 MHz- 6 GHz hybrid
- ADC resolution: 14 bits
- Interface: USB 3.0 or higher
- SDR software: none specific to the radio, extio.dll for SDR# and HDSDR.

# GLOSSARY OF ABBREVIATIONS AND ACRONYMS

| ABBREVIATIONS, ACRONYMS, AND OTHER TERMS | |
|---|---|
| *Term* | *Description* |
| 2m, 6m, 10m | 144 MHz, 50 MHz, and 28 MHz amateur radio bands |
| 15m, 20m, 40m, 80m | 21 MHz, 14 MHz, 7 MHz, and 3.5 MHz amateur radio bands |
| 59 | Standard (default) signal report for amateur radio voice conversations. A report of '59' means perfect readability and strength. |
| 599, 5NN | Standard (default) signal report for amateur radio CW conversations. A 599 report means perfect readability, strength, and tone. The 599 signal report is often used for digital modes as well. The 5NN version is faster to send using CW. It is often used as a signal report exchange when working contest stations. |
| 73 | Morse code abbreviation 'best wishes, see you later.' It is used when you have finished transmitting at the end of the conversation. |
| A/D | Analogue to digital. |
| ADC | Analogue to digital converter or analogue to digital conversion. |
| AF | Audio frequency - nominally 20 to 20,000 Hz. |
| AGC | Automatic gain control. A system where the circuit, often a receiver I.F. stage, but sometimes microphone level or another function, can reduce the signal level to prevent it from overloading the stage. This increases the dynamic range of input signal that the circuit can handle before distortion occurs. AGC is used to extend the dynamic range of superheterodyne receivers. It is used in SDR receivers to limit the increase in audio output level as the level of RF input increases. |
| algorithm | A process or set of rules to be followed in calculations or other problem-solving operations, especially by a computer. In DSP it is a mathematical formula, code block, or process that acts on the data signal stream to perform a particular function, for example, a noise filter. |
| ANF | Automatic notch filter – a filter that eliminates the effect of long-term interference signals such as carrier signals that are close to the wanted receiving frequency. Not effective against impulse noise. |
| API | Application Programming Interface. Is a set of instructions that are built into a computer program that lets a second program control aspects of the first. For example, you might write a digital voice program that takes and sends audio to SmartSDR. |
| ARM | Advanced RISC Machine. A type of microprocessor that works faster because it has a reduced list of possible instructions. |
| Bandscope | A bandscope is a spectrum display of the frequencies above and below the frequency that the radio is tuned to. The centre of the display is generally the frequency that you are listening to. This is different from a panadapter where you can listen to any frequency across the display. |

| | |
|---|---|
| BDR | Blocking dynamic range, a receiver test. |
| Bin | A narrow band of frequencies. The FFT process creates thousands of narrow bins. |
| Bit | Binary value 0 or 1. |
| Bus | The lines on a computer mainboard that connect key devices together. You can think of a bus as a data highway. It carries parallel data signals usually 16, 32, or 64 bits wide. |
| Byte | Eight binary bits. |
| C# | C Sharp – computer language (a development of C, C+, and C++). |
| Carrier | Usually refers to the transmission of an unmodulated RF signal. It is called a carrier because the modulation process modifies the un-modulated RF signal to carry the modulation information. A carrier signal can be amplitude, frequency, and/or phase modulated. Then it is referred to as a 'modulated carrier.' An oscillator signal is not a carrier unless it is transmitted. |
| CAT | Computer-aided transceiver. A computer sends predefined text strings to the radio to control and read information from the radio.<br>Icom calls it CI-V. |
| CC | Command and Control bytes. |
| CFIR | Compensating finite impulse response – a filter type with a rise at the high-frequency end to compensate for the roll-off created by the preceding CIC filters in the FPGA. |
| CI-V | Icom version of CAT control. A computer sends predefined text strings to the radio to control and read information from the radio. |
| CIC | Cascaded integrated comb filters are used in the FPGA to remove alias frequencies during the decimation process. |
| CISC | Complex instruction set computing – Standard microprocessor. |
| CODEC | Coder/decoder - a device or software used for encoding or decoding a digital data stream. Often used to encode a voice or video signal into a format suitable for digital transmission. |
| COTS | Commercial off the shelf – a component you can buy from a dealer. For DSP this means that the chips are provided already configured for digital signal processing applications. |
| CPU | Central processing unit - usually the microprocessor in a PC or other computer. |
| CQ | 'Seek You' is an abbreviation used by amateur radio operators when making a general call that anyone can answer. |
| CW | Continuous Wave – the transmission mode used to transmit Morse Code. |
| DAB | Digital audio broadcasting. The DAB standard is dominant in Europe, It is also used in Australia, and in parts of Africa and Asia. |
| D/A | Digital to analogue. |
| DAC | Digital to analogue converter or digital to analogue conversion. |
| dB, dBm, | The Decibel is a way of representing numbers using a logarithmic scale. |

| | |
|---|---|
| dBc, dBV | dB is used to describe a ratio, i.e. the difference between two levels or numbers. Decibels are often referenced to a fixed value such as a Volt (dBV), a milliwatt (dBm), or the carrier level (dBc). Decibels are also used to represent logarithmic units of gain or loss. An amplifier might have 3 dB gain. An attenuator might have 10 dB loss. |
| dBFS | Signal level compared to the full-scale level, expressed in dB. |
| DDC | Digital down conversion - the receiver part of an SDR. The incoming radio or I.F. signals are converted to digital samples with an ADC. Frequency translation to audio or the 24 kHz or 32 kHz input to a DSP chip is done by a computer device (FPGA or microcontroller) manipulating the digital signals. |
| DDS | Direct digital sampling is used by an SDR that performs digital down conversion at the incoming RF frequency. |
| DDS | Direct digital synthesis. An oscillator that uses DSP software (firmware) code to generate frequencies. |
| DHCP | Dynamic Host Configuration Protocol (DHCP) is a client/server protocol that automatically provides an Internet Protocol (IP) host with its IP address and other related configuration information such as the subnet mask and default gateway. It is the easiest way to connect an Ethernet-equipped radio to your home network. |
| digital modes | Amateur radio transmission of digital information rather than voice. It can be text, or data such as video, still pictures, or computer files. |
| digital voice | Amateur radio mode where speech is coded into a digital format and sent as tone sequences or phase shift keying. |
| direct sampling | Direct sampling or direct digital sampling (DDS) refers to an SDR that uses an ADC sampling at the incoming RF frequency. Not at an I.F. or at audio frequencies. |
| .dll | Dynamic link library – reusable software block - can be called from other programs. |
| DR3 | 3$^{rd}$ order dynamic range receiver test - needs to be tested differently for an SDR. |
| DRM | Digital Radio Mondiale is a digital broadcast radio standard used by the BBC World Service, Voice of America, Deutsche Welle, and other shortwave broadcasters. |
| DSP | Digital signal processing – a dedicated integrated circuit chip usually running internal firmware code, or a software program running on a computer. DSP uses mathematical algorithms in computer software to manipulate digital signals in ways that are equivalent to functions performed on analogue signals by hardware mixers, oscillators, filters, amplifiers, attenuators, modulators, or demodulators. |
| DUC | Digital up-conversion - the transmitter part of an SDR. Microphone audio, music, Morse code, or digital text tones are converted to digital samples with an ADC. Frequency translation up to an I.F. or the final RF frequency is done by a computer device (FPGA or microcontroller) manipulating the digital signals. It is then converted to an analogue signal using a DAC and amplified by an RF amplifier. |

| | |
|---|---|
| DX | Long distance, or rare, or wanted by you, amateur radio station. |
| EME | Earth - Moon – Earth, communications using signals 'bounced' off the moon back to Earth. These days usually using the Q65 digital mode. |
| ENOB | The effective number of bits of an ADC. The ADC can't describe very small input signals due to its internal noise performance. ENOB is the number of bits that contain usable information. |
| FFT | Fast Fourier Transformation – conversion of signals from the time domain to the frequency domain (and back). |
| FFTW | Fastest Fourier Transformation in the West – a software library used by many SDR programs. |
| FIFO | First In First Out – a type of data buffer. |
| FIR | A Finite Impulse Filter is a DSP filter (software algorithm) designed so that the output signal will always return to zero if a pulse or 'impulse' is applied at the input. |
| FPGA | Field Programmable Gate Array – a chip that can be programmed to act like logic circuits or a CPU. |
| Frame | Data packet structure – usually includes synchronisation, control, and payload information in bytes. |
| Fs | Signal frequency or sampling frequency |
| Gbit | High-speed data rate 1 Gbit = 1,000,000,000 bits per second. |
| GHz | Gigahertz – a unit of frequency = 1,000,000,000 cycles per second. |
| GPU | Graphics Processing Unit – a specialized microprocessor designed to manage data for display on a computer monitor. They were introduced to speed up the display of high-resolution images such as movies or computer games. Using a GPU reduces the processing load on the computers main processor, the CPU. |
| GUI | Graphical user interface – the form on the PC software that allows you to interact with the radio. The GUI is the knobs and display of a black box SDR. |
| HF | High Frequency (3 MHz - 30 MHz). |
| Hybrid | A radio that uses a hardware oscillator and mixer to convert the incoming received signals down to a frequency that can be sampled by an ADC. Often a superheterodyne front end followed by an SDR. |
| Hz | Hertz – a unit of frequency = 1 cycle per second. |
| I data or stream | The non-phase-shifted part of the IQ audio signal or data stream. |
| I.F. | Intermediate frequency i.e. the Fs – LO or Fs + LO output of a mixer. (Fs = Signal frequency LO = local oscillator frequency.) |
| Imaginary | Part of the complex signal in a buffer - often holds the Q signal or frequency bins below the LO frequency. |
| IMD | Intermodulation distortion - interference/distortion caused by non-linear devices like mixers. There are IMD tests for receivers and transmitters. IMD performance of linear amplifiers can also be tested. |

| | |
|---|---|
| IQ | Refers to the I and Q data streams treated as a pair of signals. |
| JT65 | A slow digital mode used for extremely weak signal communications, especially EME. Part of the WSJT-X software suite by Joe Taylor K1JT |
| kHz | Kilohertz – a unit of frequency = 1 thousand cycles per second. |
| KK | Kiss Konsole – SDR application for Windows used as an example of PC software. |
| ksps | Thousands of samples per second. |
| L band | Amateur radio 1296 MHz band (1,240 to 1,300 MHz) also known as the 23cm band. |
| latency | Latency is the delay caused by using computer software or DSP chips to process the audio signal. In the SDR receiver, it is the delay between the time the receiver detects a signal and the time that you hear it from the speaker or see it on the panadapter. In the SDR transmitter, it is the delay between your speech going into the microphone or the Morse key or bug is depressed until the signal is transmitted. It is a particular issue for CW operators where it can cause an annoying delay before the sidetone is heard. |
| LF | Low Frequency (30 kHz - 300 kHz). |
| LO | Local Oscillator frequency of a superheterodyne receiver. In an SDR with a 'zero I.F.', it is the same as the frequency that the SDR is tuned to. It is usually the frequency that is displayed at the centre of the panadapter spectrum display. |
| LSB | Lower sideband transmission. When talking about data streams or frames it is the least significant byte – lower 8 bits of a 16-bit word. |
| Mbit | A data rate in millions of bits per second |
| MDS | Minimum discernible signal – the measurement of receiver sensitivity. |
| MF | Medium Frequency (300 kHz – 3 MHz). |
| MHz | Megahertz – a unit of frequency = 1 million cycles per second. |
| MIMO | Multiple input multiple output |
| MOX | Microphone-operated-switch (can be a PTT from the PC software). |
| MSB | When talking about data streams or frames it is the most significant byte – higher 8 bits of a 16-bit word. |
| Msps | Millions of samples per second. |
| Multicore | A cable containing several wires. |
| mW | Milliwatt = 1/0000$^{th}$ watts |
| m, 2m, 6m | Meter (US) or Metre. Often used to denote an amateur radio or shortwave band; e.g. 2m, 6m, 30m, 10m, where it denotes the approximate free space wavelength of the radio frequency. Wavelength = 300 / frequency in MHz. A frequency of 30 MHz has a corresponding wavelength of 10m. |
| NB | Noise blanker – filter used to eliminate impulse noise. |
| NCO | Numerically controlled oscillator – in the case of SDRs it is a digital oscillator that outputs data samples representing a sine wave at a nominated frequency. In an SDR it usually also outputs a cosine wave, |

| | |
|---|---|
| | which is a sine wave delayed by 90°. |
| Net | An on-air meeting of a group of amateur operators. |
| Nonlinearity | Is when the relationship between the input to a device and the output signal is not linear. For example, an amplifier that outputs less power at full drive because the supply voltage droops. |
| NPR | Noise power ratio or 'white noise' receiver test. |
| NR | Noise Reduction – filter used to eliminate continuous background noise |
| Nyquist zone | A Nyquist zone is the range of frequencies that the ADC can sample without interference from unwanted alias signals. It sets the maximum frequency coverage of a direct sampling radio and ½ of the maximum panadapter size on a hybrid radio.<br>The 1st Nyquist zone is a range of frequencies equal to ½ of the sample rate. The 2nd Nyquist zone extends from ½ the sample rate up to the sample rate. Some ADCs can work up to the 11th Nyquist zone. |
| Panadapter | Panadapter is short for 'panoramic adapter.' It allows us to see a panoramic display of the band. The panadapter is sometimes confusingly referred to as a 'receiver.' It often can display a spectrum display and/or a waterfall picture. You can usually adjust the size, speed, and width of the display with a mouse or touchscreen device. You may be listening to one or more signals anywhere within the displayed spectrum of frequencies. This is different from a bandscope. |
| Payload | The part of a digital data frame that contains the wanted data signal. The rest of the bits in the data frame are used to carry synchronisation and control information. |
| PC | Personal computer - usually with a Windows, Mac, or Linux operating system. In this book, it could include a Raspberry Pi, a tablet, or even a cell phone. |
| PCB | Printed circuit board |
| PSK, BPSK, QPSK | Phase shift keying, binary phase shift keying, and quadrature phase shift keying are some of the amateur radio and commercial digital modes used for sending text. |
| PTT | Press to talk - the transmit button on a microphone – The PTT signal sets the radio and software to transmit mode. |
| Q data or stream | The 90-degree phase shifted part of the IQ audio signal or data stream. |
| Q65 | A slow digital mode used for extremely weak signal communications, especially EME. Part of the WSJT-X software suite by Joe Taylor K1JT. |
| QAM | Quadrature amplitude modulation is used by most digitally modulated point-to-point microwave radio systems. Created by combining two multi-level data streams in quadrature. The signal is both phase and amplitude-modulated. |
| QRP | Q code meaning, low-power amateur radio operation. Less than 10 watts for SSB and less than 5 watts for CW or digital modes. |
| QSD | Quadrature sampling detector – An SDR receiver front end (often a Tayloe |

| | |
|---|---|
| | detector). Single conversion with RF input and audio output. |
| QSE | Quadrature sampling exciter - SDR transmitter front end (often a Tayloe detector working in reverse). Single conversion with an audio input and an RF output. |
| QSO | Q code - amateur radio conversation or "contact". |
| QSY | Q code - request or decision to change to another frequency. |
| RDS | Radio data system – a system for sending brief test strings over the FM broadcast radio. It usually includes a station ID and sometimes the song title and artist currently playing. |
| Real | Part of the complex signal in a buffer - often holds the 'I' signal or frequency bins above the LO frequency. |
| RF | Radio frequency. |
| RISC | Reduced instruction set computing – a special microprocessor usually used for embedded systems or fast digital processing. |
| RMDR | Reciprocal mixing dynamic range – a receiver performance test. |
| RS232 | A computer interface used for serial data communications. |
| SAM | Synchronous amplitude modulation. |
| SAW | SAW (surface acoustic wave) filters are electromechanical devices commonly used in radio frequency applications. They are usually ceramic or crystal filters. |
| SBC | Single board computer, often a Raspberry Pi. |
| SDK | Software development kit. APIs are used to communicate between different applications, whilst an SDK is a tool kit to build applications and features. In most cases, an SDK application will contain an API — or even multiple APIs depending on the purpose of the software. |
| SDR | If you don't know this one you haven't been paying attention! |
| + SDR | A conventional radio with some basic SDR functionality, usually the ability to display a bandscope of up to about 200 kHz bandwidth. |
| SFDR | Spurious-free dynamic range – a performance measurement of the dynamic range of an analogue to digital converter (or another device). |
| SHF | Super High Frequency (3 GHz - 30 GHz). |
| SINAD | Signal-to-Noise Plus Distortion Ratio in dB. |
| Sked | A pre-organized or scheduled appointment to communicate with another amateur radio operator. |
| Slice | The part of the spectrum that is within the selected audio bandwidth i.e. the part that you can listen to. |
| SMD | Surface mounted device. Very small integrated circuit packages, chip capacitors, or resistors that are designed to sit on top of the circuit board and are not through-hole mounted. |
| SNR | Signal-to-Noise Ratio in dB. |
| Softcore | An emulation of a microprocessor running as a firmware program on another device such as an FPGA. Can be used as a hardware microprocessor. Not a 1970s movie. |

| | |
|---|---|
| SSB | Single sideband transmission mode. |
| Superhet | Superheterodyne receiver – conventional, non-SDR, receiver architecture that uses hardware oscillators and mixers to convert ('heterodyne') the incoming RF signal to an intermediate frequency. |
| SWL | Shortwave listener. |
| TR | Transmit / Receive. Usually refers to the antenna change over relay, or the signal that controls the, transmit to receive switching. |
| TU | Morse code abbreviation meaning 'to you.' Used when you have finished transmitting and wish the other station to respond. |
| UHF | Ultra High Frequency (300 MHz - 3000 MHz). |
| USB | Universal serial bus – serial data communications between a computer and other devices. USB 2.0 is fast, USB 3.0 is very fast. |
| USB | Upper sideband transmission. |
| VHF | Very High Frequency (30 MHz - 300 MHz). |
| VNA | A vector network analyser is a test instrument, or a mode using an SDR transceiver as a tracking spectrum analyser to test antennas, coax feeder cables, or RF filters. |
| W | watts – a unit of power (electrical or RF). |
| Word | A digital word is one or more bytes used to carry a particular piece of information. For example, the Q signal between the ANAN or Hermes radio and the PC software is a stream of 24-bit (3-byte) words. The microphone signal uses 16-bit (2-byte) words. |
| Zadig | Zadig is a program that installs Windows USB drivers, which allow an RTL dongle to communicate with the SDR software. In some cases, you have to run Zadig as well as installing the SDR program. SDR# installs Zadig as a part of its installation. Your RTL dongle might not work on a different USB port. You might have to run Zadig for each USB port that you plan to use. It is easier if you use the same SDR port every time you use your RTL dongle. |

# DRAWINGS AND IMAGES

# INDEX

## 1



## 3



## A



## B



## C



## D



## E



## F

## G



## H



## I



## J



## K



## L



## M



## N



## O



## P



## Q

# R



# S



# T



# U



# V



# W



# Y